12th International Conference on Prognostics and System Health Management (PHM 2021)

IOP Conference Series: Materials Science and Engineering
Volume 1207

Zhengzhou, China
22-24 October 2021

Editors:

Wenliao Du
Diego Cabrera
Yun Bai

ISBN: 978-1-7138-4666-6
ISSN: 1757-8981

Printed from e-media with permission by:

Curran Associates, Inc.
57 Morehouse Lane
Red Hook, NY 12571

Some format issues inherent in the e-media version may also appear in this print version.

Institute of Physics
Dirac House, Temple Back
Bristol BS1 6BE UK

Phone: 44 1 17 929 7481
Fax: 44 1 17 920 0979

techtracking@iop.org

Additional copies of this publication are available from:

Curran Associates, Inc.
57 Morehouse Lane
Red Hook, NY 12571 USA
Phone: 845-758-0400
Fax: 845-758-2633
Email: curran@proceedings.com
Web: www.proceedings.com

TABLE OF CONTENTS

IOP Conf. Series: Materials Science and Engineering 1207 (2021) 011001 doi:10.1088/1757-899X/1207/1/011001

Preface

THE MESSAGE FROM THE GENERAL CHAIRS

It is our great pleasure to welcome all of you to the PHM 2021 Zhengzhou International Conference. In the past few years, this annual conference represented one of the largest gatherings of researchers and professionals to discuss most recent advancements in prognostics and system health management. This year, however, the ongoing COVID-19 pandemic and the resulting travel restrictions in many regions impose unprecedented challenges to the organization of the conference. Instead of meetings fully in person, the Conference Organization Committee has decided to shift the conference to a hybrid mode, with which some authors attended this conference in person while others participated in this conference virtually via a shared Zoom link. We hope this shift in conference mode will not affect your enthusiasm in participating in the conference and discussions.

We are pleased to invite two well-known experts in the field to deliver keynote speeches at this conference: Professor Wilson Wang, Professor and University Research Chair in Intelligent Diagnostics and Prognostics of Engineering Systems, Lakehead University, Canada, and his talk is on "Fault diagnosis in gearboxes"; Dr. Yiqiang Chen, Deputy Chief Engineer of National Key Laboratory, China Electronic Product Reliability and Environmental Testing Research Institute, China, and his talk is about "Prognostic and health management technology for electric system based on PoF and data-driven method". The conference program is also offering three paper oral presentation sessions and one poster session, arranged in the late morning and entire afternoon on October 23.

This conference could not be organized without the earnest efforts from all our organization committee members. The conference is also indebted to dozens of volunteers who contributed to the various processes of the conference. It has been a great privilege for both of us to serve as the General Chairs of PHM 2021 Zhengzhou and it is our hope that you find the conference fulfilling and enjoyable. We thank you for your support and wish you a pleasant experience at PHM 2021 Zhengzhou.

Sincerely,

Rongjie Huang

Wenliao Du

General Chairs, PHM2021 Zhengzhou International Conference
Professors, Zhengzhou University of Light Industry

IOP Conf. Series: Materials Science and Engineering 1207 (2021) 011001 doi:10.1088/1757-899X/1207/1/011001

KEYNOTE SPEAKERS

Keynote 1: Fault Diagnosis in Gearboxes

Dr. Wilson Wang, Professor and University Research Chair in Intelligent Diagnostics and Prognostics of Engineering Systems, Lakehead University, Canada

Abstract:
Gearboxes are commonly used in rotating machinery for power transmission and speed modification. As a matter of fact, most machinery imperfections are related to defects in gearboxes, including gears and rolling element bearings. However, reliable fault detection in gears and bearings in a gearbox still remains a challenging task in this R&D field, due to reasons such as signal complexity, feature modulation and coupling. This talk will address some recent research and development in gear and bearing fault detection. The covered research aspects include signal processing for fault detection in gear systems and rolling element bearings, feature extraction, and automatic fault diagnostics in gearboxes. This speech will also discuss the related challenges and possible solution strategies.

Speaker's Bio:
Wilson Wang received his MSc in Industrial Engineering from the University of Toronto (Ontario, Canada), and PhD in Mechatronics Engineering from the University of Waterloo (Ontario, Canada) in 2002. From 2002 to 2004, he was a Senior Scientist/Project Manager at Mechworks Systems Inc. He joined Lakehead University (Thunder Bay, Canada) in 2004, where he is currently a Professor in the Department of Mechanical Engineering. His research interests include mechatronics, diagnostics, prognostics, signal processing, artificial intelligence, and machine learning. Dr. Wang is a Lakehead Research Char in Intelligent Diagnostics and Prognostics of Engineering Systems. He is the founder/director of the Research Center for Intelligent Mechatronics Systems. He has supervised about 80 highly qualified personnel (PDF, PhD and MSc students), and published about 200 peer reviewed articles. He has also

successfully developed a series of smart systems for online condition monitoring of mechanical and electrical machines.

Keynote 2: Prognostic and Health Management Technology for Electric System based on PoF and Data-Driven Method

Dr. Yiqiang Chen, Deputy Chief Engineer of National Key Laboratory, China Electronic Product Reliability and Environmental Testing Research Institute, China

Abstract:
The reliability and safety of electrical products has always been a concern in many industries like aerospace, offshore, and transportation. At present, existing prognostics techniques or approaches can generally be divided into three categories: (1) physics-based; (2) data-driven; and (3) hybrid-based. Physics-based approaches incorporate prior knowledge of physical and/or analytical models/laws with measured data, to predict the degradation behavior of a system and its RUL. Data-driven approaches, rely on historically collected data and attempt to derive models directly from the data for the RUL prediction. The hybrid-based approaches attempt to make use of the strengths of both approaches, by combining the knowledge about the physical process as well as information from the observed data together to improve the prediction performance. The use of both physics-based and data-driven approaches might be one of the suitable solutions. The report will introduce the background of the PHM for the Electric System at first. And then share our works about the chip-level PHM for the SoC, the board-level PHM for the module, and the system-level PHM for the servomechanism and so on.

Speaker's Bio:
Yiqiang Chen, Ph. D., research fellow, deputy chief engineer, team leader, Principal investigator, has long been engaged in basic theory, method and application

research of microwave, power, information processing and reliability of electronic products, progress has been made in failure mechanism and model of device, PHM of power conversion system and system on chip (SoC). About six SoC failure monitoring and warning chips have been successfully developed and verified, and a series of instruments such as the semiconductor device quantitative evaluation system and power cycle test system have been developed, and it provides a technical support for the reliability and security of electronic products. He has won the China Electronic Information Science and Technology Innovation Team Award, the Science and Technology Innovation Team Award of the Ministry of Industry and Information Technology, the working skills competition award of Guangdong Province, and the first, second and third prizes of provincial and ministerial level science and technology progress. He was selected as one of the outstanding young talents in science and technology innovation of Guangdong Province, the outstanding natural science foundation of Guangdong Province, the special support project, and the Pearl River Science and Technology Star. He is the experts on national Science and Technology Achievements Evaluation Center, Ministry of Science and Technology of the People's Republic of China, National Natural Science Foundation of China, a department of the central government, Guangdong Province and Guangzhou City Science and technology project evaluation. He is responsible for more than 20 projects, such as national key R & D Program, National Natural Science Fund, Major International Cooperation Projects, high quality projects of MIIT, pre-research on installation and development. As the first author or corresponding author, has published more than 70 SCI papers in IEEE EDL, IEEE TED, APL and other journals, and applied for 69 international and domestic invention patents, of which 7 U.S. invention patents and 35 Chinese invention patents have been granted, publishing reliability editing 3 books, 5 standards, 2 sets of software, the joint cultivation of doctoral, master more than 20 students.

CONFERENCE ORGANIZATION COMMITTEE

Honorary General Chair
Jie (Peter) Liu, Carleton University, Canada

General Chairs
Rongjie Huang, Zhengzhou University of Light Industry, China
Wenliao Du, Zhengzhou University of Light Industry, China

Steering Committee Chairs
Chuan Li, Chongqing Technology and Business University, China
Guofu Luo, Zhengzhou University of Light Industry, China

Program Chairs
Guoliang Lu, Shandong University, China
Hao Li, Zhengzhou University of Light Industry, China

Datong Liu, Harbin Institute of Technology, China
Yiming Deng, Michigan State University, USA
Xiaojian Yi, Beijing Institute of Technology, China
Fangyi Wan, Northwestern Polytechnic University, China
Mariela Cerrada, Universidad de Los Andes, Venezuela
Yan Su, Nanjing University of Aeronautics and Astronautics, China
Yaguo Lei, Xi'an Jiaotong University, China
Yixiang Huang, Shanghai Jiaotong University, China
Qiang Miao, Sichuan University, China
Bin Zhang, University of South Carolina, USA

Program Co-Chairs
Shaowei Chen, Northwestern Polytechnical University, China
Ming J. Zuo, University of Alberta, Canada
Liansheng Liu, Harbin Institute of Technology, China
Jaya Sil, Indian Institute of Engineering Science and Technology, India
Zhongkui Zhu, Soochow University, China
Qiang Han, North Minzu University, China
Lixiang Duan, China University of Petroleum (Beijing), China
Youmin Zhang, Concordia University, Canada
Jiayang Huang, Shanghai Aircraft Customer Service Corporation, China
Jian Wang, Chinese Academy of Sciences, China
Guigang Zhang, Chinese Academy of Sciences, China
Jaehwa Jeong, Korea University, South Korea
Madhav Mishra, Luleå University of Technology
René-Vinicio Sanchez, Universidad Politécnica Salesiana, Ecuador
Zhinong Li, Nanchang Hangkong University, China
Yanjun Yan, Western Carolina University, USA
Shisheng Zhong, Harbin Institute of Technology, China
Rado Zimroz, Wroclaw University of Science and Technology, Poland
Wilson Wang, Lakehead University, Canada
Agnieszka Wylomanska, Wroclaw University of Science and Technology, Poland
Zhigang Tian, University of Alberta, Canada
Yu Peng, Harbin Institute of Technology, China
Baoping Tang, Chongqing University, China
Zhiliang Liu, University of Electronic Science and Technology of China, China
Zheng Liu, University of British Columbia (Okanagan), Canada
Jiawei Xiang, Wenzhou University, China
Janice Duan, London South Bank University, UK
Yi Qin, Chongqing University, China
Christopher Mechefske, Queen's University, Canada
Aitor J. Garrido, University of the Basque Country, Spain
Markus Timusk, Laurentian University, Canada
Nagarajan Raghavan, Singapore University of Technology & Design, Singapore

IOP Conf. Series: Materials Science and Engineering 1207 (2021) 011001 doi:10.1088/1757-899X/1207/1/011001

Jiufei Luo, Chongqing University of Posts and Telecommunications, China

Arrangements Chairs
Fai Ma, University of California at Berkeley, USA
Xiaoyun Gong, Zhengzhou University of Light Industry, China

Paper Review Chairs
Haining Liu, University of Jinan, China
Qinghua Luo, Harbin Institute of Technology, China
HengYangCan (Alexander) Gu, Carleton University, Canada

Keynote Chair
Yongqiang Liu, Shijiazhuang Tiedao University, China

Tutorial Chair
Li Wang, Nanjing University of Aeronautics & Astronautics, China

Poster Chair
Xuyun Fu, Harbin Institute of Technology, China

Proceeding Chairs
Chunsheng Yang, National Research Council of Canada, Canada
Gianluca Valentino, University of Malta, Malta

Publicity Chair
Rui Zhao, National Research Council Canada, Canada

Finance Chairs
Grover Zurita Villarroel, Universidad Privada Boliviana, Bolivia
Wei Guo, University of Electronic Science and Technology of China, China

LIST OF EDITORS
Wenliao Du, Zhengzhou University of Light Industry, dwenliao@zzuli.edu.cn, China
Diego Cabrera, Universidad Politecnica Salesiana, dcabrera@ups.edu.ec, Ecuador
Yun Bai, Universidade do Algarve, yunbai@ualg.pt, Portugal

IOP Conf. Series: Materials Science and Engineering 1207 (2021) 011002 doi:10.1088/1757-899X/1207/1/011002

Peer review declaration

All papers published in this volume of *IOP Conference Series: Materials Science and Engineering* have been peer reviewed through processes administered by the Editors. Reviews were conducted by expert referees to the professional and scientific standards expected of a proceedings journal published by IOP Publishing.

- **Type of peer review: Single-blind**
- **Conference submission management system:** SoftConf.com (https://www.softconf.com/l/phm2021)
- **Number of submissions received: 203**
- **Number of submissions sent for review: 121**
- **Number of submissions accepted: 25**
- **Acceptance Rate (Number of Submissions Accepted / Number of Submissions Received X 100): 12.3%**
- **Average number of reviews per paper: 2**
- **Total number of reviewers involved: 46**
- **Any additional info on review process: Papers out of the scopes of the conference had been rejected in an initial paper screening process. Papers passing the initial screening process had been further evaluated by a group of experts in the field to make sure they meet the necessary standards for acceptance and publication. Each paper underwent formal paper review by at least two reviewers in terms of appropriateness, clarity, originality/Innovativeness, and Overall Recommendation. We used professional conference paper management system "SOFTCONF" in coordinating the paper review process and communicating with the authors. A paper was accepted for presentation and publication only if the comments from both reviewers are positive. The reviewers' comments had been provided to the authors so that they could follow the suggestions from the reviewers to further improve their papers. The authors were required to highlight the revisions in red in the final manuscript so that the Paper Review Committee could check to ensure the reviewers' comments were accommodated in the revised manuscript.**
- **Contact person for queries: Jie (Peter) Liu, Professor, Carleton University; Email: jie.liu@carleton.ca**

A distributed localization method for mobile nodes

Chao Liu[1,2,3], Qinghua Luo[1,2,3,*], Xiaozhen Yan[1,2,3], Yang Shao[1,2,3], Kexin Yang[1,2,3], Chunyu Ju[1,2,3]

[1]School of Information Science and Engineering, Harbin Institute of Technology, Weihai, China
[2]Shandong Institute of Shipbuilding Technology, Weihai, China
[3]Shandong New Beiyang Information Technology Co., Ltd. Weihai, China

*Corresponding author: luoqinghua081519@163.com

Abstract. In Wireless Sensor Networks(WSNs), the location services are the basis of many application scenarios. However, for the range-based localization method, the localization accuracy and the system robustness of the distributed localization system are difficult to guarantee, due to the uncertainty of the distance estimation and position calculation are affected by the node state and communication uncertainty. In this paper, we propose the distributed localization method based on anchor node selection and Particle Filter optimization. In this method, we analyze the uncertainty of error propagation in the Least-squares method and find that there is a proportional relation between localization error and uncertainty propagation. According to this relationship, we propose the corresponding optimization criterion methods of anchor nodes. To optimize the initial localization results, we present the distributed localization method based on anchor node optimal selection and Particle Filter. Simulation results show that the methods we proposed could effectively improve the localization accuracy of the mobile nodes and the robustness of the system.

Keywords: Distributed localization, anchor node optimization, particle filter, Wireless Sensor Networks.

1. Introduction

Wireless Sensor Networks(WSNs) are extensively used in smart cities, disaster relief, and other fields. Presently, the most universal methods of providing location services are the Global Positioning System (GPS) and the Beidou system. Nevertheless, the localization accuracy is greatly reduced in buildings, indoors, or in canyons[1], which makes reliable localization information cannot be obtained.

For the range-based localization method, the accuracy of localization depends on the distance estimation of anchor nodes[2]. Due to the complex transmission environment such as environmental interference, reflection, refraction, multi-path, and NLOS transmission[3] in the process of wireless signal transmission, the distance estimation results have a different extent of the error. At the same time, the distributed localization method[4] depend on the information exchange and coordination between the nodes. Wireless Sensor Networks have inevitable defects: the anchor nodes may have problems such as communication uncertainty[5], limited processing capacity, insufficient power supply[6], etc, which cause the input error to enlarge. Therefore, it is necessary to select reliable anchor nodes for distance estimation and localization calculation in distributed mobile localization.

The rest of this paper is structured as follows: we review related works in Chapter 2; Chapter 3 illustrates the uncertainty propagation, and propose the MSDO-PF and MEPO-PF; In Chapter 4, we evaluate the performance of the proposed methods; Finally, we give the conclusion of our work.

2. Related works

The localization accuracy of WSN is affected by factors such as communication uncertainty, nodes' processing capabilities, etc. Scholars have considered different methods to improve the location results.

Some scholars have made improvements in localization calculation methods.[5] proposed an adaptive fading factor to compensate for the inconsistency and error of the estimation according to the communication uncertainty caused by the failures of sensor nodes and mobile targets; The application of Unscented Kalman Filter (UKF), and Particle Filter (PF)[7] in WSN are also studied. Some scholars put forward methods with considering the uncertainty of anchor nodes' position. Considering both Range-Based and Range-Free node localization methods with errors in anchor nodes, The algorithm proposed by [8] calculates the similarity between nodes according to the location information and hops of anchor nodes, and uses the K most similar anchor nodes to calculate the coordinates.

In the above methods, the improvement on localization accuracy by the improved location estimation methods is limited, and the analysis after error propagation is lacking. In this paper, we analyze the error propagation mechanism of the distributed mobile nodes, and propose the minimum standard deviation optimization(MSDO) criterion and the minimum error propagation optimization(MEPO) criterion. With using the Particle Filter algorithm, we propose the distributed location method based on anchor node optimal selection and particle filter(MSDO-PF and MEPO-PF).

3. Proposed algorithm

3.1. System structure

The framework of the distributed localization method based on anchor node selection and particle filter optimization is illustrated in Fig.1, which is composed of four sub-modules.

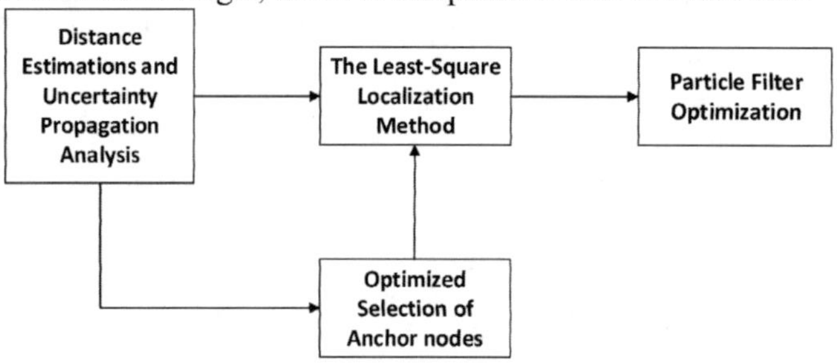

Figure 1. The framework of MSDO-PF and MEPO-PF localization method.

3.2. Uncertainty propagation analysis and Optimal selection of anchor nodes

We assume k known anchor nodes with corresponding coordinates: $(x_1, y_1), \cdots (x_i, y_i), \cdots (x_k, y_k)$, And the coordinate of the unknown node moving non-linearly is (x, y), the corresponding distance estimated by anchor nodes are $d = \{d_1, d_2, \cdots d_i, \cdots, d_k\}$.

The localization equations can be formed as followed:

$$\begin{cases} \sqrt{(x-x_1)^2+(y-y_1)^2} = d_1 \\ \sqrt{(x-x_2)^2+(y-y_2)^2} = d_2 \\ \vdots \\ \sqrt{(x-x_k)^2+(y-y_k)^2} = d_k \end{cases} \tag{1}$$

In the form of matrix equality: $AX = B$, and based on Least Square criterion, we can get the solution for the location equations:

$$X = (A^T A)^{-1} A^T B \tag{2}$$

When calculating the coordinates of moving nodes, one of the variables with uncertainty is the anchor node coordinates (x_i, y_i) $(i = 1, 2, \cdots, k)$, whose size is the sum of the real value and a neighbourhood, and the neighbourhood is decomposed into $(\delta_{x_i}, \delta_{y_i})$; The other uncertainty is the distance estimation d_i of the corresponding anchor node to the moving node, and the error is δ_{d_i}. The sensitivity coefficients of anchor node coordinate and distance estimation are defined as follows:

$$S_{x_i} = \frac{\partial((A^T A)^{-1} A^T B)}{\partial x_i} \quad S_{y_i} = \frac{\partial((A^T A)^{-1} A^T B)}{\partial y_i} \quad S_{d_i} = \frac{\partial((A^T A)^{-1} A^T B)}{\partial d_i} \quad (i = 1, 2, \cdots, k) \tag{3}$$

According to the total differentiation formula (TDF), the positioning error can be obtained:

$$\begin{pmatrix} \delta x \\ \delta y \end{pmatrix} = S_{x_1}\delta_{x_1} + S_{y_1}\delta_{y_1} + S_{d_1}\delta_{d_1} + \cdots + S_{x_i}\delta_{x_i} + S_{y_i}\delta_{y_i} + S_{d_i}\delta_{d_i} + \cdots + S_{x_k}\delta_{x_k} + S_{y_k}\delta_{y_k} + S_{d_k}\delta_{d_k} \tag{4}$$

We can get the standard deviation of localization result through (3) according to the square root rule:

$$\begin{aligned} \begin{pmatrix} \sigma_x \\ \sigma_y \end{pmatrix} &= \sqrt{\left(\frac{\partial((A^T A)^{-1} A^T B)}{\partial d_1}\right)^2 \sigma_{d_1}^2 + \cdots + \left(\frac{\partial((A^T A)^{-1} A^T B)}{\partial d_i}\right)^2 \sigma_{d_i}^2 + \cdots \left(\frac{\partial((A^T A)^{-1} A^T B)}{\partial d_k}\right)^2 \sigma_{d_k}^2} \\ &= \sqrt{\left(2(A^T A)^{-1} A^T \begin{pmatrix} 1 \\ 0 \\ \vdots \\ 0 \\ 0 \end{pmatrix} d_1 * \sigma_{d_1}\right)^2 + \cdots + \left(2(A^T A)^{-1} A^T \begin{pmatrix} 0 \\ \vdots \\ 1 \\ \vdots \\ 0 \end{pmatrix} d_i * \sigma_{d_i}\right)^2 + \cdots + \left(2(A^T A)^{-1} A^T \begin{pmatrix} 0 \\ 0 \\ \vdots \\ 0 \\ 1 \end{pmatrix} d_k * \sigma_{d_k}\right)^2} \end{aligned} \tag{5}$$

From (5), the error of the positioning result is proportional to the standard deviation of the distance estimation, the product of the estimated distance and the standard deviation of the corresponding estimation. Therefor, the smaller the range standard deviation and error propagation factor, the smaller the localization error. According to this relationship, we propose the minimum standard deviation optimization criteria(MSDO) and minimum error propagation optimization criteria(MEPO).

3.3. Improvement of the localization results with particle filter algorithm

In this section, the optimized coordinate is obtained by Particle Filter algorithm to track the motion state of the moving node. The MSDO-PF and MEPO-PF have been proposed, and the following is a detailed illustration.

Suppose that the state equation and observation equation are as follows:

$$\begin{cases} X_k = f_k(X_{k-1}, \delta_k) \\ Y_k = h_k(X_k, \gamma_k) \end{cases} \tag{6}$$

Where k denotes the motion time of mobile nodes, the random variable X_k denotes the predicted value of target location, and Y_k denotes the observed value of the target position. In this method, we suppose the $Y_k = d\_s = \{d_1\_s, d_2\_s, \cdots d_i\_s, \cdots, d_k\_s\}$ is the preliminary result of positioning after the optimization of anchor nodes, f_k and h_k are Non-linear function. δ_k denotes the system noise and γ_k denotes the observation noise, they are independent.

Construct a set $\{X_k^{(i)}, W_k^{(i)}\}_{i=1}^{N}$ containing N particles, where $X_k^{(i)}$ represents the state of the ith particle at the moment, $W_k^{(i)}$ represents the weight of this particle, and the weights satisfied that

$\sum_{i=1}^{N} W_k^{(i)} = 1$ · $f(X_k)$ denotes the real coordinates of the target at time k, and $p(X_k | Y_{1:k})$ is the posterior probability density of X_k at this time. Then the final positioning result is expressed as:

$$E[f(X_k)] = \int f(X_k)\, p(X_k | Y_{1:k})\, dX_k \tag{7}$$

In practical application, for avoiding directly extracting samples of the posterior probability distribution $p(X_k | Y_k)$, the importance sampling is introduced to improve sampling efficiency. the importance sampling method extracts samples of the known importance sampling density $q(X_k | Y_k)$. Then equation (7) can be expressed as:

$$E[f(X_k)] = \int f(X_k) \frac{p(X_k | Y_{1:k})}{q(X_k | Y_{1:k})} q(X_k | Y_{1:k})\, dX_k \tag{8}$$
$$= \int f(X_k)\, w_k(X_k)\, q(X_k | Y_{1:k})\, dX_k$$

$$w(X_k) = \frac{p(X_k | Y_{1:k})}{q(X_k | Y_{1:k})} \tag{9}$$

According to the Sequential Importance Sampling (SIS), and corresponding steps, equation (7) can be expressed as:

$$E[f(X_k)] = \sum_{i=1}^{N} W_k^{(i)} f(X_k) \tag{10}$$

Where the particle weight is:

$$W_k^{(i)} = \frac{w(X_k^{(i)})}{\sum_{i=1}^{N} w(X_k^{(i)})} \tag{11}$$

According to the theory of particle filter, resampling is added to reduce the degradation of the particle filter. With the combination of Particle Filter, the MSDO-PF and MEPO-PF can get better localization results.

4. Simulation And Analysis

We compare the localization results of MSDO-PF and MEPO-PF, the trajectory tracking results are shown in Fig.2 and Fig.3 respectively.

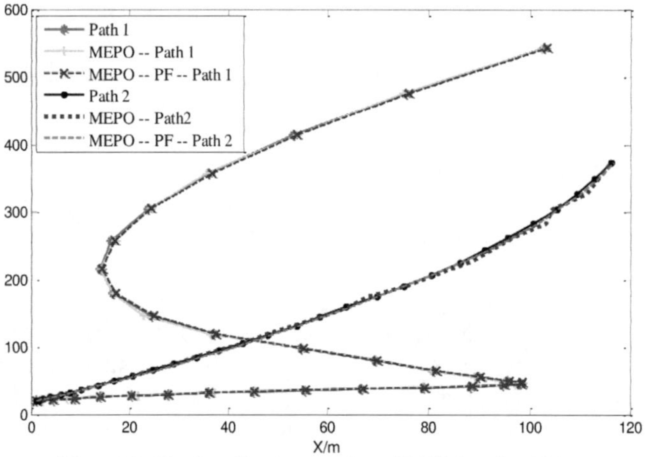

Figure 2. The localization results of MEPO using PF

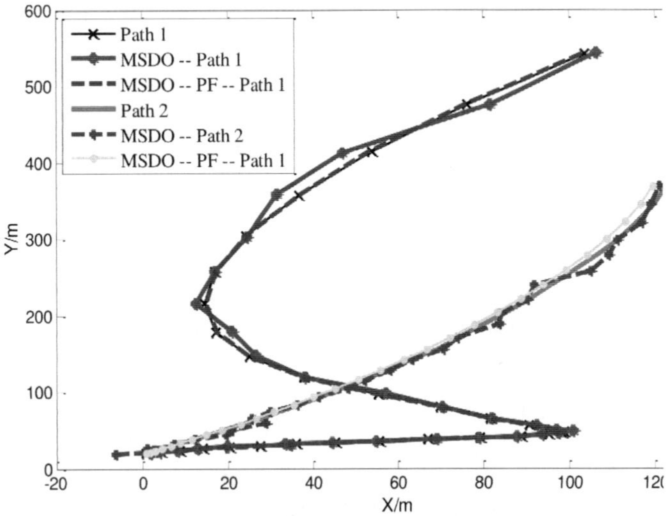

Figure 3. The localization results of MSDO using PF

We independently conducted five times localization simulations, and statistically calculated the localization effects of Path 1 and Path 2 after using MSDO method and Particle Filter(MSDO-PF) or MEPO method and Particle Filter(MEPO-PF), the experiment dates are as shown in Table 1.

Table 1. The localization effects of using three anchor nodes selection methods and Particle Filter

Methods	Average positioning error e_{MSE} (m)	Error variance e_{VAR} (m^2)
Path1 + RS-PF	0.33	0.12
Path2 + RS-PF	1.34	0.51
Path1 +MSDO-PF	0.35	0.13
Path2 +MSDO-PF	1.26	0.47
Path1 +MEPO-PF	0.28	0.05
Path2 +MEPO-PF	0.05	0.48

According to the statistical calculation, when the MSDO-PF is used to optimize the anchor node localization, compared with the RS-PF, the localization accuracy and the error variance of path 1 are equivalent, the localization accuracy of path 2 is improved by 5.9%, and the error variance is reduced by 7.8%. When MEPO-PF is used to optimize localization, compared with RS-PF, the localization accuracy of path 1 is improved by 14.8%, the error variance is reduced by 56.3%; the localization accuracy of path 2 is the same, but the error variance is reduced by 5.5%.

5. Conclusions

Considering the negative factors during distance estimation of anchor nodes, we comprehensively analyze the error propagation mechanism of the distributed mobile nodes through WSN, and propose the minimum standard deviation optimization(MSDO) criterion and the minimum error propagation optimization(MEPO) criterion. According to the utilization of the Particle Filter to optimize the localization results, we also present the distributed location method based on anchor node optimal selection and particle filtering(MSDO-PF and MEPO-PF). We experiment to evaluate the accuracy and robustness of our proposed method, and the experiment results illustrate that the proposed algorithm is accurate and stable.

Acknowledgments

Major Scientific and technological innovation project of Shandong Province of China(2020CXGC010705), China Postdoctoral Science Foundation funded project(2020M672123),

Post-doc Creative Funding in Shandong Province(244312), Post-doc Funding in Weihai, science and technology development plan project of Weihai.

References

[1] A. Maddumabandara, H. Leung and M. Liu, "Experimental Evaluation of Indoor Localization Using Wireless Sensor Networks," in IEEE Sensors Journal, vol. 15, no. 9, pp. 5228-5237, Sept. 2015.

[2] X. Yan, Q. Luo, Y. Yang, S. Liu, H. Li and C. Hu, "ITL-MEPOSA: Improved Trilateration Localization With Minimum Uncertainty Propagation and Optimized Selection of Anchor Nodes for Wireless Sensor Networks," in IEEE Access, vol. 7, pp. 53136-53146, 2019.

[3] Vempaty A, Ozdemir O, Agrawak K, Chen H. Localization in wireless sensor networks: byzantines and mitigation techniques. IEEE Trans. on Signal Processing, 2013,61(6):1495–1508.

[4] P. Zhu and W. Ren, "Fully Distributed Joint Localization and Target Tracking With Mobile Robot Networks," in IEEE Transactions on Control Systems Technology,2020.2991126.

[5] H. Zhu and M. Luo, "Hybrid Robust Sequential Fusion Estimation for WSN-Assisted Moving-Target Localization With Sensor-Node-Position Uncertainty," in IEEE Transactions on Instrumentation and Measurement, vol. 69, no. 9, pp. 6499-6508, Sept. 2020.

[6] M. I. M. Ismail, R. A. Dziyauddin, R. Ahmad, N. Ahmad, N. A. Ahmad and A. M. A. Hamid, "A Review of Energy Harvesting in Localization for Wireless Sensor Node Tracking," in IEEE Access, vol. 9, pp. 60108-60122, 2021.

[7] Pak, Jung Min et al. Improving Reliability of Particle Filter-Based Localization in Wireless Sensor Networks via Hybrid Particle/FIR Filtering[J]. IEEE transactions on industrial informatics, 2015, 11(5) : 1089-1098.

[8] H. WANG, K. LYU, H. JIANG, Y. WU, Q. YUE and Q. ZHAO, "Location Algorithm for WSNs with Mobile Anchor Node Based on Optimzed K-Nearest Neighbers," 2019 Eleventh International Conference on Advanced Computational Intelligence (ICACI), 2019, pp. 190-195.

A multi-AUV cooperative navigation method

Yang Shao[1,2,3], Qinghua Luo[1,2,3*], Chao Liu[1,2,3], Xiaozhen Yan[1,2,3], Kexin Yang[1,2,3]

[1]School of Information Science and Engineering, Harbin Institute of Technology, Weihai, China
[2]Shandong Institute of Shipbuilding Technology Weihai, China;
[3]Shandong New Beiyang Information Technology Co., Ltd. Weihai, China

*Corresponding author: luoqinghua081519@163.com

Abstract. Cooperative navigation is one of the key methods for multiple autonomous underwater vehicles (AUVs) to obtain accurate positions when performing tasks underwater. In the realistic state-space model of the multi-AUV cooperative navigation system, where the system noise does not satisfy the additivity, it is necessary to augment the dimension of the state variables before nonlinear filtering. Aiming at the problem that the error of traditional algorithms increases linearly with the dimension of state-space, a cooperative navigation method based on Augmented Embedded Cubature Kalman filter (AECKF) algorithm is proposed. The experiment results show that the AECKF cooperative navigation algorithm has better positioning accuracy and stability than the traditional algorithm.

Keywords: Autonomous Underwater Vehicle (AUV), cooperative navigation, state augmentation, Augmented Embedded Cubature Kalman filter (AECKF)

1. Introduction

In recent years, multi-AUV cooperative navigation has developed as a new type of underwater navigation technology, which has received extensive attention from countries all over the world[1]. To estimate and correct the cooperative navigation parameters of AUV, many different filtering algorithms are used based on different state-space models[2]. Using higher precision sensors or a more realistic motion model can improve navigation accuracy. However, the former will increase costs and the latter may cause the system noise not to meet the additivity.

Based on the above analysis, this paper proposes a multi-AUV cooperative navigation method based on the AECKF algorithm. Firstly, we add the system noise to the state variable to augment its dimension to solve the problem that the system noise cannot be added. Then, a multi-AUV cooperative navigation method based on the AECKF algorithm is proposed, which overcomes the shortcoming that the positioning error of traditional algorithms increases linearly with the dimension of state-space. The simulation results show that the positioning accuracy and stability of the proposed multi-AUV cooperative navigation method are better than the traditional algorithm.

2. Related Works

Among the Multi-AUV cooperative navigation methods, the algorithm based on Bayesian estimation is the widely used. Reference[3] uses the Cross Entropy algorithm combined with the Extended

Kalman filter (EKF) to effectively reduce the observation error of the following AUV. Reference[4] shows that Unscented Kalman filter (UKF) has a better approximation effect than EKF for nonlinear systems, but the sigma points used for approximation have negative values. For this reason, the Cubature Kalman filter (CKF) algorithm is proposed. However, the inherent defects make its filtering error increase linearly with the augmentation of state dimension[5]. In reference[6], the Embedded Cubature Kalman filter (ECKF) algorithm is proposed to achieve higher accuracy of filtering estimation and has better high-order expansibility.

Comprehensively considering the problems that the noise of the system does not satisfy the additivity and the increase of positioning error of the traditional algorithms after the augmentation of the state dimension during the multi-AUV cooperative navigation process, this paper proposes a multi-AUV cooperative navigation method based on the AECKF algorithm. In terms of verification methods, in addition to the simulation to verify the navigation and positioning accuracy of the algorithm.

3. Multi-AUV cooperative navigation state-space model

In actual underwater cooperative positioning, the research on the AUV positioning problem can be directly mapped to the two-dimensional space. The state equation of single AUV is as follow:

$$\begin{cases} x_{k+1} = x_k + \Delta t \bullet V_k \cos\varphi_k \\ y_{k+1} = y_k + \Delta t \bullet V_k \sin\varphi_k \\ \varphi_{k+1} = \varphi_k + \Delta t \bullet \omega_k \end{cases} \tag{1}$$

Among them, x_k and y_k are the AUV's dead reckoning positioning information; Δt is the dead reckoning cycle; V_k is the forward speed; φ_k represents the heading angle, and ω_k represents the angular velocity, where V_k and φ_k are interfered by zero-mean and independent Gaussian white noise.

The relative distance measurement information between the pilot and following AUVs is Z_k. Then the coordinate position relationship can be obtained:

$$Z_k = \sqrt{(x_k^M - x_k^S)^2 + (y_k^M - y_k^S)^2} + v_k \tag{2}$$

Where x_k^M, y_k^M and x_k^S, y_k^S are the east and north positions of the pilot AUV and the following AUV at t_k, respectively, and v_k is the Gaussian white noise.

Define $X_k = [x_k, y_k, \varphi_k]^T$ as the state variable of the following AUV, $w_k = [w_k^V, w_k^\phi]^T$ as the process noise vector, and $v_k = [v_k]^T$ as the measured noise vector. Then we can get the following equation:

$$Q_k = E\left[w_k w_k^T\right], R_k = E\left[v_k v_k^T\right] \tag{3}$$

Where Q_k and R_k are the assumed known system process noise covariance matrix and the measured noise covariance matrix, respectively.

4. State augmentation and Embedded Cubature Kalman filter

The framework of the multi-AUV cooperative navigation method based on the AECKF algorithm is shown in Figure 1.

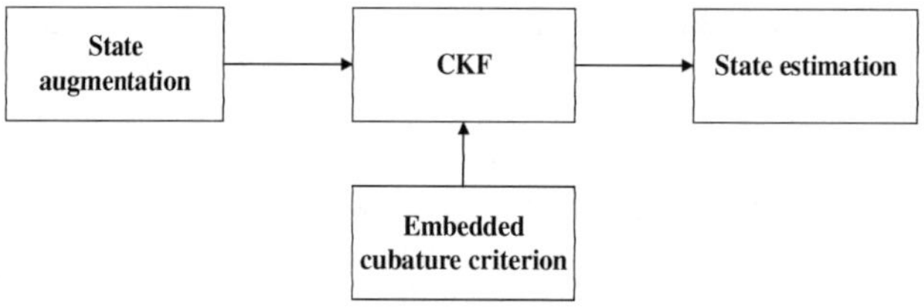

Figure 1. The framework of the AECKF algorithm.

4.1. Embedded Cubature Kalman filter(ECKF)
For n - dimensional state variables, use $N = 2^n + 1$ cubature points to achieve the numerical approximation of the integral:

$$I_N(f) = \sum_{i=1}^{N} \overline{\omega}_i^E f(\xi_i) \tag{4}$$

In equation (4), the corresponding cubature points and weight are as follow:

$$\xi_i^E = \begin{cases} [0]_i, i = 1 \\ \sqrt{2}[\delta]_i, i = 2, \cdots N \end{cases}, \overline{\omega}_i^E = \begin{cases} 1 - \dfrac{1}{2\sigma^2}, i = 1 \\ \dfrac{1}{2^{n+1}\sigma^2}, i = 2, \cdots, N \end{cases} \tag{5}$$

Where $[\delta] = [s_1\sigma^2, s_1\sigma^2, \cdots, s_1\sigma^2]^T$ is the cubature point set, and $[\delta]_i$ is the i-th point in the point set.

4.2. Algorithm steps of multi-AUV cooperative navigation based on AECKF
Consider the nonlinear system established in the previous section:

$$\begin{cases} X_{k+1} = f(X_k, u_k, w_k) \\ Z_{k+1} = h(X_k) + v_k \end{cases} \tag{6}$$

Where $X_k \in R^n$ is the n - dimensional system state variable, w_k is the n_w - dimensional system process noise, and v_k is the n_v - dimensional system observation noise.

Step1 State prediction

Replacing the cubature points set and weights of CKF with the equation (5), the Cholesky decomposition of the error covariance is performed as follow:

$$P_{k|k} = S_{k|k} S_{k|k}^T \tag{7}$$

augment the dimension of state variables and $S_{k|k}^E$:

$$\hat{X}_{k|k}^a = [\hat{X}_{k|k}^T \quad 0_{1\times n_w} \quad 0_{1\times n_v}]^T, S_{k|k}^a = \begin{pmatrix} S_{k|k} & 0 & 0 \\ 0 & \sqrt{Q_k} & 0 \\ 0 & 0 & \sqrt{R_k} \end{pmatrix} \tag{8}$$

Calculate the cubature points ($i = 1,2,\cdots,N, N = 2(n + n_w + n_v)$):

$$X_{i,k|k} = S_{k|k}^a \xi_i^E + \hat{X}_{k|k}^a \tag{9}$$

Propagate cubature points through system function:

$$X_{i,k+1|k}^* = f(X_{i,k|k}^x, X_{i,k|k}^w) \tag{10}$$

Calculate the state prediction value at time t_{k+1}:

$$\hat{X}_{k+1|k} = \frac{1}{N}\sum_{i=1}^{N} X_{i,k+1|k}^* \tag{11}$$

Calculate the prediction value of state error covariance at time t_{k+1}:

$$P_{k+1|k} = \frac{1}{N}\sum_{i=1}^{N} X_{i,k+1|k}^* X_{i,k+1|k}^{*T} - \hat{X}_{k+1|k}\hat{X}_{k+1|k}^T \tag{12}$$

Step2 Measurement prediction

Propagate the cubature points through the measurement function:

$$Z_{i,k+1|k} = f(X_{i,k+1|k}^{*x}, X_{i,k+1|k}^{*v}) \tag{13}$$

Calculate the prediction value of measurement at time t_{k+1}:

$$\hat{Z}_{k+1|k} = \frac{1}{N}\sum_{i=1}^{N} Z_{i,k+1|k} \tag{14}$$

Calculate the autocorrelation and the cross-correlation covariance matrix:

$$P_{ZZ,k+1|k} = \frac{1}{N}\sum_{i=1}^{N} Z_{i,k+1|k} Z_{i,k+1|k}^{T} - \hat{Z}_{k+1|k}\hat{Z}_{k+1|k}^{T}, P_{XZ,k+1|k} = \sum_{i=1}^{N} \overline{\omega}_i^{E} X_{i,k+1|k} Z_{i,k+1|k}^{T} - \hat{X}_{k+1|k}\hat{Z}_{k+1|k}^{T} \quad (15)$$

Step3 State update

Calculate the filter gain at time t_{k+1}:

$$K_{k+1} = P_{XZ,k+1|k} P_{ZZ,k+1|k}^{-1} \quad (16)$$

Calculate the state estimate value at time t_{k+1}:

$$\hat{X}_{k+1|k+1} = \hat{X}_{k+1|k} + K_{k+1}(Z_{k+1} - \hat{Z}_{k+1|k}) \quad (17)$$

Calculate the estimate value of the state error covariance at time t_{k+1}:

$$P_{k+1|k+1} = P_{k+1|k+1} - K_{k+1} P_{ZZ,k+1|k} K_{k+1}^{T} \quad (18)$$

5. Simulation comparison and performance evaluation

5.1. Simulation environment construction

On a two-dimensional plane, a multi-AUV cooperative navigation system based on a single pilot boat is established. The speed of AUVs are maintained at $2m/s$. The system status update cycle is $\Delta t = 10s$, and the carrier's underwater navigation time is $T = 2500s$. Set the variance of the process noises to $\sigma_v^2 = (0.1m/s)^2$ and $\sigma_\omega^2 = (1^\circ/s)^2$ and the variance of the measured noise to $\sigma_r^2 = (1m)^2$. The initial value of the state estimate after taking the augmented dimension is as follow:

$$\hat{X}_{0|0}^{a} = \begin{bmatrix} 500 & 500 & \pi/4 & 0 & 0 & 0 \end{bmatrix}^T, P_{0|0} = diag\begin{bmatrix} 1 & 1 & 0.01 & 0 & 0 \end{bmatrix} \quad (19)$$

5.2. Simulation results and analysis

Figure 2 is the comparison of the positioning accuracy based on AECKF and other cooperative navigation algorithms. As shown, the AECKF algorithm can effectively reduce the errors of the other algorithms during the co-location process in the case of the high dimensionality of the system.

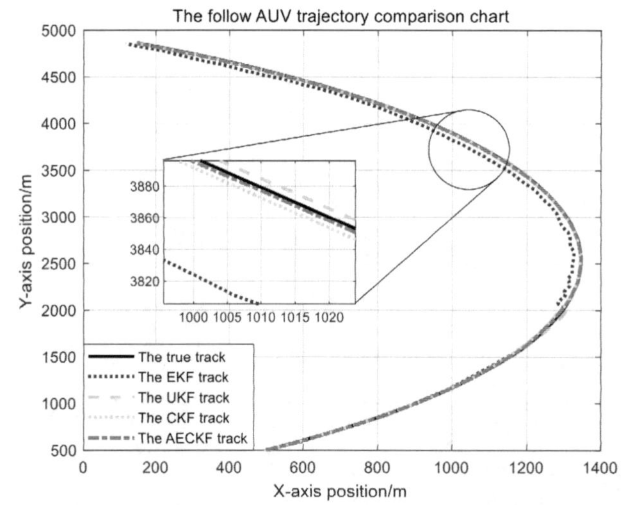

Figure 2. Different filter algorithms track and real track.

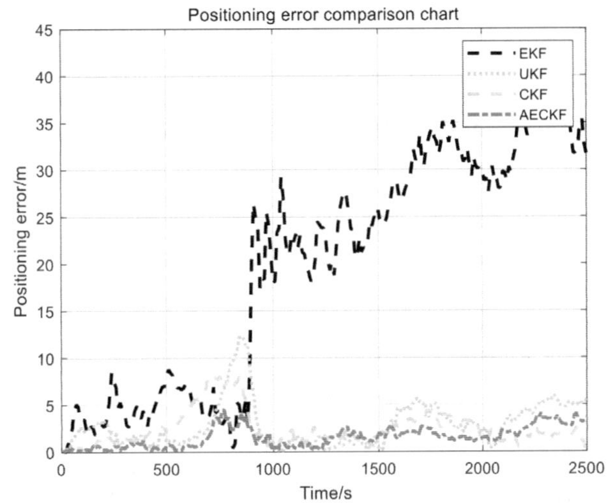

Figure 3. Positioning errors of different algorithms.

Figure 3 shows the comparison of the positioning error curves obtained by the four algorithms. It can be seen that under the simulation conditions, the maximum positioning error of the AECKF filter algorithm is less than 5m and the fluctuation range of the error curve is about 4m, indicating that it has better accuracy and robustness. The maximum error distance and error curve fluctuation degree of CKF, UKF, and EKF algorithms increase sequentially, which is difficult to maintain higher accuracy and stability compared with the AECKF algorithm.

Table 1. Performance comparison of different algorithms for cooperative navigation.

Algorithm	RMSE			
	x-axis position /m	y-axis position/m	Heading angle/rad	Distance error/m
EKF	22.494	7.688	2.343	23.772
UKF	2.531	2.771	0.747	3.753
CKF	2.009	2.250	0.612	3.016
AECKF	1.106	1.538	0.307	1.894

Table 1 lists the co-location performance indicators of four different algorithms. From the statistical data in Table 1, it can be seen that the AECKF algorithm proposed in this paper has higher positioning accuracy than other cooperative navigation methods. Compared with EKF, UKF, and CKF, the multi-AUV cooperative navigation algorithm based on AECKF improves the positioning accuracy by 92.03%, 49.53%, and 37.20%, respectively. This is mainly due to the comprehensive consideration of the increased dimension of the system after the augmentation of the state. For the adverse effects of filtering accuracy, the embedded cubature criterion is used to overcome this shortcoming, thereby obtaining higher positioning accuracy.

6. Conclusions

This paper proposes an Augmented ECKF algorithm, which adds process noise and measurement noise to state variables and uses the ECKF algorithm to suppress the positioning estimation error that increases with the increase of the state-space dimension. The simulation results show that this algorithm can effectively overcome the shortcomings of reduced or even divergent filtering accuracy of the CKF and UKF algorithms when the system state dimension is relatively high. The accuracy and stability of the AECKF algorithm is significantly better than many other traditional nonlinear filtering algorithms. But while improving the accuracy and stability of the navigation, it also increases the running time of the algorithm to a certain extent.

Acknowledgments

This work is partly supported by Major Scientific and technological innovation project of Shandong Province of China (2020CXGC010705), China Postdoctoral Science Foundation funded project (2020M672123), Post-doc Creative Funding in Shandong Province (244312), and Post-doc Funding in Weihai, science and technology development plan project of Weihai.

References

[1] Xu B, Li S, Razzaqi A A, et al. Cooperative Localization in Harsh Underwater Environment based on the MC-ANFIS[J]. IEEE Access, 2019, PP(99):1-1.
[2] Xu Bo, Li Shengxin, Jin Kunming, Wang Lianzhao. Multi-autonomous underwater vehicle cooperative positioning method based on radial basis function neural network-assisted Cubature Kalman filter[J]. Acta Armamentarius, 2019, 40(10): 2119 -2128.
[3] Zhang L, Li Y, Liu L, et al. Cooperative Navigation Based on Cross Entropy: Dual Leaders[J]. IEEE Access, 2019, 7:1-1.
[4] Allotta B, Chisci L, Costanzi R, et al. A comparison between EKF-based and UKF-based navigation algorithms for AUVs localization[C]// Oceans. IEEE, 2015:1-5.
[5] Jiang L, Cai B, Tao T, et al. A CKF based GNSS/INS train integrated positioning method[C]// International Conference on Mechatronics & Automation. IEEE, 2010.
[6] Liu Hua, Miao Chen. Square root embedded cubature Kalman particle filter algorithm[J]. Journal of Nanjing University of Science and Technology: Natural Science Edition, 2015(4):471-476.

Fault diagnosis of rolling bearing based on multi-scale one-dimensional convolutional neural network

Xukun Hou[1], Pengjie Hu[1], Wenliao Du[1,*], Xiaoyun Gong[1], Hongchao Wang[1], Fannian Meng[1]

[1]School of Mechanical and Electronic Engineering, Zhengzhou University of Light Industry, Zhengzhou, China

*Corresponding author: dwenliao@163.com

Abstract. Aiming at the typical non-stationary and nonlinear characteristics of rolling bearing vibration signals, a multi-scale convolutional neural network method for bearing fault diagnosis based on wavelet transform and one-dimensional convolutional neural network is proposed. First, the signal is decomposed into multi scale components with wavelet transform, and then each scale component is reconstructed. The reconstructed signal is subjected to the Fourier transform to obtain the frequency spectrum representation, which is used as the input of the one-dimensional convolutional neural network. Finally, one-dimensional convolution neural network is used to learn the features of the input data and recognize the bearing fault. The performance of the model is verified by using data sets of rolling bearing. The results show that this method can intelligent feature extraction and obtain 99.94% diagnostic accuracy.

1. Introduction

Rolling bearing is one of the most vulnerable parts of rotating machinery, and about 1/3 of the mechanical faults in rotating machinery are caused by rolling bearings [1]. In order to avoid the economic losses and security risks caused by bearing faults, it is particularly important to monitor, analyze and diagnose the rolling bearing faults in time.

At present, the process of bearing fault diagnosis mainly includes three aspects: data preprocessing, feature extraction and feature classification [2]. Traditional fault diagnosis mainly uses time domain, frequency domain and time-frequency domain analysis to obtain features. Multi-dimensional fault features are extracted by time-frequency domain analysis methods such as local feature scale decomposition, wavelet transform (WT), empirical mode decomposition (EMD) and so on, which are used as the basis of pattern recognition [3]. Among them, wavelet transform (WT) has solid mathematic foundation, and can decompose the vibration signals into multi-scale. Wang et al. [4] used discrete wavelet transform and sparse representation to obtain new sparse wavelet energy features to identify rolling bearing faults. Shen et al. [5] developed a bearing intelligent fault diagnosis method based on wavelet transform and support vector regression. All of the above methods need to extract features manually, which depends on professional experience.

Recently, deep learning is widely used in pattern reorganization, especially in image processing, computer vision[6]. Deep learning model can directly extract abstract features from the original signal without manual feature extraction. With its powerful nonlinear modeling and feature representation ability, deep learning shows great potential in any aspects, many scholars have applied deep learning to mechanical equipment fault diagnosis[7][8].

Deep convolution neural network (CNN) is a typical deep learning model. Compared with other models, CNN has good generalization ability and robustness [9]. At present, CNN has been applied in bearing fault diagnosis methods, mainly including one-dimensional and two-dimensional convolutional neural network. In the fault diagnosis of bearing, because the input signal is one-dimensional, the classical two-dimensional CNN model needs to convert one-dimensional signal into two-dimensional image, so the one-dimensional convolution neural network model is used to diagnose the one-dimensional vibration signal of bearing. In order to avoid the influence of learning only one-dimension feature on the diagnosis accuracy, it is necessary to use wavelet transform to decompose the original signal into multi-scale components, reconstruct each scale component of the obtained signal, and construct a spectrum representation conducive to feature learning, so as to learn more useful fault feature information.

To sum up, aiming at the non-stationary and non-linear characteristics of rolling bearing vibration signal, wavelet transform is used for multi-scale analysis to accurately extract the different fault state feature information contained in time-frequency domain, and then it is brought into the one-dimensional convolution neural network model for feature learning and classification to realize intelligent fault diagnosis.

2. Theoretical Basis

2.1 Wavelet Transform

In the squared integrable real number space $L^2(R)$, if the function $\psi(t)$ satisfies the admissible condition:

$$C_\psi = \int_{-\infty}^{\infty} \frac{|\psi(t)|^2}{t} dt < \infty \tag{1}$$

Then $\psi(t)$can be used as the basic wavelet.$\psi(t)$ generates a family of functions $\{\psi_{a,b}(t)\}$ through scaling and translation, which is called wavelet function:

$$\psi_{a,b}(t) = \frac{1}{\sqrt{a}} \psi\left(\frac{t-b}{a}\right) \tag{2}$$

where, a is the scale factor and b is the displacement factor wavelet transform of signal $x(t)$ is defined as:

$$WT_x(a,b) = \frac{1}{\sqrt{a}} \int_{-\infty}^{\infty} x(t) \times \psi^* \left(\frac{t-b}{a}\right) dt \tag{3}$$

where, $\psi^*(t)$ is a conjugate of the wavelet function $\psi(t)$.

2.2 One-dimensional convolutional neural network

The network is composed of input layer, convolution layer, batch normalization (BN), pooling layer, full connection layer, dropout layer and output layer. The training process can be divided into two parts. The forward propagation is a neural network to calculate the predicted value according to the samples; Back propagation uses gradient descent method to update the weights.

1) Forward propagation calculation process

One-dimensional convolution process can be expressed:

$$x_j^l = F(\sum_N Conv1D(x_i^{l-1}, k_{ij}^l) + b^l) \tag{4}$$

where k is the convolution kernel, j is the number of convolution kernels, N is the convolution calculation area, i is the it data, x^{l-1} is the convolution input of the first l layer, x^l is the convolution output, $Conv1D(\ )$ is one-dimensional convolution calculation, b^l is the bias, $F(\ )$ is the activation function. The activation function is Relu function.

2) Error reverses propagation process

The chain derivation rule is used to update the parameters. Assuming that the error of convolution transfer is known to be ∇x^l , then the error ∇x^{l-1} of the upper layer and the error ∇w^l of the parameter to be updated are as follows:

$$\nabla x^{l-1} = \frac{\partial L}{\partial x^{l-1}} = \nabla x^l \bullet \frac{\partial x^l}{\partial x^{l-1}} \tag{5}$$

$$\nabla w^l = \sum_i \frac{\partial L}{\partial w^l} = \sum_i (\nabla x_i^l w_{i+p-1}^{l-1}) \tag{6}$$

where, L is the objective function, x^{l-1} is the original value of the upper layer, w is the parameter to be updated.

3. WT-1DCNN diagnosis model and process

The intelligent diagnosis of rolling bearing using the diagnosis model based on multi-scale one-dimensional convolutional neural network can be divided into the following 8 steps, as is shown in Figure1.

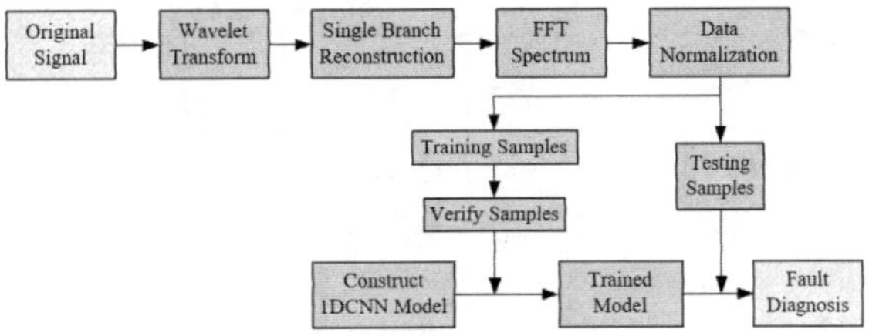

Figure1. Fault diagnosis process

1) The time domain signals of rolling bearing working state are collected by sensors;

2) The multi-scale representation of the signal is obtained by wavelet transform;

3) Each layer of wavelet signal is reconstructed by a single branch;

4) The amplitude spectrum of the reconstructed signal is obtained by Fast Fourier Transform;

5) Normalize the obtained amplitude data, and construct each scale data into a one-dimensional feature vector;

6) The feature samples are divided into training samples, verification samples and testing samples according to a certain proportion;

7) Construct a one-dimensional convolutional neural network model, use training samples to train the model, and verify with verification samples, and finally get a diagnosis model;
8) Input the testing sample into the model for verification, and get the final diagnosis accuracy.

4. Experimental verification and analysis

4.1 Data Description

The experimental data comes from CWRU [10]. The experimental platform consists of a motor, a coupling and a load motor, the bearing to be tested is the motor drive end bearing. The failure of the bearing is a single point damage simulated by EDM. The state of the bearing can be divided into normal (N), inner ring failure (IF),rolling element failure (BF) and outer ring failure (OF), according to the damage degree of the failure (fault diameter), each type of failure can be divided into mild (fault diameter 0.18mm), moderate (fault diameter 0.36mm) , Severe (fault diameter 0.53mm) three different degrees of failure. The motor runs under four loads of 0, 1, 2, and 3 hp, and the speed is 1800 rpm. The experiment contains two sampling frequency: 12 kHz and 48 kHz. This article uses the data collected under the sampling frequency of 12 kHz as the experimental data. In this experiment, the data set is divided into 10 health states; each health state has a total of 400 data samples, a total of 4000 data samples. 70% of the characteristic samples are randomly selected as training samples, the remaining 30% as testing samples, and 30% of the training samples as verification set.

4.2 Model building

According to the 10 state data samples of rolling bearings, db5 wavelet is used for 3-layers decomposition, then the wavelet signal of each layer is reconstructed into a single branch, and the reconstructed signal is subjected to fast Fourier transform to obtain a spectrum representation. The obtained spectrum data is normalized and constructed into a one-dimensional feature vector, which is used as the input of the 1DCNN model.

The neural network framework is keras. When constructing 1DCNN model, there are 11 layers of network, including input layer, 3 convolution layers, 3 BN layers, pooling layer, dropout layer, full connection layer and output layer, as shown in Table 1.

Table 1. One-dimensional convolutional neural network structure

Layer type	Output Shape	Param
Input	(None, 2048)	/
Conv1D_1	(None, 128,128)	8320
Batch Normalization_1	(None, 128,128)	512
Conv1D_2	(None, 128,64)	24640
Batch Normalization_2	(None, 128,64)	256
Conv1D_3	(None, 128,32)	6176
Batch Normalization_3	(None, 128,32)	128
Max Pooling1D	(None, 64,32)	/
Flatten	(None, 2048)	/
Dropout	(None, 2048)	/
Output	(None, 10)	/

4.3 Result and discussion

For the testing samples, in order to avoid the error caused by one experiment data, 10 experiments were carried out. The accuracy of the test set is 99.74% to 100%, and the average accuracy of 10 experiments is 99.94%, which indicates that the model selected in this experiment has good stability.

In order to show that wavelet transform can accurately extract fault features for non-stationary vibration signals, the original signal is directly input into 1DCNN, and the feature samples after wavelet transform are input into the same 1DCNN model for comparison. After 10 experiments, as shown in Figure 2 the average accuracy of 1DCNN is 98.02%, while the average accuracy of WT-1DCNN is 99.94%.

In order to verify the superiority of the method proposed in this paper, the relative data sets are compared with different methods. Du et al [11]. developed a method for diagnosing rolling bearing faults (Wavelet Leaders) based on the multifractal features of wavelet leadership, and got 88.9% accuracy. Li et al [12]. proposed a fault diagnosis method based on short-time Fourier transform and CNN (STFE-CNN), which realized end-to-end fault pattern recognition, and the accuracy rate was 97.93%. Min et al [13][13]. proposed a CNN-based multi-sensor fusion method (MCNN), which fused the signals of multiple sensors and input them into CNN to diagnose the 10 health states of the bearing, with a 99.41% correct rate. Ding et al [14]. proposed a new energy fluctuation multi-scale feature mining method (WPE-2DCNN) based on wavelet packet energy (WPE) images and two-dimensional deep convolutional networks, which was used for bearing fault diagnosis, and the accuracy rate reached 98.8. %. Comparing the five methods, the results are shown in Figure 3. It can be seen that the WT-1DCNN method used in this article has the highest fault diagnosis accuracy.

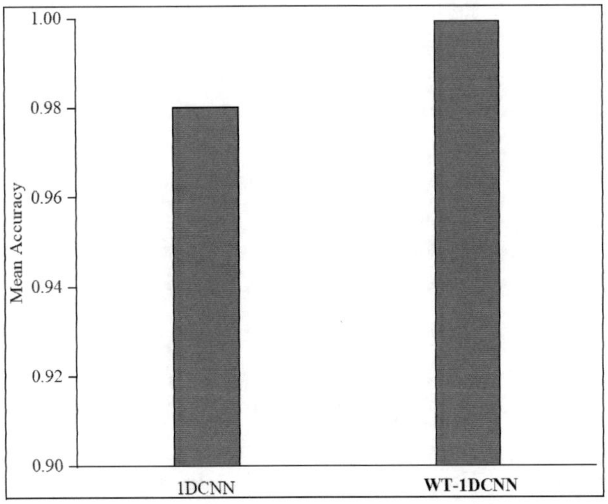

Figure2. 1DCNN and WT-1DCNN

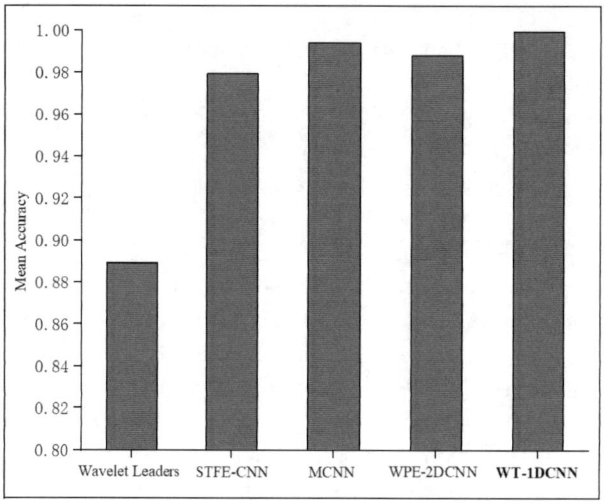

Figure3. Accuracy of different methods

5. Conclusion

In order to solve the problem that the artificial feature extraction in bearing fault diagnosis depends on experience and has weak generalization and adaptability, a multi-scale one dimensional convolution neural network is proposed for bearing fault diagnosis. Compared with other diagnosis methods, the method is as follows advantage.

Compared with traditional intelligent diagnosis methods, the use of wavelet transform to perform multi-scale analysis of signals can more effectively dig out the hidden feature distribution in fault data, and establish a complex nonlinear relationship between fault features and bearing faults.

One dimensional convolution neural network is used to train the input one-dimensional signal, which effectively integrates feature automatic learning and fault classification, avoids the dependence on complicated signal processing technology and rich diagnosis experience, simplifies the actual diagnosis process and achieves good diagnosis effect.

Acknowledgments

This work was supported by the National Nature Science Foundation of China (Grant No. U1804141), the Key Scientific Research Projects of Colleges and Universities in Henan Province (Grant No. 21A460033), the Foundation for University Key Teacher of Henan Educational Department（Grant No.2018GGJS091）, the Program for Innovative Research Team in University of Henan Province (Grant No. 20IRTSTHN015),and the key specialized and development projects in Henan province(No.202102310257).

Reference

[1] Y. Lei, F. Jia, et al. Opportunities and Challenges of Mechanical Intelligent Fault Diagnosis under Big Data, J. Journal of Mechanical Engineering, 2018, 54(05): 94-104.

[2] L.Wen, X. Li, Liang G, et al. A new convolutional neural network-based data-driven fault diagnosis method, J. IEEE Transactions on Industrial Electronics, 2018(7): 5990-5998.

[3] M. Zhao, M. Kang, B. Tang et al. Deep residual networks with dynamically weighted

wavelet coefficients for fault diagnosis of planetary gearboxes, J. IEEE Transactions on Industrial Electronics, vol. 65, no. 5, pp. 4290–4300, 2017.

[4] S. Yin, S. Ding, X. Xie, et al. A review on basic data-driven approaches for industrial process monitoring, J. IEEE Transactions on Industrial Electronics, vol. 61, no. 11, pp. 6418-6428, 2014.

[5] C. Wang, M. Gan, C. Zhu. Intelligent fault diagnosis of rolling element bearings using sparse wavelet energy based on over complete DWT and basis pursuit, J. Intell. Manuf. (2015)1-15.

[6] C. Shen, D. Wang, F. Kong, et al. Fault diagnosis of rotating machinery based on the statistical parameters of wavelet packet paving and a generic support vector regressive classifier, J. Measurement. 46 (2013) 1551-1564.

[7] B. P. Duong and J. M. Kim. "Non-mutually exclusive deep neural network classifier for combined modes of bearing fault diagnosis," Sensors, vol. 18, no. 4, pp. 1129-1143, 2018.

[8] W. Li，W. Shan. Bearing fault classification and identification based on depth belief network, J. Journal of Vibration Engineering，2016，29 (2) :340 -347.

[9] D. Peng, Z. Liu, H. Wang, et al. A novel deeper one-dimensional CNN with residual learning for fault diagnosis of wheelset bearings in high-speed trains, J. IEEE Access, vol. 7, pp. 10278-10293, 2018.

[10] Bearing Data Center accessed on Mar. 2017. [Online].

[11] W. Du, J. Tao, Y. Li, et al. Wavelet leader multifractal features based fault diagnosis of rotating mechanism, J. Mechanical Systems and Signal Processing, 2014, 43(1-2):57-75.

[12] H. Li, H. Zhang, X. Qin, et al. Bearing fault diagnosis method based on short-time Fourier transform and convolutional neural network, J. Vibration and Shock, 2018, v.37; No.327(19) :132-139.

[13] M. Xia, T. Li, L. Xu, et al. Fault Diagnosis for Rotating Machinery Using Multiple Sensors and Convolutional Neural Networks, J. IEEE/ASME Transactions on Mechatronics, 2017, PP (99):1-1.

[14] X. Ding, Q. He. Energy-Fluctuated Multiscale Feature Learning with Deep ConvNet for Intelligent Spindle Bearing Fault Diagnosis, J. IEEE Transactions on Instrumentation and Measurement, 2017, PP (8):1-10.

Modelling dimensional change of machined parts and optimizing replacement time and initial position of cutting tool

Renyan Jiang

School of Science and Technology, Changsha University, Changsha, China

Email: jiang@csust.edu.cn

Abstract. In a manufacturing process, a key quality characteristic such as a certain dimension of machined parts can have a systematic shift toward the upper or lower specification limit due to the cutting tool wear. That is, the quality characteristic is a stochastic process with trend. As a result, the process capability and quality loss vary with time and depend on the initial position of the cutting tool. In this setting, decision problems that need to be solved are optimization of replacement time and initial position of the cutting tool. This paper proposes a non-homogenous Wiener process to model the dimensional change of machined parts; and the above-mentioned decision problems are solved through combining this model with a cost model that involves Taguchi quality loss. A real-world example is included to illustrate the appropriateness of the proposed models. The results are useful for both reliability researchers and manufacturing engineers.

1. Introduction

In a manufacturing process, the wear of a cutting tool can lead to itself failure and other problems such as nonconforming geometry and poor surface finish [1-3]. Thus, the preventive replacement of cutting tools becomes necessary for improving manufacturing productivity and quality [4].

There are two kinds of approaches to model the wear process of the cutting tool. In the first kind of approaches, the underlying variable is wear amount, which can be measured in a direct way (e.g., optical or vision system) or in an indirect way (e.g., cutting force, vibration and acoustic emission) [5]. In this case, the wear limit is determined based on the standard [6]. In the second kind of approaches, the underlying variable is a quality characteristic such as a certain dimension of produced parts [7-8]. This characteristic is highly related to the tool wear and hence reflects the consequence of the tool wear [9]. In this case, the quality failure threshold can be taken as the upper specification limit (USL) or lower specification limit (LSL) of the quality characteristic.

The tool wear results in a systematic shift of the quality characteristic toward its USL or LSL; and this shift cannot be physically removed from the process unless the cutting tool is adjusted or replaced [7]. As a result, the quality characteristic is a stochastic process with trend; and the process capability and quality loss vary with time and depend on the initial position of the cutting tool, which is described by the initial value of the characteristic [1]. In this setting, two basic decision problems are (a) optimization of preventive replacement time of the cutting tool and (b) optimization of the initial position of the cutting tool.

The first problem has attracted wide attention; and the decision models include cost models, process capability models and their combination. The cost models are usually associated with a known tool life distribution and an age or a block replacement policy [1, 10]. In a replacement cycle, periodic adjustment, resetting or grinding actions may be performed. Thus, the involved costs include replacement cost [1], resetting, adjustment or grinding cost [1, 9, 11], production capacity loss and quality loss [2, 12]. The quality loss is usually described by the Taguchi quadratic loss function [13]. As an exception, Xu and Cao present a sectional model to represent the percentage of quality failures and the quality loss is proportional to this percentage [2]. This approach actually assumes that any conforming item does not have quality loss, which is different from the Taguchi's idea.

The process capability index (PCI) depends on process mean function and process variance of the quality characteristic. The process variance is the sum of the non-random variance caused by the trend term and the random variance after removing the trend term. As a result, the PCI varies with time, and the cutting tool is replaced when the PCI drops blow a critical point [7]. Though it looks reasonable, the PCI-based replacement may be too conservative, as shown later. Deng et al. optimize the tool replacement time based on a cost model subject to a PCI constraint [11].

For the optimization problem of initial position of the cutting tool, no decision model is found. This gap needs to be filled. In addition, it is noted that the current cost models are associated with a given distribution of the tool failure time. It appears more natural to use a stochastic process model for representing the change of the quality characteristic with time. From such a model, the distribution of time to quality failure can be derived and the PCI can be evaluated. This paper aims to address the above two problems. A non-homogenous Wiener process is proposed to model the change of the quality characteristic and the decision problems are solved through combining this model with a cost model. A real-world example is included to illustrate the appropriateness and usefulness of the proposed models.

The paper is organized as follows. The proposed models are presented in Section 2 and illustrated in Section 3. The paper is concluded in Section 4.

2. Proposed models

Let X denote a quality characteristic of a machined part in a manufacturing process and x is its observation. The LSL and USL of X are known. Let T denote its target and it is usually taken as T = (LSL + USL)/2. Since the LSL is usually larger than zero so that the initial value of X (denoted as X_0) is larger than zero.

A sample with size n is drawn at discrete time points t_i (i = 1, …, m) and let x_{ij} (j = 1, …, n) denote their values. Usually, $t_i = t_{i-1} + \Delta$, where Δ is called the sampling interval. Clearly, X changes with time t and hence is a stochastic process. Let $X(t)$ denote this process, and $\mu(t)$ and $\sigma^2(t)$ denote the mean and variance (relative to the mean) of the process, respectively.

The process mean has an increasing or decreasing trend due to the tool wear. Without loss of generality, the increasing trend is assumed in the remainder of the paper. The cumulative wear curve of the cutting tool is often inverse S-shaped, as shown in figure 1, implying that the wear rate is bathtub-shaped (e.g., see [14-17]).

The process mean should have a similar trend as the wear curve of the cutting tool. Thus, the models for $\mu(t)$ need to be developed. The process variance, $\sigma^2(t)$, is the variance of residual $X(t) - \mu(t)$, and can be a constant or a function of t. For the latter case, the models for $\sigma(t)$ need to be developed. The residual is usually assumed to be normally distributed, and hence $X(t)$ is a non-homogenous Wiener process. The proposed stochastic process model and decision model are outlined as follows.

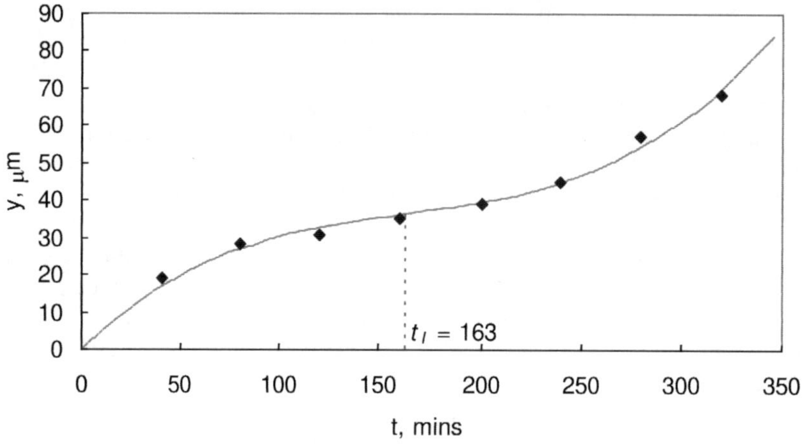

Figure 1. A wear curve at a cutting speed of 81.64 mm/min in high-speed milling [17].

2.1. Stochastic process model of the quality characteristic

Three optional models for $\mu(t)$ are shown in the second row of table 1. It is noted that all the three models have common constant and linear terms. For a given model, a_0 can be viewed as the initial position of the cutting tool. In a condition-based maintenance setting, a_0 can be viewed as an individual-oriented parameter and a_1 to a_3 as population-oriented parameters. The coefficients can be easily obtained through regression. Two regression statistics to assist model selection are R Square (denoted as R^2) and Standard Error (denoted as σ_e). The best model has the smallest value of $\rho = \sigma_e/R^2$.

The plots of Models 1-3 are all inverse S-shaped. The third row shows the horizontal coordinates of the inflection points.

The wear rate is defined as $r(t) = \mu'(t)$. It is clear that the wear rate is symmetry for Model 1 and non-symmetry for Models 2 and 3. The asymptotic wear rates are shown in the fourth and fifth rows. As seen, the models have considerably different trends for small and large t.

Three optional models for $\sigma(t)$ are shown in the last row of table 1. It is noted that the third option is the usual assumption for the Wiener and gamma processes. For a given process mean function, the plot of residuals can provide a clue to determine which is better, and the model selection can be made based on the goodness-of-fitting such as the sum of squared errors or the like.

Table 1. Optional models for $\mu(t)$ and $\sigma(t)$.

	Model 1	Model 2	Model 3
$\mu(t) =$	$\sum_{k=0}^{3} a_k t^k$	$\sum_{k=0}^{3} a_k t^{k/2}$	$\sum_{k=0}^{3} a_k t^{k/3}$
t_I	$-a_2/(3a_3)$	$a_1/(3a_3)$	$(-a_1/a_2)^3$
$r(0)$	a_1	∞	∞
$r(\infty)$	$\approx 3a_3t^2$	$\approx 1.5a_3t^{0.5}$	a_3
$\sigma(t) =$	a	$\alpha r(t)$	$\alpha[\mu(t)-a_0]^{0.5}$

All the model parameters can be jointly estimated using the maximum likelihood method (MLM). The log-likelihood function is given by

$$\ln(L) = \sum_{j=1}^{n}\sum_{i=1}^{m} \ln\{\phi[x_{ij};\ \mu(t_i),\ \sigma(t_i)]\} \tag{1}$$

where $\phi(.)$ is the normal density function. The maximum likelihood estimates (MLEs) of the parameters can be obtained through maximizing ln(L). This can be done using Solver of Microsoft Excel.

The distribution of time to the USL is given by

$$F(t) = 1 - \Phi[USL; \mu(t), \sigma(t)] \quad (2)$$

It is not the distribution of time to the hard failure of the cutting tool and can be viewed as the distribution of time to the quality failure.

2.2. Process capability index

Pearn and Hsu define the process capability index (CPI) as below [7]

$$C_{pmk}(t) = \frac{d - |T - \mu(t)|}{3\sqrt{[T-\mu(t)]^2 + \sigma^2(t)}} \quad (3)$$

where d = (USL - LSL)/2.

For the proposed mean models with LSL < a_0 < T, the CPI at t = 0 is given by

$$C_{pmk}(0) = \frac{d - |T - a_0|}{3(T - a_0)} = \frac{d}{3(T - a_0)} - 1/3 \quad (4)$$

This implies that $C_{pmk}(0)$ increases as a_0 increases. Let t_T denote the solution of $\mu(t) = T$. When $t \approx t_T$, we have

$$C_{pmk}(t_T) = \frac{d}{3\sigma(t_T)} = 2 \times \frac{d}{6\sigma(t_T)} \quad (5)$$

Generally, $d/[6\sigma(t_T)] > 1$ and hence $C_{pmk}(t_T)$ is generally larger than 2. Let t_U denote the solution of $\mu(t) = USL$. When $t = t_U$, we have

$$C_{pmk}(t_U) = \frac{d/2}{3\sqrt{(d/2)^2 + \sigma^2(t_U)}} < \frac{d/2}{3d/2} = 1/3 \quad (6)$$

According to the above analysis, the CPI is a unimodal function of time t; and the process must be stopped at a time point within (t_T, t_U).

2.3. Age replacement policy and cost model

The age replacement policy is appropriate for the cutting tool since it is relatively expensive [18]. Three cost elements are considered: preventive replacement cost c_p with a probability of 1 – $F(t)$, failure replacement cost c_f with a probability of $F(t)$ and a Taguchi quality loss, which is given by

$$l(t) = KV(t) = K\{[T - \mu(t)]^2 + \sigma^2(t)\} \quad (7)$$

Let c_q denote that the quality loss when x = USL. Using this to Equation (7) yields

$$l(t) = KV(t) = c_q\{[T - \mu(t)]^2 + \sigma^2(t)\}/d^2 \quad (8)$$

Let p denote that the production rate, i.e., the number of produced parts per unit time. The total quality loss over (0, t) is given by

$$L(t) = \int_0^t l(\tau) p d\tau \approx 0.5p \sum_{k=1}^{N(t)} [l(k\Delta t) + l(k\Delta t - \Delta t)]\Delta t \quad (9)$$

where Δt is a small time interval and $N(t) = t/\Delta t$.

The cost rate model is given by [18]

$$J(t) = [c_f F(t) + c_p R(t) + L(t)] / \int_0^t R(\tau) d\tau \quad (10)$$

The optimal replacement age corresponds to the minimum of $J(t)$. Let t^* denote the optimal replacement age and J^* denote the corresponding cost rate.

2.4. *Optimal initial position*

It is noted that t^* and J^* are functions of a_0. That is, J^* can be written as

$$J^*(a_0) = \varphi(a_0) \quad (11)$$

The optimal initial position of the cutting tool corresponds to the minimum of $J^*(a_0)$. For a set of given values of a_0, compute the corresponding values of J^*. Fitting these points into a smooth curve yields $\varphi(a_0)$, and the optimal value of a_0 can be obtained from the fitted curve.

3. Illustration

3.1. *Data and earlier results*

Pearn and Hsu present a case study, which deals with a particular type of aluminium lid product produced on a press [7]. The key characteristic is lid height, and its USL, LSL and T are 68.4 mm, 64.65 mm and 66.525 mm, respectively. The data collected over one-week period are shown in table 2 in terms of values of x_{ij} ($i = 1, 2, \ldots, 7; j = 1, 2, \ldots, 15$).

Table 2. Values of x_{ij} in mm at t_i in day.

$j \backslash t_i$	1	2	3	4	5	6	7
1	66.100	66.335	66.470	66.542	66.670	66.722	66.872
2	66.261	66.295	66.387	66.551	66.665	66.722	66.931
3	66.147	66.335	66.456	66.501	66.684	66.715	66.860
4	66.214	66.361	66.402	66.504	66.644	66.777	66.836
5	66.133	66.314	66.468	66.568	66.689	66.724	66.922
6	66.223	66.335	66.430	66.546	66.715	66.770	66.943
7	66.216	66.428	66.480	66.470	66.695	66.803	66.907
8	66.288	66.337	66.428	66.572	66.732	66.770	66.900
9	66.159	66.397	66.413	66.618	66.665	66.753	66.929
10	66.252	66.337	66.499	66.625	66.606	66.789	66.919
11	66.288	66.418	66.387	66.599	66.717	66.758	66.862
12	66.242	66.416	66.504	66.656	66.675	66.805	66.922
13	66.297	66.423	66.432	66.596	66.727	66.774	66.836
14	66.304	66.361	66.516	66.594	66.708	66.800	66.929
15	66.221	66.435	66.546	66.665	66.739	66.781	66.950

Figure 2 shows the plot of data. As seen, $X(t)$ has an increasing trend due to the tool wear. Therefore, some minimum level of process capability such as $C_{pmk} > 1$ can be maintained through monitoring the PCI.

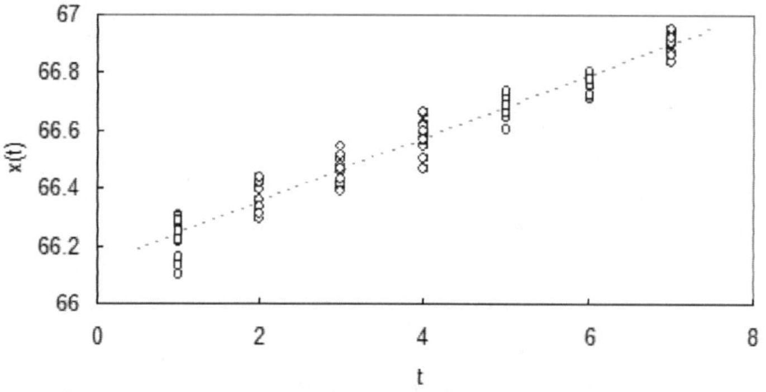

Figure 2. Plot of data.

Pearn and Hsu derive the distribution of C_{pmk} [7]. According to the derived distribution, the critical value of C_{pmk} associated with a type-I error of 0.05 and sample size $n = 15$ is 1.55. When the estimated process capability is smaller than the critical value, the process is stopped and the tool is reset or replaced. For the current example, C_{pmk} is computed using the sample mean and sample variance [7]. The results are displayed in figure 3. Since C_{pmk} is smaller than the critical value at $t = 7$, Pearn and Hsu recommend that the tool should be replaced at $t = 7$. This replacement time is too conservative as illustrated later.

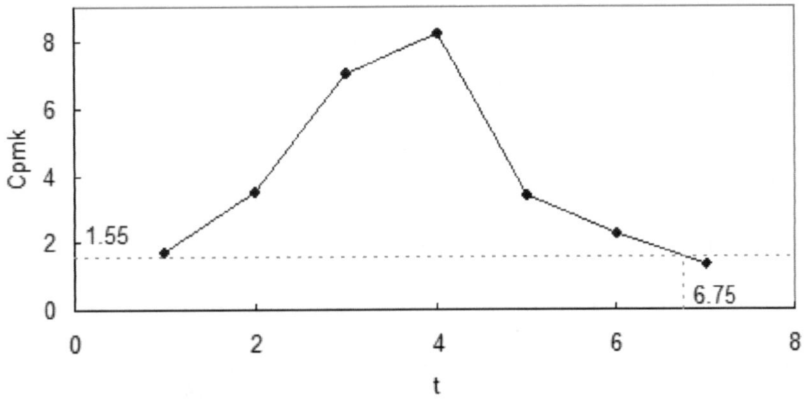

Figure 3. Plot of Cpmk.

3.2. Results obtained from the proposed models

3.2.1. Process mean function. Figure 4 shows the plot of process mean function (i.e., those dots), which looks linear. Considering the fact that the wear rate of the cutting tool is bathtub-shaped, it is more reasonable to assume that the process mean function is inverse S-shaped.

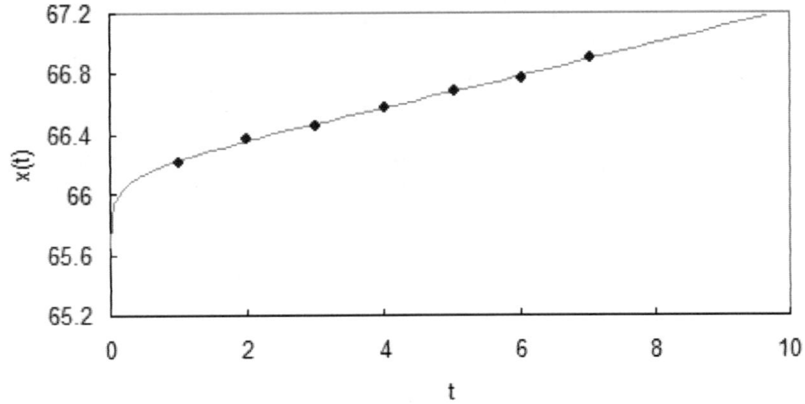

Figure 4. Plot of empirical mean degradation function

The regression coefficients of the optional models and the corresponding values of ρ are shown in table 3. In terms of ρ, Model 3 is slightly better than the other three models. This will be further confirmed in Section 3.2.3.

Table 3. Regression parameters and performances of the process mean functions.

	Linear	Model 1	Model 2	Model 3
a_0	66.131	66.067	65.770	65.051
a_1	0.1093	0.1742	0.6221	2.2242
a_2		-0.01733	-0.2251	-1.4806
a_3		0.001341	0.05698	0.4298
ρ	0.049969	0.049068	0.049032	0.049016

3.2.2. Process variance function. Let σ_i denote the sample standard deviation at t_i. The empirical wear rate is computed using difference method given by

$$r(t_{i-1} + \Delta t/2) = [x(t_i) - x(t_{i-1})]/(t_i - t_{i-1}) \tag{12}$$

where $\Delta t = 1$. Define

$$x(t_{i-1} + \Delta t/2) = [x(t_i) + x(t_{i-1})]/2 \tag{13}$$

Thus, using (12) and (13), one can compute correlation coefficient between the sample standard deviation and the empirical wear rate. Similarly, the correlation coefficients between the sample standard deviation and $[x_i\text{-}a_0]^q$ with $q = 10^{-5}$ and 0.5 can be computed. Here, $q = 10^{-5}$ corresponds to Model 1 and $q = 0.5$ corresponds to Model 3. The best model has the largest correlation coefficient. For the current example, Table 4 shows the correlation coefficients of Models 1-3. As seen, Model 2 is the best, implying that the standard deviation function is approximately proportional to the wear rate.

Table 4. Correlation coefficients between $\sigma(t)$, $r(t)$ and $[\mu(t)\text{-}a_0]^q$.

Model 1 ($q = 10^{-5}$)	Model 2	Model 3 ($q = 0.5$)
-0.8061	0.3900	-0.8028

3.2.3. Maximum likelihood estimates of model parameters.
For the standard deviation model given by Model 2, table 5 shows the MLEs of parameters of the fitted Wiener process model.

Table 5. Maximum likelihood estimates of the Wiener process model parameters.

a_0	a_1	a_2	a_3	α
65.934	0.3031	-0.0332	0.02085	0.4075

3.2.4. Distribution of time to the USL. The distribution of time to the USL is given by equation (2), which can be well approximated by a normal distribution with parameters $\mu = t_U = 19.36$ and $\sigma = 0.4075$. Clearly, the distribution is highly concentrated.

3.2.5. Process capability index. Figure 5 shows the plot of PCI. As seen, the maximum of PCI occurs at $t = 3.54$, which corresponds to $\mu(t) = T$. When $t = 7.95$, $C_{pmk} = 1$. It seems that the cutting tool should be replaced $t = 8$. However, the value of F(8) is nearly equal to zero, implying that replacing the cutting tool at $t = 8$ is too early. In other words, the PCI-based replacement decision is too conservative.

It should be noted that the third process mean model has $r(0) = \infty$ so that $\sigma(0) = \infty$ according to the second standard deviation model. As a result, $C_{pmk}(0) = 0$, which appears unreasonable. Therefore, the value of $C_{pmk}(t)$ inferred from the fitted model for small t is generally unreliable.

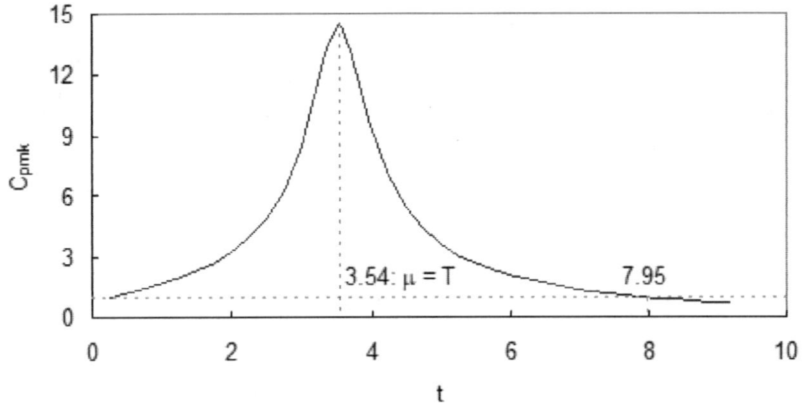

Figure 5. Plot of $C_{pmk}(t)$.

3.2.6. Cost model for optimizing tool replacement time. Assume that production rate per unit time is p = 1 and cost parameters are (c_p, c_f, c_q) = (1, 4, 0.5). Since the distribution is highly concentrated, it is reasonable to assume that the optimal replacement time t^* is much smaller than $\mu - 3\sigma$ so that $F(t^*) \approx 0$. In this case, Equation (10) can be simplified as

$$J(t) = [c_p + L(t)]/t \, , \; t \le \mu - 3\sigma \tag{14}$$

Figure 6 shows the plot of of $J(t)$. As seen, the optimal solution is $t^* = 11.81$ ($< \mu - 3\sigma = 18.97$), which is much larger than the PCI-based solution (= 7.95). This confirms the conclusion that *the PCI-based replacement decision is too conservative*.

It is noted that the cost rate curve is flat in the adjacent region of t^* so that the relative error of the cost rate within (10.87, 12.80) is smaller than 1%. This interval can be used as the opportunistic replacement interval.

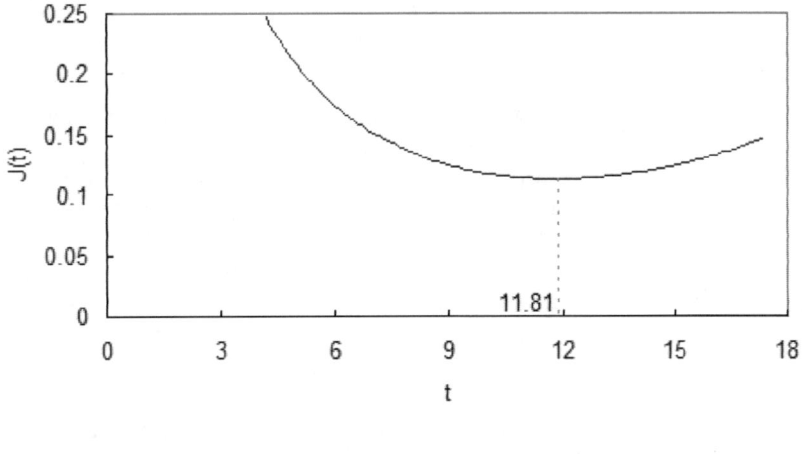

Figure 6. Plot of $J(t)$.

3.2.7. Optimal initial position of the cutting tool. As mentioned earlier, parameter a_0 represents the initial position of the cutting tool. Different value of a_0 leads to different values of t^* and J^*. Figure 7 shows the plot of J^* vs. a_0. As seen, there exists an optimal value of a_0, where J^* achieves its minimum. For the current example, the optimal value of a_0 is 65.229.

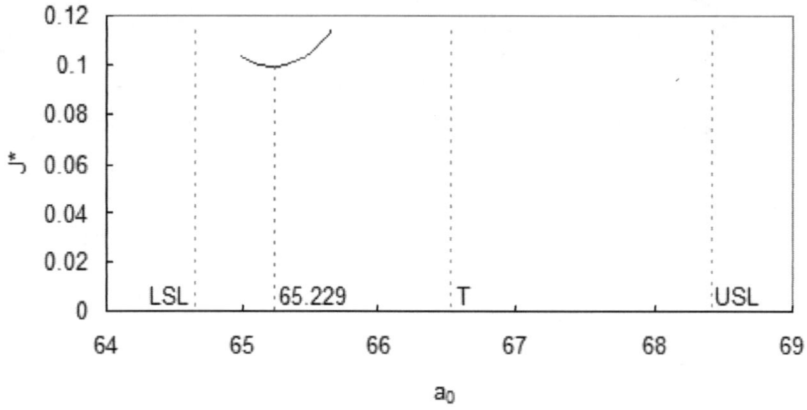

Figure 7. Plot of J^* vs. a_0.

Table 6 shows the optimal solutions before and after the optimization of initial position. As seen, through the optimization of initial position, the useful life of cutting tool increases by 27.13% and the cost rate decreases by 12.60%. The effect of the optimization is very obvious.

Table 6. Current and optimal initial positions and corresponding optimal solutions.

	a_0	t^*	$J(t^*)$
Before optimization	65.655	11.81	0.1138
After optimization	65.229	15.01	0.0994
Relative error, %		27.13	12.60

4. Conclusions

Dimensions of a machined part change with time due to the cutting tool wear. The dimension can be modelled by a stochastic process and the resulting model can be used to optimize the replacement time and initial position of the cutting tool. The main contributions of the paper have been:

- the Wiener process has been proposed to model the dimensional change in a manufacturing process; three process mean models and three process variance models have been proposed, and the method for model selection has been developed;
- a method to optimize the initial position of the cutting tool has been proposed and its effectiveness has been illustrated; and
- the CPI-based replacement decision is found to be too conservative. As a result, the paper is useful for both reliability researchers and manufacturing engineers.

Two topics for future research are:

- to develop the inverse S-shaped process mean models, whose initial wear rates are finite, and
- to develop the cost models that include the possible resetting, adjustment and/or grinding costs.

Acknowledgment

The research was supported by the National Natural Science Foundation of China (No. 71771029).

References

[1] Sheikh A K, Raouf A, Sekerdey U A and Younas M 1999 Optimal tool replacement and resetting strategies in automated manufacturing systems *Int. J. Prod. Res.* **4** 917–37.

[2] Xu W and Cao L 2015 Optimal tool replacement with product quality deterioration and random tool failure *Int. J. Prod. Res.* **53** 1736-45.

[3] Tian W, Wells L J and Camelio J A 2016 *ASME 2016 11th International Manufacturing Science and Engineering Conference* pp 1-10.

[4] Li Z, Liu R and Wu D 2019 Data-driven smart manufacturing: tool wear monitoring with audio signals and machine learning *J. Manuf. Processes* pp 66-76.

[5] Ambhore N, Kamble D, Chinchanikar S and Wayal V 2015 Tool condition monitoring system: A review *Materials Today: Proceedings* **2** 3419-28.

[6] ISO 1993 *ISO3685: Tool-life testing with single-point turning tools* (International Standard)

[7] Pearn W L and Hsu Y C 2007 Optimal Tool replacement for processes with low fraction defective *European Journal of Operational Research* **180** 1116-29.

[8] Zhang M, He Z and Liu Z 2007 *2007 2nd IEEE Conference on Industrial Electronics and Applications* pp 516–21.

[9] Hsu B M, and Shu M H 2010 Reliability assessment and replacement for machine tools under wear deterioration *Int. J. Adv. Manuf. Technol.* pp 355-65.

[10] Vagnorius Z, Rausand M and Sorby K 2010 Determining optimal replacement time for metal cutting tools *European Journal of Operational Research* pp 407-16.

[11] Deng Y, Zhu H , Zhang G and Yin H 2012 Optimal tool replacement decision method based on cost and process capability *Mechanical Engineering and Technology* (Advances in Intelligent and Soft Computing vol 125) ed Zhang T(Berlin: Springer Berlin Heidelberg) pp 9-14.

[12] Cao L, Xu W, Li H, Chen J and Chen P 2014 Tool replacement decision-making method for automatic production line *Int. J. Mech. Eng. Educ.* **50** pp 131-5.

[13] Genichi T 1986 *Introduction to quality engineering: designing quality into products and processes* (New York: White Plains).

[14] Sheikh A K and Ahmad M 1987 A reliability model for a nonlinear damage process *Reliability Engineering* **18** pp 73-99.

[15] Prasad K N and Ramamoorthy B 2001 Tool wear evaluation by stereo vision and prediction by artificial neural network *J. Mater. Process. Technol.* **112** pp 43-52.

[16] Su J C, Huang C K and Tarng Y S 2006 An automated flank wear measurement of microdrills using machine vision *J. Mater. Process. Technol.* **180** pp 328-35.

[17] Lei X, Liao W, Xie F, Zheng K and Zhao J 2013 Predict on cutter wear and life in high-speed milling of pre-sintered zirconia used for denture *Journal of Nanjing University of Science and Technology* **37** pp 567-72.

[18] Murthy D N P and Jiang R 2008 *Maintenance: decision models for management* (Beijing: Science Press).

Research on DC-DC power health management technology based on degradation test and deep learning

Linlin Shi[1], Pengfei Yu[1,2,*], Shilie He[1], Zhenwei Zhou[1], Linghui Meng[1], Junbin Liu[1], Liye Cheng, Yun Huang

[1]The Fifth Electronics Research Institute of Ministry of Industry and Information Technology, Guangzhou, China
[2]South China University of Technology, Guangzhou, China

*Corresponding author: pfyuhust@163.com

Abstract. The health state of DC-DC power supply is the key factor to determine whether the electronic equipment can operate normally. The failure and deterioration of the power supply system will lead to the collapse of the entire electronic system. The research in this paper is based on the long-term high temperature degradation test data of a certain type of DC-DC power supply. The degradation law of power supply is studied by data preprocessing and noise reduction of sensitive parameters such as input current, output current, input voltage and output voltage. On this basis, we use the method of deep learning to model the efficiency of power supply in the process of degradation test. The experimental results show that the efficiency time series modeling of power supply degradation using LSTM method can effectively reflect the law of power supply efficiency degradation. Based on the DC-DC power health management technology combining degradation test and deep learning, the advanced fault prediction model is used to reflect the change law of power supply in the real degradation process. This method has certain theoretical and engineering value for power PHM modeling and application.

Keywords: degradation, DC-DC, Health Management, deep learning.

1. Introduction

With the progress and development of modern science and technology, switching power supply in power electronics, aerospace, ships, computers, electric vehicles, industrial control, communication equipment and other fields have been a big rule of application, in the development of the national economy where people's lives are indispensable. In recent years, with the significant increase of power consumption of system equipment, the improvement of load has put forward higher requirements for the performance and reliability of power supply system. As we know, switching power supply usually works in harsh environment, the environment including thermal stress, humidity, shock vibration, vacuum, salt spray and other stress environment, but also includes electromagnetic radiation, higher particles, strong electromagnetic interference and other harsh working conditions. Power supply as the core component of equipment, its importance is self-evident. Switching power supply often plays the role of the "heart" of the electronic system, and the failure and deterioration of the power supply system can lead to the collapse of the entire electronic system. Whether the power supply works normally or not determines the stability, safety and reliability of the electronic system. Prognostic and Health Management (PHM) technology was first developed by the U.S. Military to improve equipment safety and mission success, and is an important means to improve support effectiveness and reduce support costs. The research contents include condition monitoring, fault diagnosis, fault prediction, health management and decision, data communication, transmission and expression, etc.

This technology has been promoted and applied by researchers at home and abroad, and is becoming an important part of the development and application stages of new-generation aircraft, ships, vehicles, electronic systems, mechanical systems and other equipment or systems [1]-[2]. With the rapid development of electronic products, it is necessary to introduce state-based maintenance method into the field of power supply reliability assurance. At present, the research on intelligent maintenance of power supply equipment at home and abroad mainly focuses on fault diagnosis and life prediction of key components of power supply. Zhou Shihong [3] based on the analysis of the main degradation mode of aluminum electrolytic capacitor and the influence of degradation on the output signal of DC-DC converter, and taking a specific DC-DC converter as the research object, preprearily constructed an experimental system for predicting the remaining service life of DC-DC power supply. Yu Shan et al. [4] analyzed and studied the influence of ambient temperature and ripple current on the service life of electrolytic capacitors from the internal structure and failure mechanism of electrolytic capacitors, and predicted and analyzed the capacitor life of switching power supply by measuring the internal temperature rise of the capacitor. Li Xiaohong et al. [5] used the accelerated degradation test method to evaluate the reliability and predict the life of a highly reliable DC-DC hybrid power supply. From the perspective of health management of power supply, Wang Fengru et al. [6] studied the method of fault feature extraction, condition monitoring and fault prediction related technologies of intelligent secondary power system level. The premise of realizing intelligent maintenance of module power based on health management is to accurately obtain the health status of power supply.

In this paper, the DC-DC power supply was subjected to high temperature degradation test for 9 months, and the health sensitive monitoring parameters during the degradation test were collected. Through feature analysis, data preprocessing, efficiency analysis and deep learning modeling of monitoring data, the variation rule of power degradation test data is analyzed, which provides experimental analysis and data demonstration for power failure prediction and health management engineering application.

2. High temperature degradation test of power supply

In order to study the degradation law of DC-DC power supply at high temperature, we carried out high temperature degradation test of switching power supply. The hardware and software of the whole test included high temperature box, power supply, DC-DC module power supply, load, current sensor, data acquisition card, temperature acquisition card, acquisition software, etc., as shown in Figure 1. The power supply test lasted from April 7, 2020 to December 4, 2020, and the temperature box was set at a constant temperature of 80 degrees Celsius. Six DC-DC power supplies were used in the high temperature degradation test of power supply. The power supply is supplied to a power system manufacturer with specifications of 18-75VDC input and 5VDC output. In the test, the input voltage is 28V and the output voltage is 5V. Through self-developed data acquisition software, input current, input voltage, output current, output voltage four parameters are collected for each power supply. In fact, the acquisition software can support 56 channels of data acquisition at the same time, here the power data only used 24 channels. All the test data were covered from April 21, 2020 to December 4, 2020.Adopt the method of random or fixed selection, select one of the 20 folders in a certain time as the data representative of one day.1 folder contains about 3 minutes of sampling data, 24 parameters of 6 power supplies are collected and stored in DAT format. Through continuous power supply test and data collection analysis, the health status of power supply in power supply test can be evaluated and analyzed in real time, and the method and model of power supply health evaluation can be established.

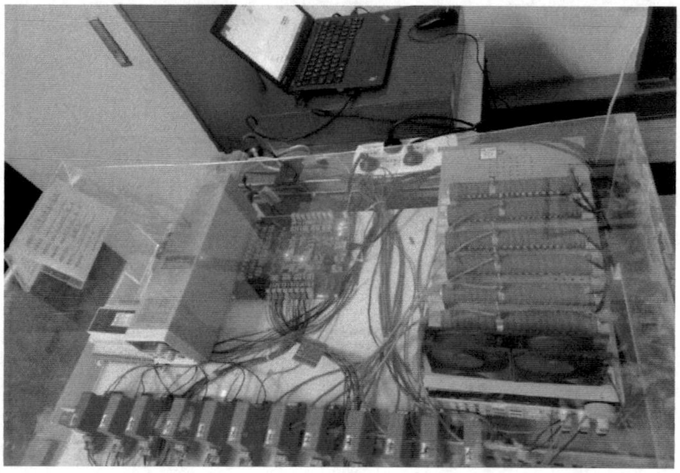

Fig. 1 High temperature test of DC-DC power supply: There are 6 power supplies in the high temperature box.

3. Power failure prediction based on Deep-learning

Deep neural network (DNN) has attracted more and more attention from researchers in various fields in recent years [7]-[9]. It employs a multi-neural layer hierarchy that extracts information from input data by processing it layer by layer, a "deep" layer structure that allows it to learn the representation of complex raw data with multiple levels of abstraction. Starting with raw input, DNN automatically discovers complex structures in large data sets and learns useful features layer by layer. LSTM(Long Short-Term Memory)[10]-[12] is a deep learning method, which is a kind of neural network used to process sample time series data. It has obvious advantages in solving the gradient disappearance and gradient explosion problems in the process of long sequence training.

3.1. Power degradation modeling method based on LSTM

Recurring Neural Network (RNN) can realize long-term dependence in learning. Different from BP neural network, every time RNN network is trained, the hidden layer will be affected by the previous state. The RNN network can be expanded according to the time series, as shown in Figure 2.Long Short Temp Memory Network (LSTM) is a special structure of RNN, which was proposed by Hochreiter& Schimidhuber (1997) and has been continuously improved since then. Figure 3 shows the network structure of LSTM. The key position of LSTM lies in the unit, which runs from the beginning to the end along the transfer line. There is a linear interaction in the middle, and information can be transferred directly along it without changing.

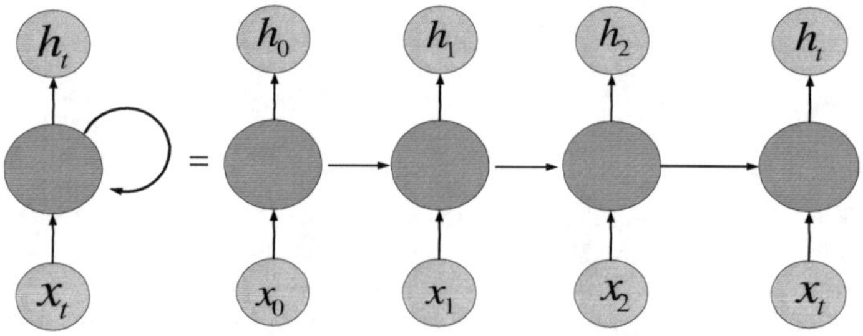

Fig. 2 RNN expansion diagram of cyclic neural network

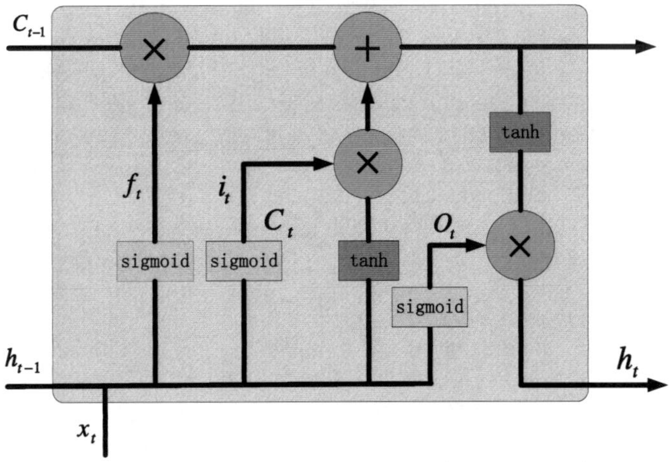

Fig. 3 LSTM hidden layer structure of cyclic neural network

LSTM uses the structure of gates to add or drop information to each cell. The units of the LSTM network have three types of gates: 1. Forgetting gate 2. Input gate 3.State the door. One of the most critical LSTM gates is the forgetting gate, which controls the flow of information in and out (or through and out) through the following three functions:

$$f_t = \sigma(W_f X_t + b_f)$$
$$i_t = \sigma(W_i X_t + b_i)$$
$$f_t = \sigma(W_o X_t + b_o)$$

Where is the network output SIGMOID activation function. $\sigma(\cdot)$ The activation function selects Sigmoid to make it impossible to pass the gate when approaching 0, but it can pass the gate when approaching 1. Therefore, the output of the network depends on the current information itself.

Set the current time as t, and after the state expression of LSTM network is expanded, we can get:

$$S_t = \sigma(W_f X_t + b_f)S_{t-1} + \sigma(W_i X_t + b_i)X_t$$

After the activation function, there are:

$$S_t = \tanh[\sigma(W_f X_t + b_f)S_{t-1} + \sigma(W_i X_t + b_i)X_t]$$

By calculating the partial derivative, we can know that the LSTM network has a term similar to the above RNN, but in LSTM, it is:. $\prod_{j=k+1}^{t} \frac{\partial S_j}{\partial S_{j-1}} = \prod_{j=k+1}^{t} \tanh' \sigma(W_f X_t + b_f)$

In the process of long-term use, the conversion efficiency of power supply will degrade, so the degradation data of long-term power supply forms the time series of conversion efficiency. When the efficiency of the power supply degenerates to a certain threshold, it indicates that the fault occurs. Therefore, the memory characteristics of LSTM network can be used to better predict the trend of power degradation, and the accurate prediction of power conversion efficiency plays an important role in the health management of power supply.

3.2. Power data preprocessing and power conversion efficiency calculation

All types of switching power supply involve conversion efficiency, which is defined as the ratio of output power to input power, for DC/DC switching power supply. The power supply with 100% efficiency is the ideal power supply, and the conversion efficiency of the actual switching power supply is less than 1. This is because the power loss will be generated inside the switching power supply. The greater the power loss, the lower the conversion efficiency of the switching power supply. The loss of switching power supply is mainly caused by on-off loss, switching loss and hysteresis eddy current. For the non-isolated DC/DC switching power supply such as Buck, Boost and Buck-Boost, there is no transformer inside, so it is mainly the conduction loss and switching loss. For forward, flyback, half bridge, full bridge and other isolated DC/DC switching power supply, its internal transformer, so the three kinds of losses have. For the sake of generality, only the influence of on-off loss and switching loss on the switching power supply conversion efficiency is considered here. Switching power supply in the process of long-term use of the on-off loss will increase, the change of switching loss is temporarily unknown. Considering that most of the loss of the switching power supply is the on-loss, the power loss of the switching power supply will gradually increase and the conversion efficiency will gradually decrease in the process of use. Therefore, the health assessment of DC/DC switching power supply can be realized by monitoring and calculating the conversion efficiency of the switching power supply.

In view of the degradation situation of the 6 power supply, we choose the test data of the 3# power supply for analysis. The experiment lasted from 2020.04.21 to 2020.12.04. Figure 4 shows the changes of input current, input voltage, output current and output voltage of 3# power supply. Obviously, the data we collected contains outliers and noise. In order to analyze the hidden rules in the data, we carried out a cleaning operation on the data. Fig. 5 shows the characteristic parameter curve of power supply after cleaning operation. It can be seen that the output voltage changes very little during the degradation test, while the other three parameters have certain amplitude changes. In order to further analyze the performance variation of the power supply in the degradation test, we calculated the conversion efficiency of the power supply 3#. Fig. 6 and Fig. 7 respectively show the change curve of power supply efficiency of No. 3 in the high temperature degradation test of power supply after the abnormal data were not cleaned and the abnormal data were cleaned. Data unwashed, that power the degradation law of hidden in the noise and outliers, and after cleaning the rule of data is more noticeable after data processing and analysis of the 3 # power supply, can find the following test: phase 1 test (2020.05.22-2020.07.02) the efficiency of the power supply change first appeared certain amplitude periodic increase, this is caused by faults test load in the early set, then the efficiency of power supply 3 changes tend to be stable, but not obvious degradation trend. By adding the data of 2020.07.02-2020.12.04 in the second stage, after cleaning the data, it can be found that the power performance degradation in this stage is obvious, which is manifested as a significant decline in the power conversion efficiency.

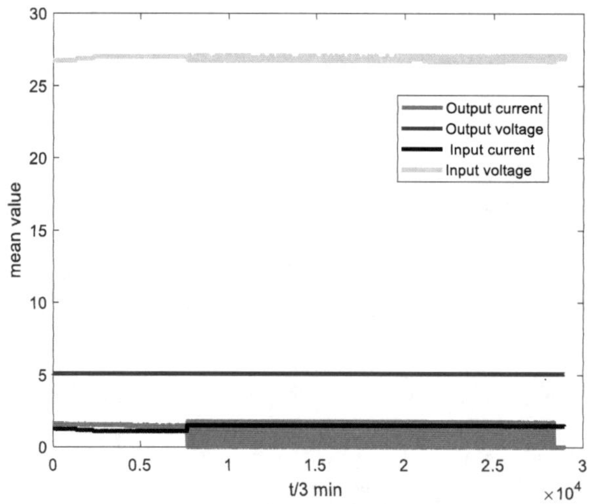

Fig. 4 Variation of the four monitoring parameters of power No. 3 in the high-temperature degradation test of power supply without cleaning abnormal data.

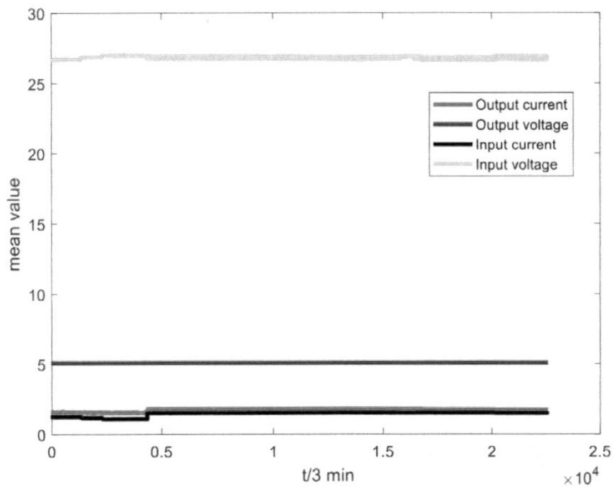

Fig. 5 Variation of the four monitoring parameters of No.3 power supply in the high-temperature degradation test under the condition of cleaning abnormal data.

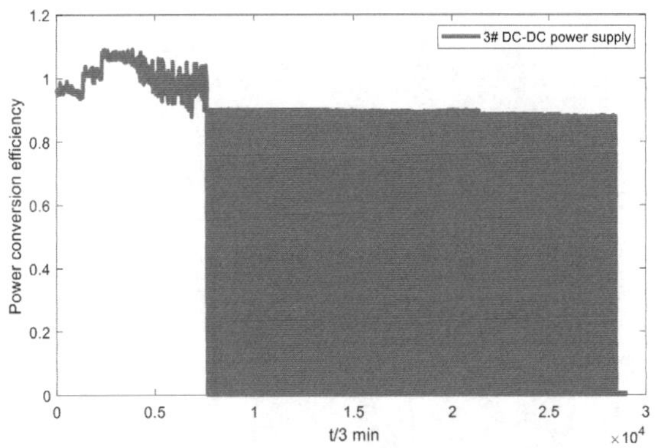

Fig. 6 Change curve of power efficiency of No. 3 in high temperature power supply degradation test without cleaning abnormal data.

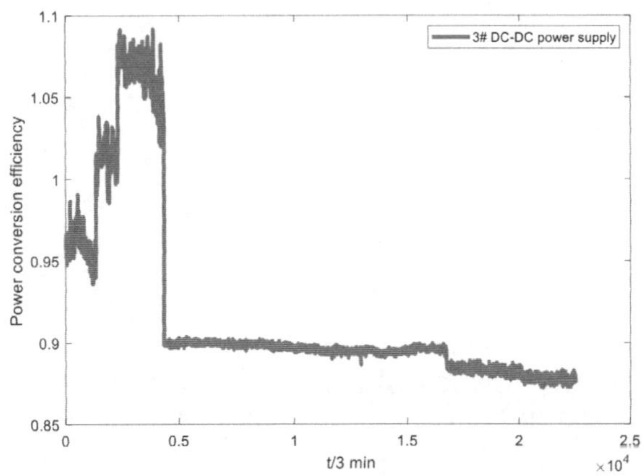

Fig. 7 Variation curve of power efficiency of # 3 in high temperature power supply degradation test under the condition of cleaning abnormal data.

3.3. Power failure prediction model and analysis based on LSTM

In order to establish a fault prediction model for power degradation, a machine learning model can be established to predict the time prediction of efficiency in the process of power degradation based on the time series of power efficiency, to judge the trend of efficiency change and to predict the degree of efficiency decrease in the future.

Therefore, on the basis of the research in Section 3.2, we build a LSTM prediction model for the degradation test data of 3# power supply. The test time range covered by the data we intercepted is from May 22 to June 25. Based on the conversion efficiency calculation method of power supply, the efficiency change curve of 3# power supply during May 22 solstice and June 25 can be calculated, as shown in Figure 8 below.

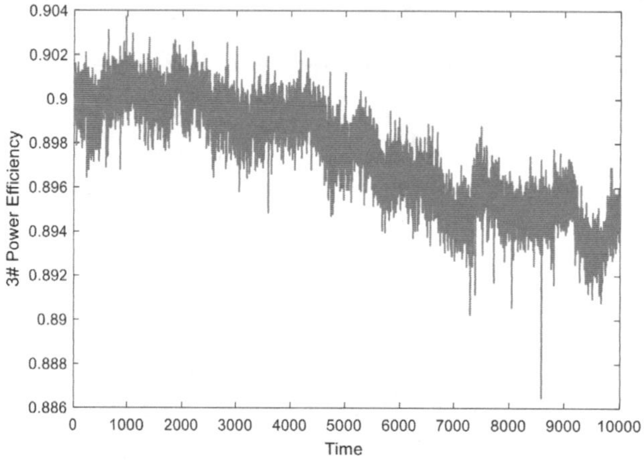

Figure 8 #3 Power Supply May 22 solstice June 25 Device conversion efficiency change.

Table 1 Power efficiency LSTM prediction network training parameter Settings.

The parameter name	Parameter Setting	Parameter interpretation
The objective function optimizer	Adam	Adaptive moment estimation
Max Epochs	120	Maximum number of iterations
Gradient Threshold	1	The gradient threshold
Initial Learn Rate	0.005	Initial learning rate
Learn Rate Schedule	piecewise	Learning rate format
Learn Rate Drop Period	125	Learning rate decline cycle
Learn Rate Drop Factor	0.2	Learning rate attenuation factor

The time series length of the conversion efficiency of the power supply is 10000, and the power efficiency value of 500 time-steps in the future is predicted. In order to predict the future trend of power conversion efficiency, the deep learning LSTM model, namely the long and short memory network, is used here. The model training data is 95% of the power efficiency sequence length, that is, it contains 9500 data samples. The length of the model test set or prediction set is 500, that is, to predict the change of the power conversion efficiency at 500 steps in the future. In order to obtain better fitting and prevent training divergence, the training data are normalized to have zero mean value and unit variance. In the prediction, we used the same parameters as the training data to standardize the test data. The established LSTM network architecture contains one input feature dimension and one output feature dimension. The network structure of this deep learning model is composed of four layers of network, namely, the input layer, the LSTM layer, the full connection layer and the regression layer. The LSTM hidden layer contains 200 neural network units. LSTM training parameters are set as follows: the maximum number of parameter iterations is 120, and the gradient threshold is 1 to prevent gradient explosion. The initial learning rate is set as 0.005, the learning rate framework is set as a segmented form, the fading cycle of the learning rate is 125, and the fading factor is 0.2. The parameter settings are shown in Table 1.

On the basis of setting relevant network parameters, the LSTM network training process is started. The error convergence and loss curves of the training process are shown in Fig. 9. The whole training session lasted seven minutes and 25 seconds. After obtaining the trained LSTM model, we predict the power efficiency data of 500 time-steps in the future based on the model. Fig. 10 and Fig. 11 respectively show the future 500-step conversion efficiency of power supply 3# predicted by LSTM model and the comparison curves between the predicted and observed values of power supply 3# based on power efficiency LSTM model. It can be found that LSTM has a relatively high accuracy in predicting the conversion efficiency of power supply degradation, which can reach the accuracy of 1e-2. Therefore, the Long-Short Term Memory network is adopted to predict the power supply degradation, which has a relatively good prediction effect and accuracy.

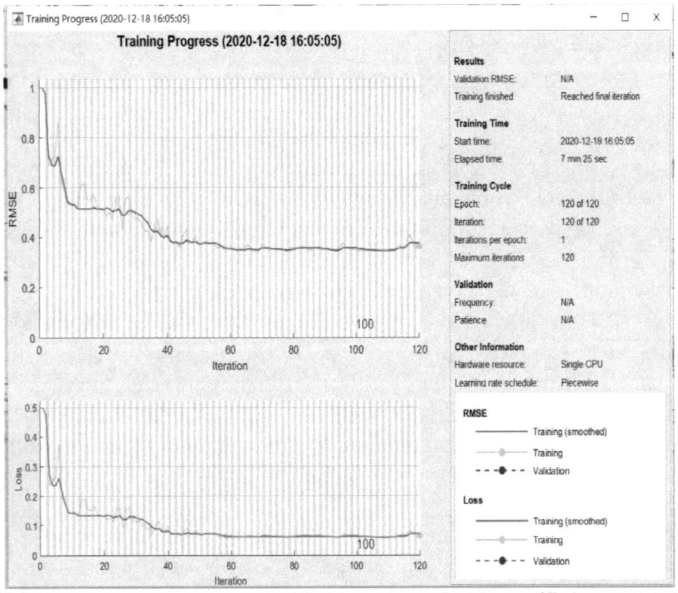

Fig. 9 Training error convergence and loss curves of power efficiency LSTM model.

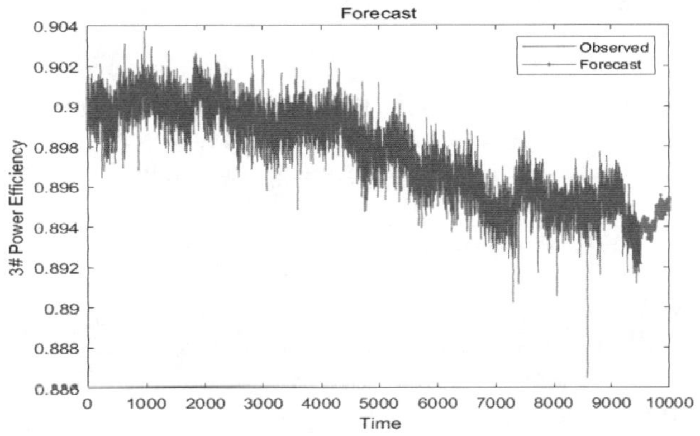

Fig. 10 Predicts the future 500-step conversion efficiency of power supply # 3 based on the power efficiency LSTM model.

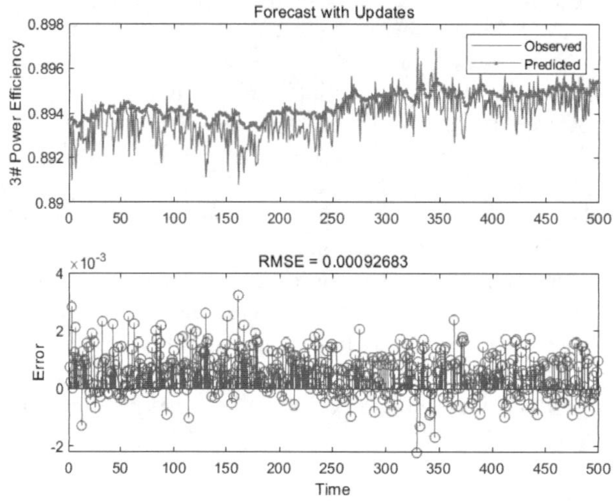

Fig. 11 Comparison of predicted and observed values of 3# power supply based on power efficiency LSTM model.

4. Conclusions

This paper carried out high temperature degradation test for DC-DC power supply, and monitored and collected the parameters related to the health state of the power supply during the degradation test. The paper analyzes the data of power supply with certain degradation characteristics in the test process, and excavates the degradation law of power supply through the pretreatment and cleaning of data characteristics. It is found that the conversion efficiency of the power supply degrades to a certain extent in the process of long-term high temperature test, which reflects the performance of the power supply is decreasing. We selected the time series of conversion efficiency in the degradation test process, and predicted the variation trend of power efficiency through the LSTM model, and achieved a good prediction effect. The modeling technology based on the combination of test and data mining model proposed in this paper has certain guiding significance and practical value for the excitation of power failure law and the establishment of power health management model.

Acknowledgments

The authors wish to thank the anonymous reviewers for their constructive comments and suggestions. This work was financially supported by the Key Laboratory Fund (JAS192800120) and the Key Laboratory Fund (JAS202800220).

References

[1] Peng Y, Liu D, Peng X.A review: Prognostics and health management[J]. Journal of Electronic Measurement & Instrument, 2010,24 (1) : 1007-1015.

[2] Lee J, Wu F, Zhao W, et al. Reviews, Methodology and Applications of Prognostics for Rotary Machinery Systems [J]. Mechanical Systems & Signal Processing, 2014,42 (1/2) : 314-334.

[3] Zhou S H, Simulations on the Physicochemical Characteristics of DC-DC Power Supply [J].Computer Engineering and Applications, 2011,47 (31): 234-237.

[4] YU S, XU Z W, DONG J Q. Analysis of electrolytic capacitor life prediction in switching power supply[J]. Journal of Power Supply, 2016,14 (6) : 87-92.

[5] Li X H, Lu J, Qiu Z W, et al. Study on Long-time Evaluation Technique for Highly Reliable DC-DC Hybrid Power Supply [J]. Microelectronics, 2012,42 (1): 138-140.

[6] Wang F R , Ping Y U. Approach of state determination for secondary electrical power source based on Mahalanobis distance[J]. Transducer and Microsystem Technologies, 2016.

[7] Hinton G , Deng L , Yu D , et al. Deep Neural Networks for Acoustic Modeling in Speech Recognition:The Shared Views of Four Research Groups[J]. IEEE Signal Processing Magazine, 2012, 29(6):82-97.
[8] Dahl G E , Yu D , Deng L , et al. Context-Dependent Pre-Trained Deep Neural Networks for Large-Vocabulary Speech Recognition[J]. IEEE Transactions on Audio Speech & Language Processing, 2011,20 (1) : 30-42.
[9] Jun, Du, Li-Rong, et al. An Experimental Study on Speech Enhancement Based on Deep Neural Networks[J]. IEEE Signal Processing Letters, 2014.
[10] Gers, Felix A , Schmidhuber, et al. Learning to Forget: Continual Prediction with LSTM.[J]. Neural Computation, 2000.
[11] IEEE Transactions on Neural Networks & Learning Systems, 2016, 28(10):2222-2232. (in Chinese with English abstract) [10] Koutnik K, Srivastava R K, Srivastava R K, et al. A spatial Odyssey based on Neural Networks and Neural Networks [J].
[12] Huang Z , Wei X , Kai Y . Bidirectional LSTM-CRF Models for Sequence Tagging[J]. Computer Science, 2015.

Vibration response of defect-ball-defect of rolling bearing with compound defects on both inner and outer races

Wei Luo[1], **Changfeng Yan**[1,*], **Junbao Yang**[1], **Yaofeng Liu**[1], **Lixiao Wu**[1]

[1]School of Mechanical & Electrical Engineering, Lanzhou University of Technology, Lanzhou, China

*Corresponding author: changf_yan@163.com

Abstract. Aiming at the problem that the existing compound defects model of rolling bearings under radial load is difficult to reflect the actual contact between rolling elements and defects. A new model is proposed to accurately reflect the simultaneous or sequential contact between inner and outer race defects and rolling elements. Considering the coupled excitation between shaft and bearing and pedestal, time-varying displacement excitation, and radial clearance, a four degree-of-freedom vibration model of rolling bearing with compound faults on both inner and outer races is built. The vibration equations are calculated by the method of numerical way, and the model is verified by experiment. The vibration response characteristics of the Defect-Ball-Defect model are studied, which renders a theoretical criterion for bearing fault diagnosis.

Keywords: Compound defects, Defect-Ball-Defect model, dynamic model, vibration response characteristics, fault diagnosis.

1. Introduction

As one of the core components of rotating machinery, the bearings are used widely in modern industrial machinery. In the long-term operation, the bearing will have some problems such as pitting, spalling, cracking, and accidental loss of materials in the contact area, threatening the safety and reliability of mechanical equipment operation. The analyze on the dynamic model and vibration characteristics of the defects bearing is beneficial for structural design and optimization, diagnose bearing faults, and life calculation of rolling element bearing [1].

The vibration signal is a powerful means to diagnose the defect of the bearing. Lots of the rolling ball bearing models with composite defects were constructed to predict this signal. Patel et al. [2, 3] established a 6-DOF dynamics model of the bearing system to explore the vibration response of the rolling element bearing with local and multiple defects. Gao et al. [4] investigated the influence of the angle between fault and fault, the amount of faults, and the position of fault on the dynamic response of the ball bearings. Cao et al. [5] illustrated the vibration response modeling ways of cylindrical roller bearings with local fault, multiple faults, and composite faults. Wang et al. [6] investigated the influence of compound faults on the dynamic vibration response under different speeds, loads and sizes of faults, respectively. A dynamic model with multiple-point defects of the inner race was found by Yaqub et al. [7] to simulate the effect of the spalling position distribution on the rolling ball bearing vibration. Zhu et al. [8] introduced the Lempel-Ziv complexity measure into the model to indicate the disorderly of vibration signals of local and compound faults. Zhang et al. [9] set up the vibration model of the rolling bearings with compound faults on the outer race and rolling element and analyzed the variation of amplitude under different speed, defective size and load.

Although many efforts have been made in the composite defects resulted in the vibration response, there is little research about the vibration mechanism of ball bearing when the inner and outer race defects contact the ball simultaneously or successively. In this article, considering the coupled excitation between the shaft and bearing and pedestal, time-varying displacement and radial clearance, a 4-DOF vibration model of the bearing is established. The vibration equations are calculated by the numerical value method, and the model is verified by the experiment on the test rig of MFS. The vibration response characteristics of the Defect-Ball-Defect model are studied, which supplied a fresh idea for exploring the vibration response mechanism of compound defects between the rolling bearing inner race and outer race.

2. Establishment of the dynamic model

2.1. Model simplification

The interaction relationship of bearings is hugely complicated, so the bearing vibration model is built according to some reasonable assumptions: (a) The contact between rolling elements and races meets Hertz's contact condition. (b) The rolling elements are distributed evenly around the shaft. (c) Ignore the influence caused by inertial force and gyro moment. (d) The speed of the inner ring rotates is the same as motor, and the outer ring is fixed on the pedestal.

Based on the above assumptions, the bearing system is further simplified as the dynamic model presented in figure 1(a). Figure 1(b) shows that the contact between the inner ring-shaft and outer ring-pedestal are simplified as spring-damping connections. The collision between the ball and raceways is equivalent to linear spring connections.

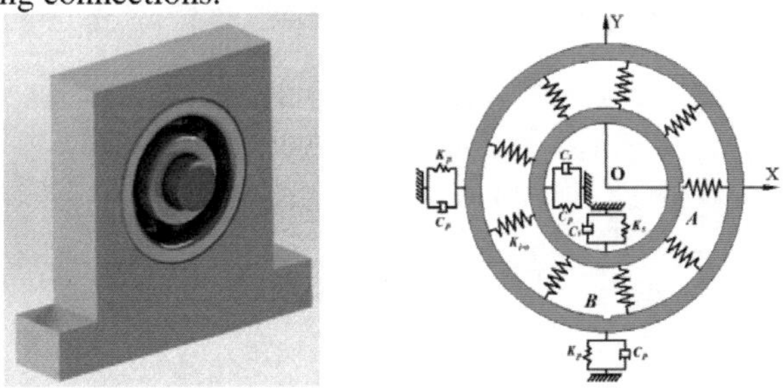

(a) (b)

Figure 1. Simplified bearing model. (a) Simplified model of outer ring-pedestal and inner ring-shaft. (b) Simplified model of the rolling elements and races.

2.2. Time-varying displacement excitation

In the compound defects model, one part of the time-varying displacement is given rise to the variable stiffness vibration of the bearing, and the other part is caused by the defects of the bearing races. So, the total radial deformation of compound defects can be given as follows:

$$\delta_{o+i} = (X_s - X_p)\cos(\theta_{bk}) + (Y_s - Y_p)\sin(\theta_{bk}) - C_r - \lambda_{o+i} \tag{1}$$

where C_r represents the radial clearance, λ_{o+i} refer to the additional displacement of rolling elements through inner and outer race defects, θ_{bk} denotes the position angle of k-th rolling element at t time, can be obtained by equation (2)

$$\theta_{bk} = 2\pi(k-1)/N_b + w_c t + \theta_0,\ k = 1,2,\cdots,N_b \tag{2}$$

where θ_0 is the position of the first rolling element at 0 s, N_b is the number of the ball, and w_c refers to the angular velocity of the cage, can be given by equation (3)

$$w_c = 0.5(1 - D_b / D_m) w_s \tag{3}$$

where D_b is the diameter of the rolling element, D_m is the pitch dimension, and w_s refers to the rotation speed of the shaft.

2.3. *Contact force*

In accordance with Hertz's contact condition, the time-varying contact force can be calculated [10].

$$F = K_b \delta^{1.5} \tag{4}$$

where δ refers to the deformation amount, and K_b references the equivalent contact stiffness coefficient, which can be written as following [10]:

$$K_b = \left[1/(1/K_i^{1.5} + 1/K_o^{1.5})\right]^{1.5} \tag{5}$$

where K_i and K_o represents the contact stiffness between the rolling elements and inner/outer races, and see Harris et al. 's research for calculation [10].

The contact force could be written by equation (6)

$$\begin{cases} F_x = K_b \sum_{i=0}^{N_b} (\eta \delta_{o+i})^{1.5} \cos\theta_{bj} \\ F_y = K_b \sum_{i=0}^{N_b} (\eta \delta_{o+i})^{1.5} \sin\theta_{bj} \end{cases} \tag{6}$$

where η is used to judge whether rolling elements enter the load area.

$$\eta = \begin{cases} 1 & \delta_{i+o} > 0 \\ 0 & \delta_{i+o} \le 0 \end{cases} \tag{7}$$

2.4. *Dynamic equation*

From the analysis in section 2.2 and 2.3, and combined with figure 1, a 4-DOF dynamic equation can be expressed as follow:

$$\begin{cases} M_s \ddot{X}_s + C_s \dot{X}_s + K_s X_s + F_x = W_x \\ M_s \ddot{Y}_s + C_s \dot{Y}_s + K_s Y_s + F_y = W_y \\ M_p \ddot{X}_p + C_p \dot{X}_p + K_p X_p - F_x = 0 \\ M_p \ddot{Y}_p + C_p \dot{Y}_p + K_p Y_p - F_y = 0 \end{cases} \tag{8}$$

where W_x and W_y are the external loads applied by the bearing system. The vibration equations are calculated by the Rung-Kutta method. The time step is 10^{-6} s, the rotation speed is 1200 r/min, the radial load is 50 N, and the axial load is 0 N. The initial displacement is $X_p=Y_p=X_s=Y_s=1\times10^{-6}$ mm, and the initial velocity is 0.

3. Defect-Ball- Defect solution model

The compound defects model can analyze the vibration signals of bearing under normal circumstances, but it can't explain the vibration response characteristics of Defect-Ball-Defect, quantitatively. Therefore, a solution model for the Defect-Ball-Defect contact condition is present.

As shown in figure 2(a), the model builds a coordinate with the center of the outer ring as the origin, setting the initial defect position of the outer raceway as 270° direction, setting the initial defect position of the inner race freely, and setting the initial position of the 1st ball as 0° . The ball rolls over the outer ring defect when the bearing starts to run. According to $w_s > w_c$, the rolling element will contact the inner and outer race defect simultaneously or after another. This model is defined as the Defect-Ball-Defect model.

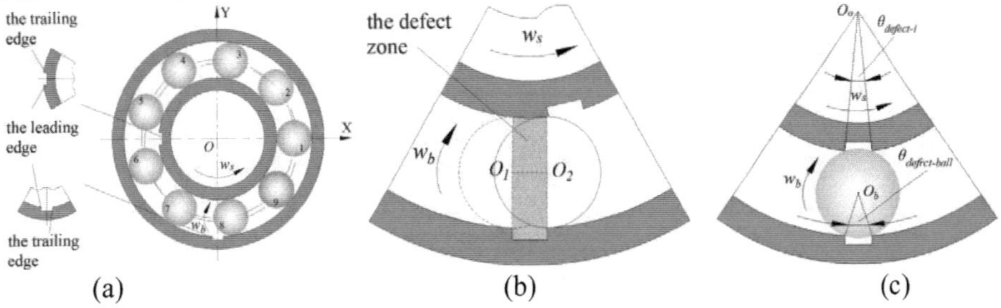

Figure 2. The solution model of Defect-Ball-Defect. (a) The position of defects and balls at 0 s. (b) Defect zone. (c) Angle is relative to the defect zone.

The half-angle of the balls in the defect area can be calculated by equation (9)

$$\theta_{defect-ball} = \arcsin(L_o / D_b) \tag{9}$$

The half-angle of the inner race in the spalling zone can be expressed as follow:

$$\theta_{defect-i} = \arcsin(L_i / D_i) \tag{10}$$

4. Experiment validation and model analysis

To further verify the rationality of the vibration model of rolling ball bearing built in this article. ER-16k bearing is selected as experimental bearing, a rectangular defect with a width of 0.6 mm and a depth of 0.25 mm was fabricated on the inner and outer races of the rolling bearing by Laser ablation to simulate the fault, and the basic parameters of the ER-16k bearing are displayed in Table.1. Figure 3 is the test rig for collecting acceleration signals of bearing.

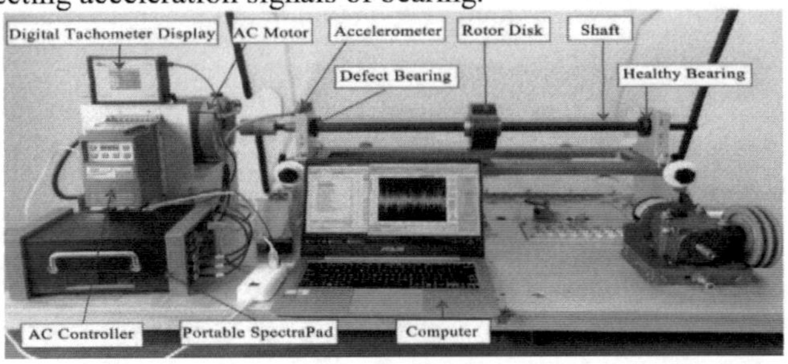

Figure 3. MFS test rig.

The rotating frequency is 20 Hz when the speed of shaft is 1200 r/min. The theoretical fault frequencies of the outer ring is 71.44 Hz and inner ring is 108.6 Hz, respectively, and see Harris et al.'s research for calculation [10].

Table 1. Parameters of the ER-16k bearing.

Parameters	Value	Parameters	Value
Inner race dimension D_i (mm)	30.59	Pitch dimension D_m (mm)	39.65
Outer race dimension D_o (mm)	46.47	Ball dimension D_b (mm)	7.94
Poisson ratio μ(/)	0.3	Number of ball N_b (/)	9

Young modulus E (GPa)	219	Radial clearance C_r (mm)	0.0055

The datas of the simulation bearing are the agree with the experimental. As shown in figure 4, the simulated signals of the rolling ball bearing with the composite defects and figure 4(b) is a time-domain amplified signal from 1.2 s to 1.3 s. The three equally spaced impulses can be seen in 1.228-1.27 s during the colldsion between the ball and outer raceway defect. Three equally spaced impulses can be seen in 1.23-1.257 s during the interaction between the rolling element and the inner race defect. Furthermore, the intervals are approximately 0.009 s and 0.014 s for the peak of each shock wave, which is equal to $1/f_{BPFI}$ and $1/f_{BPFO}$, respectively. It can be shown from figure 4(c) that the fault frequency of the shaft, outer and inner ring is 19.84 Hz, 71.41 Hz and 108.6 Hz, respectively.

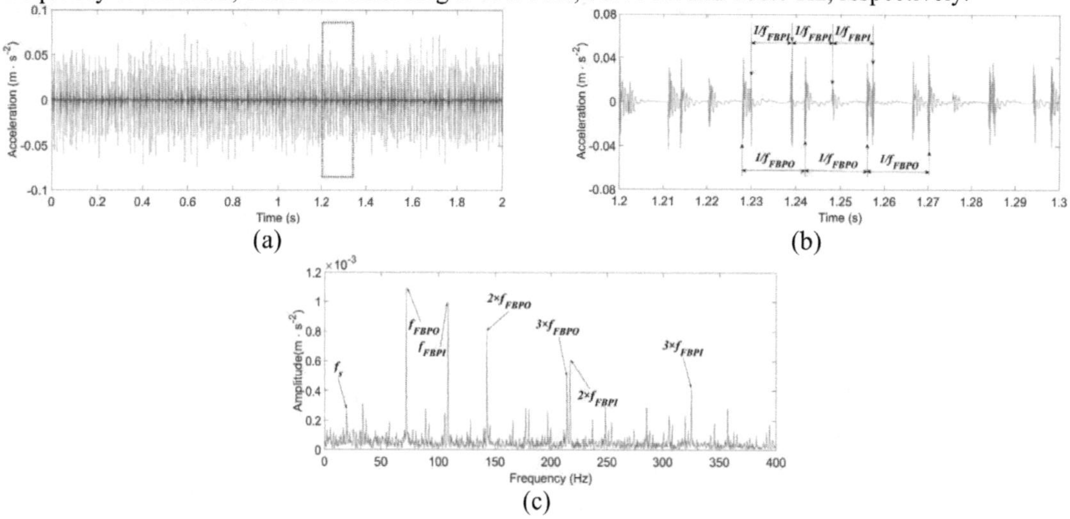

Figure 4. Simulated signals. (a) Acceleration waveform of time-domain. (b) Enlarged acceleration waveform of time-domain. (c) Acceleration waveform of frequency-domain.

Figure 5 shows the experimental signals of the rolling ball bearing with the compound defects. Figure 5(b) is a time-domain amplified signal. It can have three equally spaced impulses seen in 1.244-1.283 s during the colldsion between the ball and outer race defect. Three similarly spaced impulses can be seen in 1.253-1.28 s during the the ball impacts inner race defect. what's more, the intervals are approximately 0.009 s and 0.014 s for the peak of each shock wave, which is equal to $1/f_{BPFI}$ and $1/f_{BPFO}$, respectively.

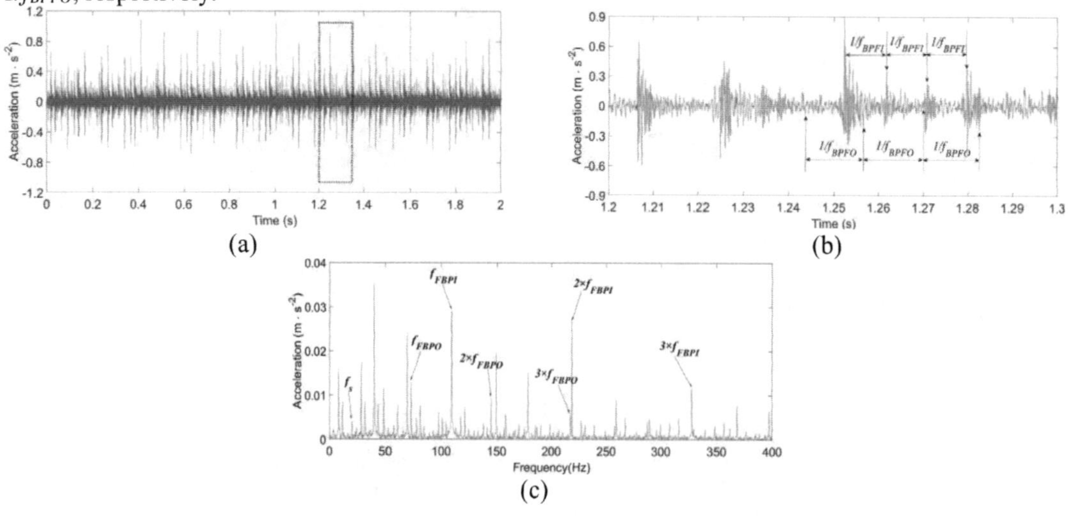

Figure 5. Experimental signal. (a) Acceleration waveform of time-domain. (b) Enlarged acceleration waveform of time-domain. (c) Acceleration waveform of frequency-domain.

In figures 4(a) and 5(a), the obvious periodic impulses appear to the experimental and simulated signal. In figure 4(c) and Figure 5(c), f_{BPFO}, f_{BPFI}, $2\times f_{BPFO}$, $3\times f_{BPFO}$, $2\times f_{BPFI}$ and $3\times f_{BPFI}$ can be seen obviously. In figure 5(c), the fault frequencies of the outer ring and inner ring are 72.42 Hz and 109.5 Hz for the experimental signals in the frequency domain, respectively. The errors of f_{BPFO} and f_{BPFI} between measured and simulated results are 1.39% and 0.82%, respectively. The main reason for the relative errors is the sliding of rolling elements and the vibrations from the shaft and pedestal.

As a result, the simulated acceleration signal agrees with experimental results and confirms the accuracy and effectiveness of the proposed fault model.

5. Vibration characteristic of Defect-Ball-Defect contact condition

The defects width of the inner and outer races is 0.5 mm, the depth is 0.25 mm, the applied load is 50 N, and the rotating speed of the shaft is 1200 r/min. The initial position of the inner race fault is 270° direction. Figure 6(a) shows the contact condition of bearing in the defect area within 0 to 0.5 s. It can be seen the Defect-Ball-Defect contact condition occurred in the running process of the 6th ball. Figure 6(b) is an enlarged view of the contact situation of the 6th ball. It can be seen from the drawing that the time of the 6th ball to roll over the fault zone is 0.1492s-0.1517 s, and the inner race defect to roll over the fault area is 0.1499-0.1501 s. Figure 6(c) is the time-domain signal, where t_{im1}, t_{im2} and t_{im3} are the impulse signals generated by the collision between the 7th ball, 6th ball and 5th ball after entering the outer race defect, respectively. t_{a1}, t_{ai1} and t_{a2} are impulse signals generated by the collision of the defective trailing edge of the inner race with the 8th ball, 7th ball and 6th ball, respectively.

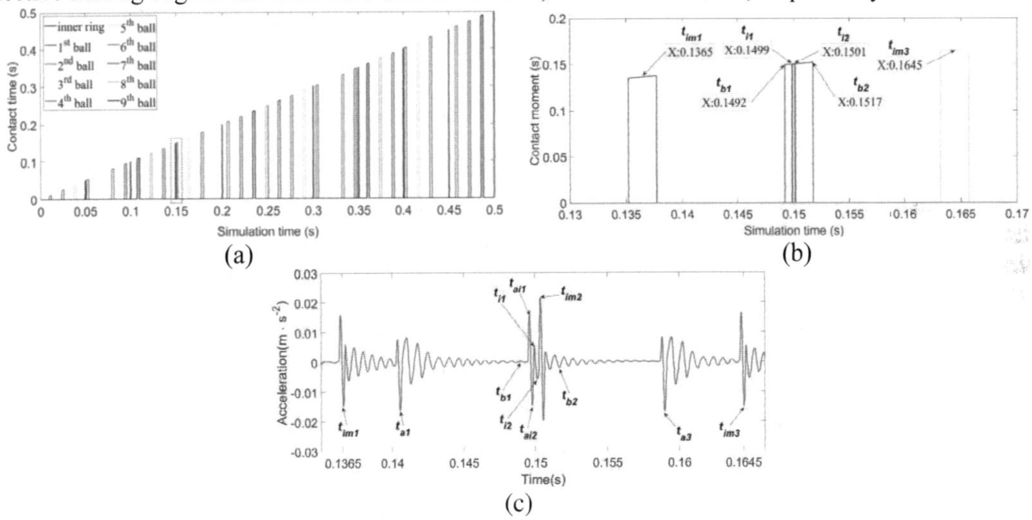

Figure 6. Defect-Ball-Defect contact condition. (a) The contact condition of the balls and inner ring defect within 0.5 s. (b) Enlarged the contact state for the 6th ball. (c) An enlarged time-domain diagram in this case.

The 6th ball starts entering the outer race defect at t_{b1}, reducing stress and showing a step signal in the time domain. Before the 6th ball collided with the defective trailing edge of the outer race, the inner race defect rolled over the 6th ball. t_{i1} to t_{i2} is the theoretical time for the inter race defect rolled over the 6th ball, but the time is t_{ai1} to t_{ai2} when the inner race defect rolled over the 6th ball in the simulation signal. It is 0.0002 s ahead compared with the theoretical value, caused by the deformation of the race caused by the variable stiffness vibration of the bearing in the load zone. The 6th ball collides the trailing edge of the defect of outer race at t_{im2}, which produced a high-frequency impulse signal in the time domain. Its acceleration amplitude is higher than other impulse signals. The reason is that the previous impulse signal had not wholly disappeared and the next impulse signal increased the acceleration amplitude. When time is t_{b2}, the 6th ball leaves the defect, regarded as a pressure recovery process, and the vibration signal gradually returns to stability.

6. Conclusions

Considering the effects of the coupled excitation between the shaft and bearing and pedestal, time-varying displacement excitation and radial clearance, a 4-DOF vibration model of rolling ball bearing with compound faults on both inner and outer races are found. The built model was validated by experiment. The results show that the simulation results agree with the test results, which proves the correctness and feasibility of the model.

Based on the compound faults model, the vibration response characteristics of the races defects and rolling elements contact simultaneously or successively are obtained by the Defect-Ball-Defect solution model. The results show that the time for the inner ring to impact the rolling element is 0.0002s earlier than the theoretical calculation value in the Defect-Ball-Defect contact condition. The coupling of vibration signals can be observed clearly, which increases the amplitude of vibration acceleration and accelerates the damage of bearings.

Acknowledgments

This work was supported by the National Natural Science Foundation of China (Grant No. 51765034) and the Hongliu First-class Disciplines Development Program of Lanzhou University of Technology.

References

[1] Alfares M, Saleem O, and Majeed M 2019 Analytical study of thermal variation impact on dynamics of a spindle bearing system *Proceedings of the Institution of Mechanical Engineers Part K Journal of Multi-body Dynamics* 233(4) 146441931984168

[2] Patel V N, Tandon N and Pandey R K 2010 A dynamic model for vibration studies of deep groove ball bearings considering single and multiple defects in races *Journal of Tribology* 132(4) 041101-0411101-10

[3] Patel V N, Tandon N and Pandey R K 2014 Vibrations generated by rolling element bearings having multiple local defects on races *Procedia Technology* 14 312-319

[4] Gao X, Yan C, Liu Y, Yan P and Wu L 2020 A 4-DOF dynamic model for ball bearing with multiple defects on raceways *Proceedings of the Institution of Mechanical Engineers Part K Journal of Multi-body Dynamics* 146441932097287

[5] Cao H, Su S, Jing X and Li D 2020 Vibration mechanism analysis for cylindrical roller bearings with single/multi defects and compound faults *Mechanical Systems and Signal Processing* 144 106903

[6] Wang K, Yan C, Wang F, Chang B and Wu L 2020 Dynamic characteristics of compound fault in deep groove ball bearing (in Chinese) *Journal of Harbin Institute of Technology* 52(1) 133-140

[7] Yaqub M, Gondal I and Kamruzzaman J 2012 Multiple-points fault signature's dynamics modeling for bearing defect frequencies *Proc.int.conf.contr* 2729-2734

[8] Zhu Y, Yuan X, Zhang Y, Zhu C and Zhou Z 2013 Vibration modeling of rolling bearings considering compound multi-defect and appraisal with lempel-ziv complexity (in Chinese) *Journal of Vibration and Shock* 32(16) 23-29

[9] Zhang X, Yan C, Liu Y, Yan P, Wang Y and Wu L 2020 Dynamic modeling and analysis of rolling bearing with compound fault on raceway and rolling element *Shock and Vibration* 2020(5) 1-16

[10] Harris T A and Kotzalas M N, Rolling bearing analysis, *Taylor & Francis Abingdon-on-Thames*, fifth ed., UK, 2006

Research on torsional vibration monitoring system of ship power shafting

Shufei Xue[1], Shuochen Wu[1], Qing Tang[1], Shulin Liu[1], Bai Liu[1]

[1] China Institute of Marine Technology and Economy, Beijing, China

Abstract. In this paper, the real-time monitoring technology of ship power system torsional vibration is studied. The photoelectric non-contact measurement method is used to measure the torsional vibration intensity of shafting, and the analysis service network management platform is established to realize the functions of real-time monitoring of shafting torsional vibration and upload and collect alarm data, through the front-end system, the torsional vibration strength and other parameters of the shafting are collected and transmitted to the background. The data of the background system is used to calculate and analyze, and the status of the shafting is alarmed. The system realizes continuous monitoring and data recording of torsional vibration index of ship power system, and ensures the operation performance and safe operation of ship power system. It provides theoretical and technical support for the future development of new technologies and related research, such as durability simulation and durability virtual test of ship power critical parts.

Keywords: Power shafting, Torsional vibration, Monitoring system, Sensor.

1. Introduction

Marine propulsion shafting is an important part of a marine power plant, which is mainly composed of power plant, transmission shaft and propeller[1]. The transmission shaft is used to transmit the power from the power plant to the propeller to make it rotate and generate propulsion. The torsional vibration of shafting is one of the important factors affecting the reliability of marine power plants. Severe torsional vibration will cause excessive alternating stress in shafting, resulting in deformation and fatigue failure of shafting, and even torsion fracture[2]. In recent years, frequent shaft breaking accidents have sounded an alarm for us. In order to improve the safety of ships in all aspects, we should pay special attention to the torsional vibration of ship propulsion shafting[3]. Therefore, this paper studies the real-time monitoring technology of ship power system torsional vibration, develops a set of ship power shafting torsional vibration monitoring system, realizes the continuous monitoring and data recording of ship power system torsional vibration index, and ensures the operation performance and safe operation of the ship power system.

2. System composition and signal acquisition

2.1. System Composition

The real-time monitoring system of shafting torsional vibration is a comprehensive system with the functions of real-time data acquisition and processing, analysis, alarm, record storage, print output and so on. It includes the front-end data acquisition and preprocessing system, including the measurement and analysis system of shafting torsional vibration stress and shaft power, shafting vibration measurement and analysis system, shafting energy efficiency monitoring and feedback system, transmission system and back-end processing system. The overall block diagram is shown in Figure 1.

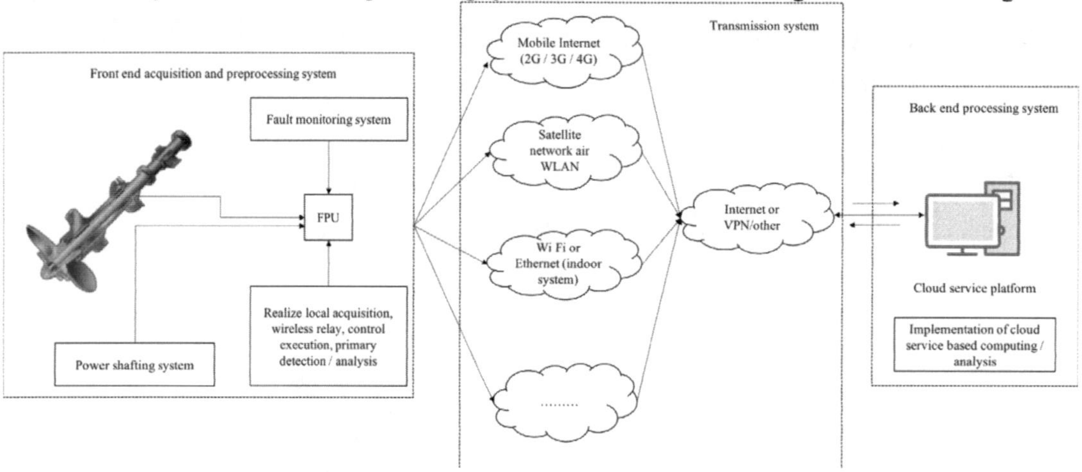

Figure 1. Detection system of marine power shafting.

2.2. *Signal Acquisition*

As shown in the figure, the measurement and analysis of ship power shafting torsional vibration mainly adopts photoelectric non-contact measurement method to measure the torsional vibration strength of shafting. The vibration signal of shafting is collected by sensors, and the amplitude of shafting vibration is judged in the monitoring and early warning unit through preprocessing technology. According to the preset alarm parameters, combined with intelligent comprehensive judgment, adaptive correction technology, fault feature model prediction, fast Fourier transform (FFT), cloud computing intelligent upgrade and other technologies, the monitoring and early warning of shafting vibration is carried out. Installation diagram of each sensor

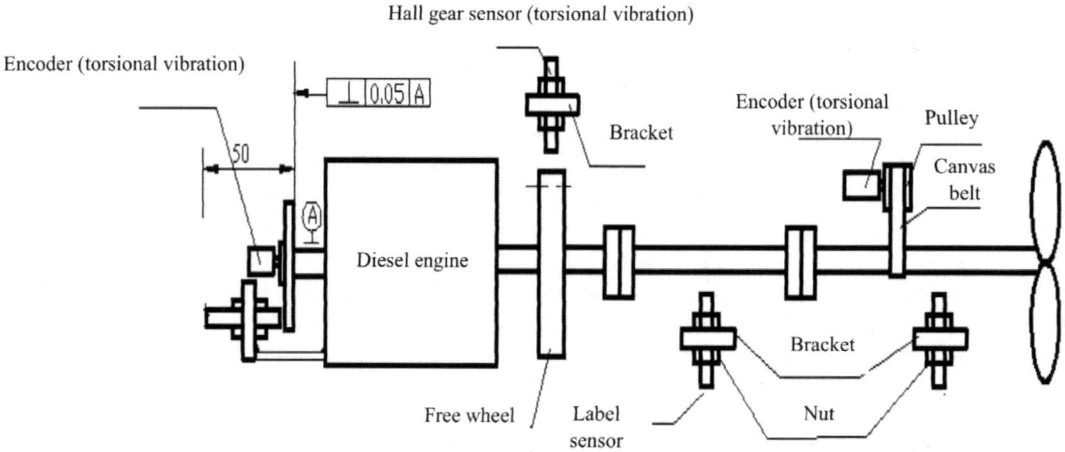

Figure 2. Torsional vibration signal acquisition.

2.3. *Torsional Angle Threshold and Actual Measured Value*

The torsional vibration monitoring system (the first generation) of marine power shafting is a tool used to continuously monitor the torsional angle of power shafting. When the amplitude reaches or exceeds the safety limit, the system will send out an alarm signal. At the same time, the system can display the engine speed, operation time and so on. The amplitude obtained by measuring torsion angle can be output to PC for result analysis. As shown in the figure, the actual measured torsion angle of 70 turns, the actual measured torsion angle of 60 turns and the torsion angle threshold are respectively. It can be seen from the threshold chart that the threshold values corresponding to the torsional vibration speed of 60 and 70 revolutions are 23 and 22 respectively. In the actual measured torsional angle values, the peak value of torsional angle at 60 revolutions is about 16, and the peak value of torsional angle at 70 revolutions is about 12, which do not exceed the corresponding threshold values. It can be concluded that the vessel is in normal operation.

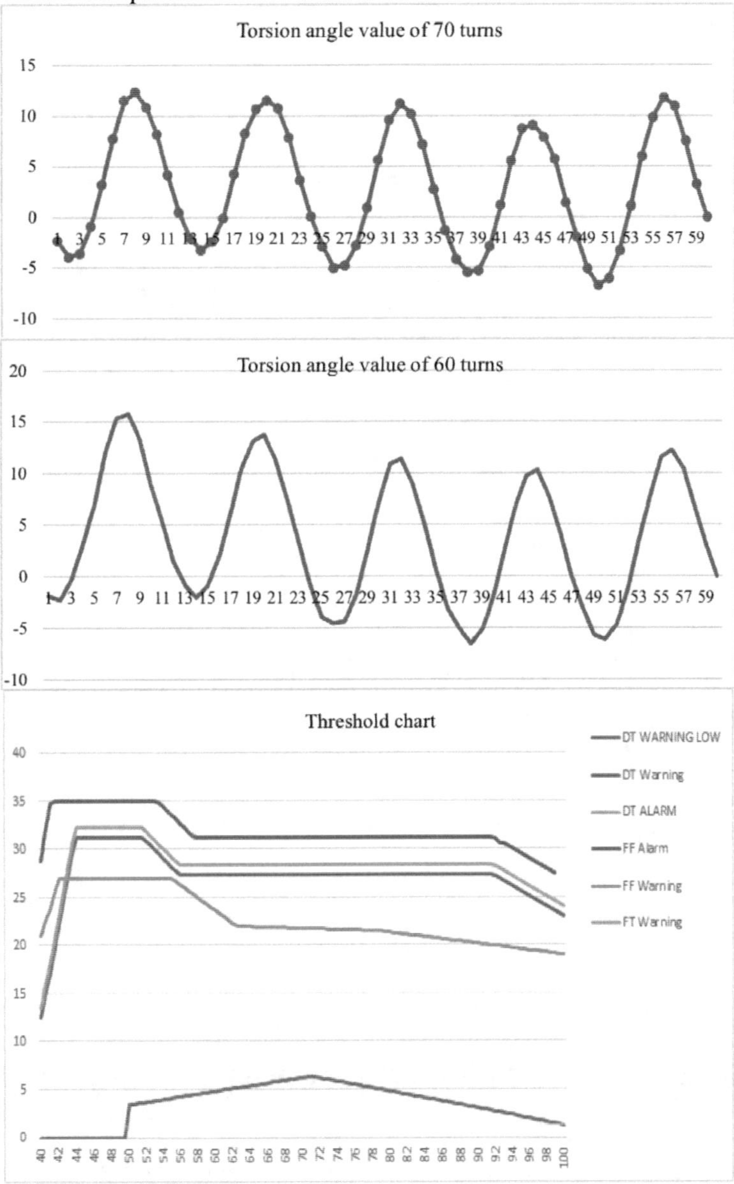

Figure 3. Torsional angle threshold and actual measured value.

3. Design of Shafting Torsional Vibration Online Monitoring System

Development of algorithm model research and embedded software of rotor vibration monitoring and fault alarm prediction analysis system: in the monitoring, alarm and early warning unit, development of various application module components based on different algorithm models, running in high-performance embedded processor system under the control of embedded real-time operating system (RTOS), and its technical architecture is shown in Figure 4. The data acquisition of torsional vibration, rotational speed, shock absorber oil temperature and pressure are carried out by the sensor module. After preprocessing the collected data, real-time data analysis, fault prediction characteristic analysis and adaptive data analysis based on cloud service after data storage can be carried out.

On the one hand, the real-time data analysis can go through the time domain amplitude calculation module and frequency domain characteristic analysis module. The analysis results can be displayed and queried in real time, and the working state of the shafting can also be comprehensively judged by the time-frequency analysis results. According to the appearance time and frequency-domain threshold data comparison, the safety alarm or limit alarm can be given, and the occurrence time of various alarms in the system can be monitored, duration and overrun data are stored and analyzed.

On the other hand, the fault prediction characteristic analysis module filters and processes the real-time monitoring data and alarm records according to the pre-stored parameters and algorithms, and gives fault warning according to the processing results. The adaptive data analysis module adjusts the alarm threshold data in time domain and frequency domain appropriately through the long period analysis of the measured historical data and the preset adaptive algorithm. In addition, the data upgrade module from the cloud can collect and upload the measured data and alarm data of the ship on the one hand, and update the fault early warning model and adaptive data analysis model upgraded by cloud computing on the other hand. The cloud service module can also update the time domain threshold data and frequency domain threshold data. The comprehensive judgment module gives the alarm signal intelligently through the preset strategy model judgment.

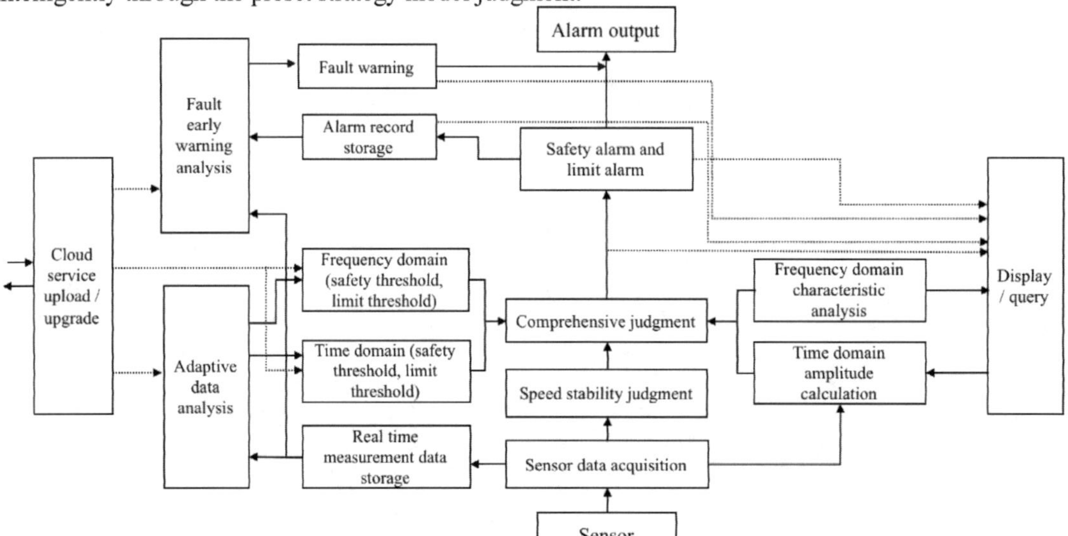

Figure 4. Technical framework of torsional vibration monitoring system.

4. Demonstration of Ship Shafting Torsional Vibration Monitoring System

4.1. Design of Calculation and Analysis Service Platform for Monitoring System

The calculation and analysis program uses the big data obtained from the real-time torsional vibration monitoring of multiple ships to analyze and process the data including the real-time torsional vibration monitoring data of multiple ship types, alarm record data, torsional vibration fault data, etc. Through various analysis programs constructed by the server, extracts the data features, establishes the torsional vibration feature database, and establishes various analysis models, the model algorithm is extracted and an embedded software package is generated to upgrade the shafting torsional vibration monitoring system.

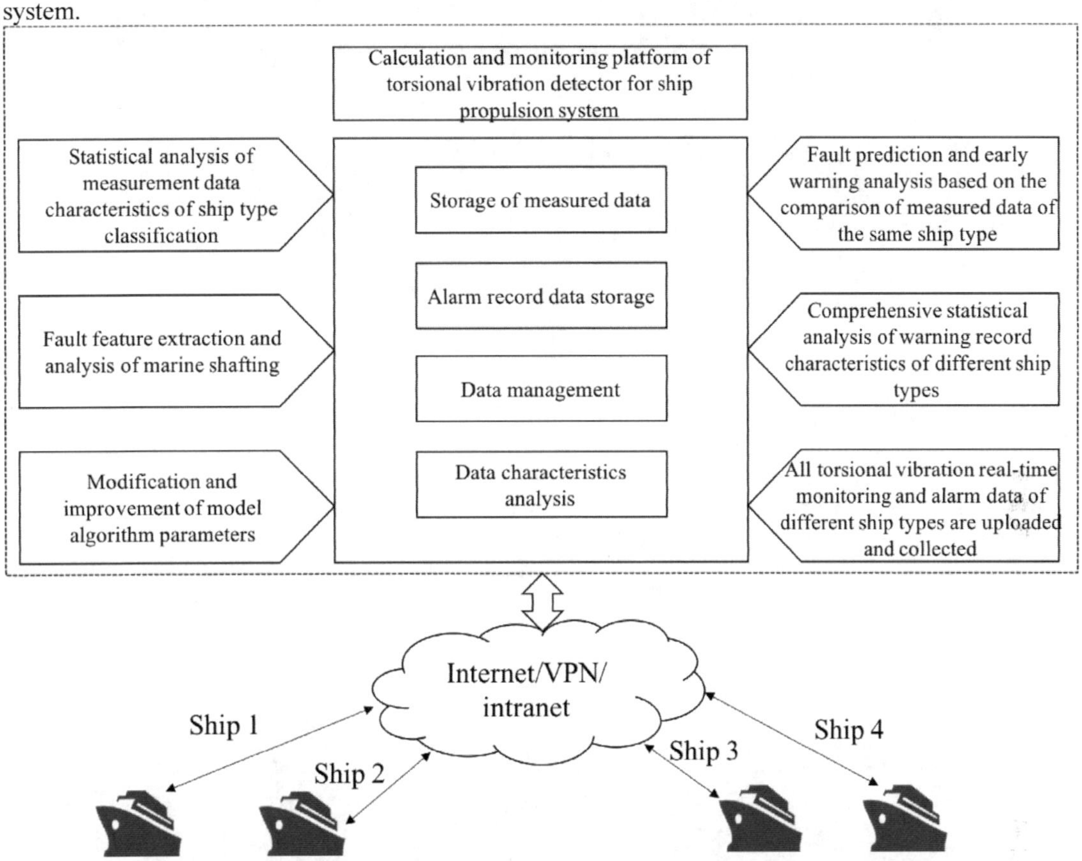

Figure 5. Block diagram of system calculation and analysis service platform.

The network management platform of calculation and analysis service as shown in Figure 5 is established to provide upload, management and upgrade service systems for all shafting torsional vibration monitoring systems of the whole ship system. Based on the above-mentioned big data analysis and processing, the system realizes various processing and analysis functions, including real-time torsional vibration monitoring and alarm data upload and collection; Statistical analysis of monitoring data characteristics combined with ship type classification; Comprehensive statistical analysis of alarm record characteristics of different ship types; Real time monitoring data and fault characteristics extraction and analysis of shafting fault ship; The structure and parameters of adaptive model algorithm are improved and modified; The improvement and modification of algorithm structure and parameters of shafting fault prediction model; Fault prediction and early warning analysis based on real-time monitoring data comparison of the same ship type.

4.2. Function Demonstration of Monitoring System

The monitoring host is usually placed in the engine control panel in the engine control room. All cable connections are located at the rear of the monitoring host, which is composed of a touch screen and printed circuit board with a microprocessor. It has the function of calculating the phase speed of rotating elements.

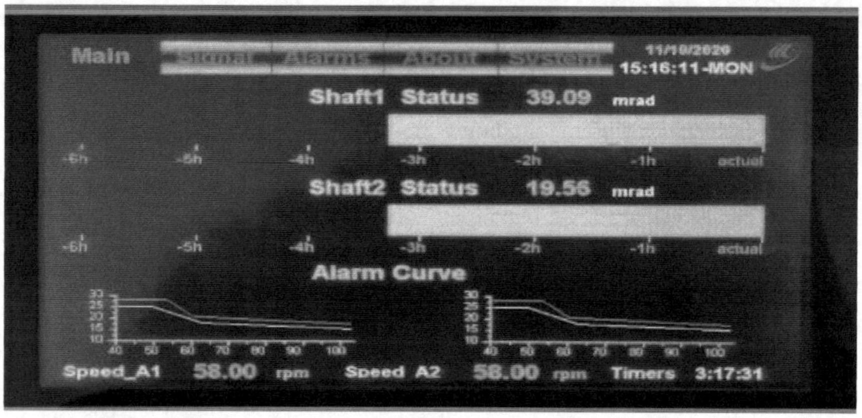

Figure 6. Torsional vibration monitoring system of ship power shafting.

The specific monitoring interface is shown in the figure below, as shown in Figure a, the main interface shows the trend of torsional vibration of the current relevant parameters of the monitored components. As shown in Figure b, the signal interface displays the real-time measured values of the relevant parameters of the torsion angle of the power shaft in one cycle, including the real-time amplitude of the rotation angle in one cycle. As shown in Figure c, the curve interface shows the specific power shaft torsion angle limit curve of the monitored component. Color: Orange: warning limit curve; Red: alarm limit curve. As shown in Figure d, the alert log display includes two parts: the current active alert and the logged alert. As shown in Figure e, the system check interface displays the information of the monitor system itself. As shown in Figure f is the warning recording interface, the system sends the alarm signal to AMS of the ship, and AMS makes a prompt or alarm sound to inform the operator. After the operator confirms the event, the monitoring system stops transmitting the alarm information, and AMS stops responding. The operator can also check all active alarms on the alarms page.

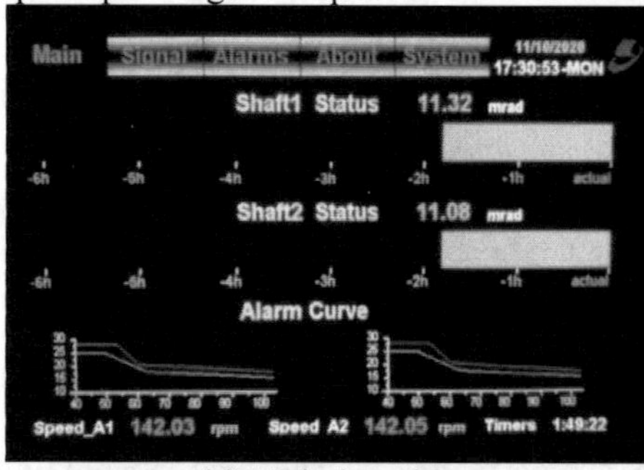

(a) Main interface.

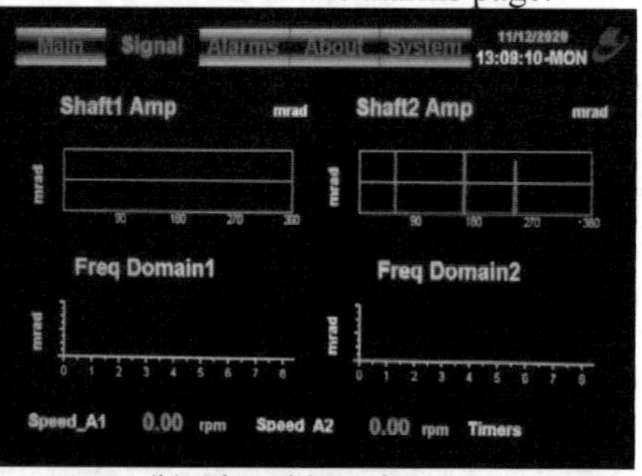

(b) Signal interface.

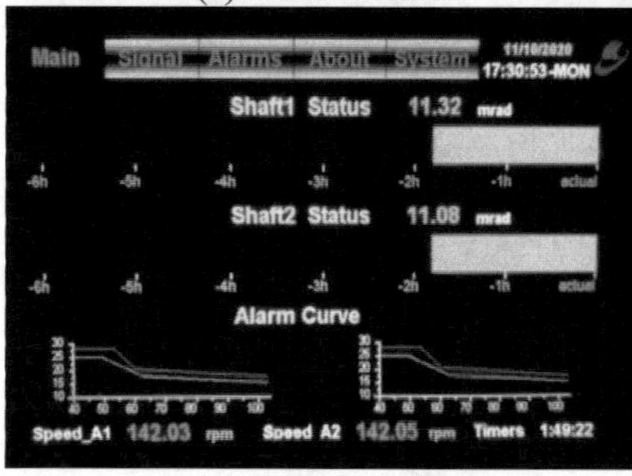

(c) Curve interface.

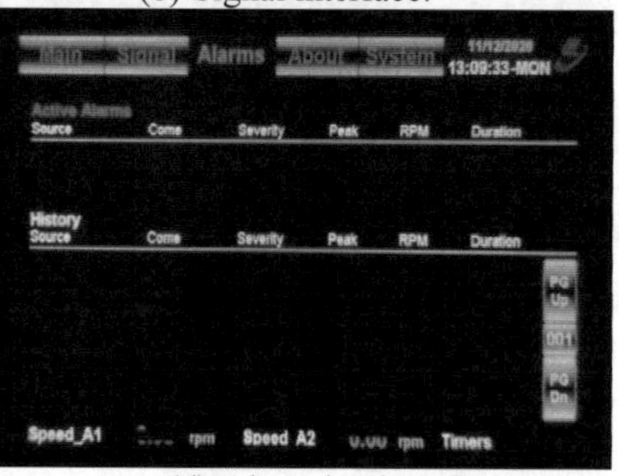

(d) Alarm log interface.

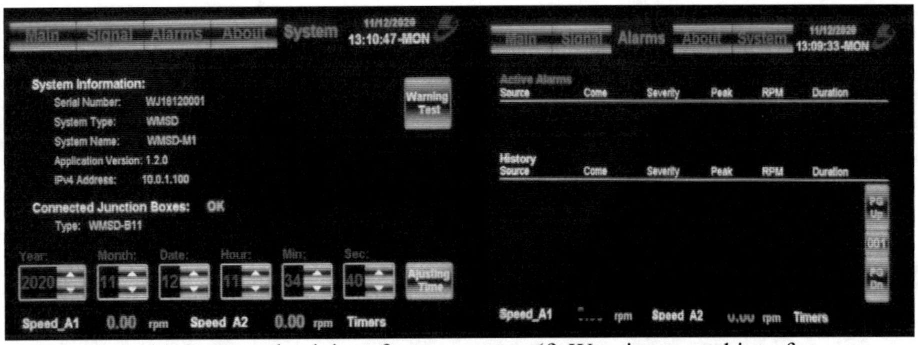

(e) System check interface. (f) Warning record interface.

Figure 7. Function demonstration of each part of the monitoring system.

5. Conclusions

In this paper, the technical research of ship power shafting torsional vibration monitoring system meets the requirements of multi working conditions, dual channel synchronous online monitoring of ship power shafting torsional vibration, and plays an important role in ensuring the main engine operation performance and safe operation of the ship:

(1) Compared with the regular monitoring of shafting torsional vibration characteristics in the past, the marine power shafting torsional vibration monitor can complete the continuous monitoring of the indicators and obtain the first-hand real-time data, so as to analyze the continuous slow change and short-term mutation of shafting characteristics, and grasp the abnormal or fault symptoms of marine propulsion shafting in time, so as to take corresponding measures, timely replacement or repair of failed parts to avoid or reduce the occurrence of major accidents.

(2) The fault prediction function of the torsional vibration monitor is the biggest characteristic of the instrument which is different from the previous torsional vibration tester. Based on a large number amount of continuous accumulated data, the multi-party early warning calculation model is designed to improve the accuracy of fault diagnosis of monitoring systems.

(3) Once a fault occurs in the propulsion shafting of a ship, the monitoring system can automatically record the complete data and information of the fault process, so as to diagnose and analyze the fault causes afterwards and avoid the occurrence of similar faults again; Through the analysis of the causes and properties of the abnormal state of ship propulsion system, the change of its operation state, the deterioration degree and development trend of the performance of main parts and subsystems are mastered, and the normal operation time is predicted, which provides the basis for the scientific decision-making of equipment maintenance time and mode.

References

[1] Han Yangquan, Fan Ming, Ji Haijun. Analysis of influencing factors on torsional vibration of ship propulsion shafting. Guangdong shipbuilding, 2014,34 (04): 65-67 + 77.

[2] Mao Lanxia, Wang Jianping. Analysis of influence factors on torsional vibration of ship propulsion shafting. Jiangsu shipbuilding, 2012, 000 (005): 26-28.

[3] Zhang wenshuai. Torsional vibration analysis and vibration avoidance of ship propulsion shafting. Harbin University of Technology, 2011.

Graph-domain features and their application in rotating machinery fault diagnosis

Yiyuan Gao[1, 2, *], Wenliao Du[2], Xiaoyun Gong[2], Dejie Yu[1]

[1]State Key Laboratory of Advanced Design and Manufacturing for Vehicle Body, Hunan University, Changsha, China
[2]Mechanical and Electrical Engineering Institute, Zhengzhou University of Light Industry, Zhengzhou, China

*Corresponding author: yygao@hnu.edu.cn

Abstract. To more effectively extract the non-stationary and non-linear fault features of mechanical vibration signals, a novel fault diagnosis method for rotating machinery is proposed combining time-domain, frequency-domain with graph-domain features. Different from the conventional time-domain and frequency-domain features, the graph-domain features generated from horizontal visibility graphs can extract the fault information hidden in the graph topology. Aiming at the problem that too many features will lead to information redundancy, the Fisher score algorithm is applied to select several of sensitive features which are then fed into the support vector machine to diagnose the faults of rotating machinery. Experimental results indicate features extracted from the three domains can be used to obtain higher diagnosis accuracy than that extracted from any single domain or dual domains.

Keywords: Rotating machinery, fault diagnosis, horizontal visibility graph, graph-domain features.

1. Introduction

In recent years, graph signal processing [1] has provided a new perspective for mechanical fault diagnosis. Based on the graph, new vertex domain, graph spectral domain and vertex-graph spectral domain are developed and applied to the analysis of mechanical vibration signals, which correspond to the traditional time domain, frequency domain and time-frequency domain respectively. The corresponding and transformation relations of these six signal domains are shown in Figure 1. The premise of graph signal processing for mechanical fault diagnosis is to convert the vibration signal into an appropriate graph. To successfully introduce the graph Fourier transform, Ou et al. [2] first mapped a bearing vibration signal into a path graph based on the structural correspondence between time series and path graph. Researches show that horizontal visibility graph (HVG) can characterize time series more exactly and has better performance for retaining the dynamic characteristics of bearing vibration signals than path graph [3, 4]. Hence, vibration signals of rotating machinery are mapped into a HVG in this paper.

Researchers have also constructed many statistical features in the graph domains to express the structural and dynamic characteristics of graphs. At present, the existing graph-domain features include the vertex-domain features and graph spectral-domain features. The vertex-domain features, such as mean vertex degree, mean path length and clustering coefficient, have been widely applied in the dynamic analysis of temperature sequence, human heartbeat and other time series [5, 6]. There are two ways to obtain the graph spectral-domain features. The first is to construct the functions of the eigenvalues of the graph matrices, which are mainly used for describing the internal structural characteristics of the chemical molecules and molecular recognition [7, 8]; the other is to analogize the conventional frequency-domain features, which have been used for diagnosing the rolling bearing faults effectively [9].

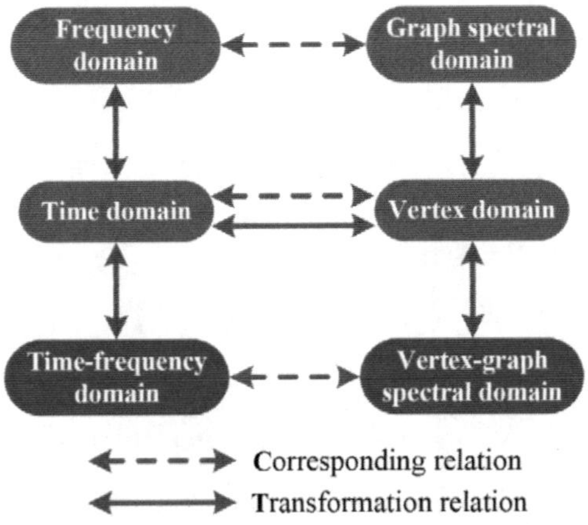

Figure 1. Corresponding and transformation relations of the six signal domains.

Feature extraction is significantly important for rotating machinery fault diagnosis. In addition to the traditional time-domain and frequency-domain features, the existing graph-domain features, including vertex-domain features and graph spectral-domain features, are also extracted after each vibration signal is mapped into a HVG. Too many features will lead to information redundancy, hence the Fisher score algorithm (FS) [10], one of the most widely used supervised feature selection algorithm, is used to measure the sensitivity of all features. Finally, several of more sensitive features are fed into the support vector machine (SVM) [11] to diagnose the rotating machinery faults. Experimental data demonstrate that the proposed method using features in time domain, frequency domain and graph domain has higher diagnostic accuracy than other methods using features in any single domain or dual domains. This is mainly because the feature of graph-domain has the ability to extract fault information hidden in graph topology effectively, and the combination of time-domain, frequency-domain and graph-domain features can provide more fault information.

The remainder of the paper is arranged as follows. The related theory is introduced in Sections 2 and 3. Then the proposed method is described in Section 4. In Section 5, the experimental results are analyzed. Finally, the conclusions of the paper are performed in Section 6.

2. Horizontal visibility graph

The horizontal visibility algorithm is proposed by Luque et al. [12], which can accurately map the dynamics characteristics of a random time series into the topological properties of a HVG. Generally, a graph consists of a finite amount of vertices and edges connecting the vertices. Let $\{x_i\}_{i=1,2,\cdots,N}$ be a time series with N any real values, each vertex in the HVG corresponds to each value of the time series in turn. If the subsequent geometric criterion is satisfied in the time series, the vertices i and j in the HVG will be connected by an undirected edge.

$$x_i, x_j > x_n, \ \forall n \,|\, i < n < j. \tag{1}$$

A graph can be expressed by its adjacency matrix **A**. HVG is an undirected and unweighted graph, and hence HVG's adjacency matrix is a symmetric 0-1 matrix. If there are an edge between any two vertices, the corresponding element in the adjacency matrix is set to 1, otherwise, 0.

Another commonly used matrix for graphs is the Laplacian matrix **L**. The Laplacian matrix **L** is expressed as **L=D-A**, where **D** is a diagonal matrix whose ith diagonal element equals the number of edges connected to the vertex i, that is, the sum of all the elements in the ith row or ith column of **A**.

In addition, the graph signal is interpreted as a mapping from a set of vertices in the HVG to a set of real numbers in the time series. The graph signal f can be written in the form of a vector as Eq. (2).

$$f = [f_1\ f_2 \cdots f_{N-1}\ f_N]^{\mathrm{T}} \in \mathbb{R}^N. \quad (2)$$

where the signal value f_i in the graph signal (the ith data value of the time series) is indexed by the vertex i in the HVG.

3. Graph-domain features

In this paper, the extracted graph-domain features include the vertex-domain features and graph spectral-domain features, which are described below.

3.1. Vertex-domain features

Degree, the number of edges connected to a vertex, is a simple and important concept among the attributes of individual vertices in graphs. The larger the degree of a vertex, the more important it is. If k_i represents the degree of the vertex i, the mean vertex degree of the graph is defined as Eq. (3).

$$K = \frac{1}{N}\sum_{i=1}^{N} k_i \quad (3)$$

The distance d_{ij} between two vertices i and j in the graph is set to the amount of edges connecting the shortest path between these two vertices, and the mean path length shown in Eq. (4) represents the mean of the distances between any two vertices.

$$P = \frac{1}{N(N-1)}\sum_{i\neq j} d_{ij} \quad (4)$$

The clustering coefficient reflects the tightness between the vertices of a graph. The local clustering coefficient C_i is the clustering coefficient of each vertex in the graph, and the global clustering coefficient C shown in Eq. (5) is the mean of the local clustering coefficients of all vertices.

$$C_i = \frac{2E_i}{k_i(k_i-1)} \qquad C = \frac{1}{N}\sum_{i=1}^{N} C_i \quad (5)$$

where E_i is the number of the closed triangle structures connected to the vertex i, and $k_i(k_i-1)/2$ is the total amount of the triples connected to vertex i.

In the field of graph signal processing, there is also an important vertex-domain feature, namely the graph Laplacian quadratic form shown in Eq. (6), which provides a measure of the global smoothness of a graph signal.

$$S = f^{\mathrm{T}}\mathbf{L}f \quad (6)$$

3.2. Graph spectral-domain features

At present, there are two ways to obtain the graph spectral-domain features. The first is to construct the functions of the eigenvalues of the graph matrices. **A** and **L** are the two most commonly used matrices. For the HVG mapped from a mechanical vibration signal, both **A** and **L** are real symmetric matrices. Therefore, **A** and **L** can be directly subjected to the standard orthogonal decomposition to obtain their eigenvalues and eigenvectors. λ_i and μ_i represent the eigenvalues of **A** and **L** respectively, and the common graph spectral-domain features are defined in Table 1.

Table 1. Common graph spectral-domain features.

Feature name	Feature expression	Feature name	Feature expression

Graph energy	$A_1 = \sum_{i=1}^{N} \lvert \lambda_i \rvert$	The sum of powers of the Laplacian eigenvalues of graphs	$L_3 = \sum_{i=1}^{N} \lvert \mu_i \rvert^{1/3}$
Estrada index	$A_2 = \sum_{i=1}^{N} e^{\lambda_i}$		$L_4 = \sum_{i=1}^{N} \mu_i^2$
Laplacian energy	$L_1 = \sum_{i=1}^{N} \lvert \mu_i - 2M/N \rvert$		$L_5 = \sum_{i=1}^{N} \lvert \mu_i \rvert^3$
Laplacian-energy-like invariant	$L_2 = \sum_{i=1}^{N} \sqrt{\mu_i}$		

The other is to analogize the conventional frequency-domain features. Similar to the definition of the traditional Fourier transform, the graph Fourier transform of a graph signal f shown in Eq. (7) can be expressed as f expanded with the eigenvectors of **L** [1].

$$\hat{f}(\mu_i) := \langle f, \upsilon_i \rangle = \sum_{i=1}^{N} f_i \upsilon_i^* \tag{7}$$

where $\hat{f}(\mu_i)$ and υ_i represent the graph spectrum coefficient and the eigenvector corresponding to the eigenvalue μ_i of **L**, respectively. By analogizing with the frequency-domain features in the reference [13], the corresponding 13 graph spectral-domain features are defined in Table 2.

Table 2. Thirteen graph spectral-domain features corresponding to frequency-domain features.

Feature expression	Feature expression	Feature expression
$G_1 = \frac{1}{N}\sum_{i=1}^{N} \hat{f}(\mu_i)$	$G_2 = \frac{1}{N-1}\sum_{i=1}^{N} (\hat{f}(\mu_i) - G_1)^2$	$G_3 = \frac{\sum_{i=1}^{N} (\hat{f}(\mu_i) - G_1)^3}{N(\sqrt{G_2})^3}$
$G_4 = \frac{\sum_{i=1}^{N} (\hat{f}(\mu_i) - G_1)^4}{N G_2^{\,2}}$	$G_5 = \frac{\sum_{i=1}^{N} \mu_i \hat{f}(\mu_i)}{\sum_{i=1}^{N} \hat{f}(\mu_i)}$	$G_6 = \sqrt{\frac{1}{N}\sum_{i=1}^{N} (\mu_i - G_5)^2 \hat{f}(\mu_i)}$ $G_7 = \frac{G_6}{G_5}$
$G_8 = \frac{\sum_{i=1}^{N} (\mu_i - G_5)^3 \hat{f}(\mu_i)}{N G_6^{\,3}}$	$G_9 = \frac{\sum_{i=1}^{N} (\mu_i - G_5)^4 \hat{f}(\mu_i)}{N G_6^{\,4}}$	$G_{10} = \frac{\sum_{i=1}^{N} \sqrt{(\mu_i - G_5)} \hat{f}(\mu_i)}{N G_6}$
$G_{11} = \sqrt{\frac{\sum_{i=1}^{N} \mu_i^2 \hat{f}(\mu_i)}{\sum_{i=1}^{N} \hat{f}(\mu_i)}}$	$G_{12} = \sqrt{\frac{\sum_{i=1}^{N} \mu_i^4 \hat{f}(\mu_i)}{\sum_{i=1}^{N} \mu_i^2 \hat{f}(\mu_i)}}$	$G_{13} = \frac{\sum_{i=1}^{N} \mu_i^2 \hat{f}(\mu_i)}{\sqrt{\sum_{i=1}^{N} \hat{f}(\mu_i) \sum_{i=1}^{N} \mu_i^4 \hat{f}(\mu_i)}}$

4. The proposed fault diagnosis method for rotating machinery

Based on the time-domain, frequency-domain and graph-domain features, a novel fault diagnosis method is proposed for rotating machinery. The proposed method is described as follows:

(1) Signal collection. The vibration signals of rotating machinery under different conditions are collected. The vibration signal of each condition is directly separated into two portions: the first portion is used to randomly and uninterruptedly capture the training samples; the second portion is used to randomly and uninterruptedly capture the testing samples.

(2) Feature extraction. First, 11 time-domain features and 13 frequency-domain features in the reference [13] are directly extracted. Next, each sample is mapped into a HVG, and then the corresponding graph signal indexed by the HVG can be obtained. Finally, the 24 graph-domain features including the 4 vertex-domain features (K, P, C, S) and the 20 graph spectral-domain features ($A_{1\text{-}2}$, $L_{1\text{-}5}$, $G_{1\text{-}13}$) are also extracted. For each sample, a total of 48 features are extracted from the time domain, frequency domain and graph domain. Notably, all features are normalized to eliminate the dimensional influence between feature data.

(3) Feature selection. The FS algorithm is applied to calculate the sensitivity of all features, and several of more sensitive features are chosen as the fault features of rotating machinery. The training feature subsets and testing feature subsets are established using those selected features.

(4) Pattern recognition. After feeding the training feature subsets into the SVM for training, the testing feature subsets are fed into the trained SVM model, and then the state of each testing sample is identified according to the output results.

Figure 2 illustrates the flow chart of the proposed method.

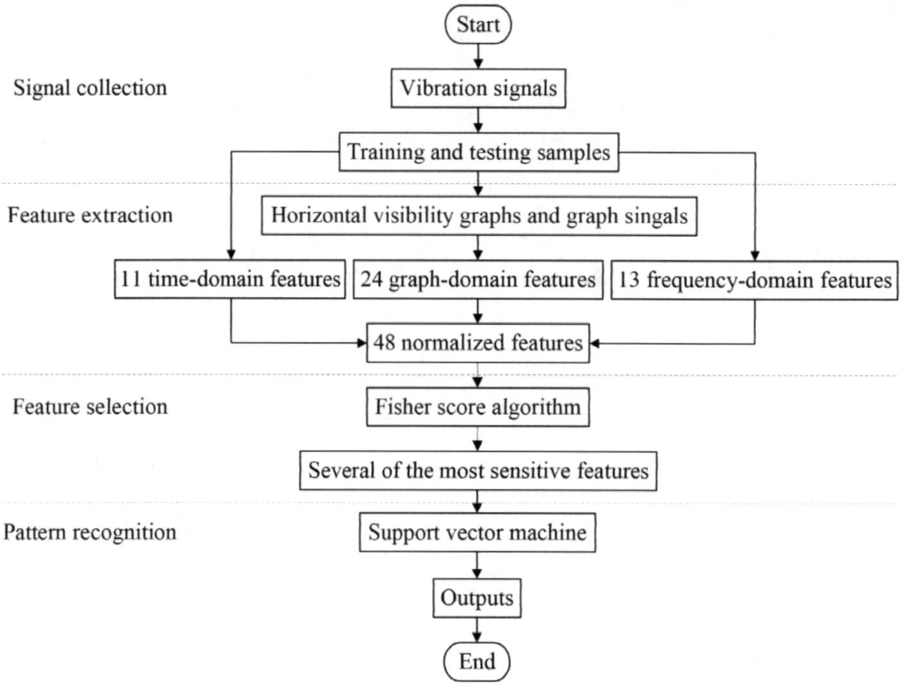

Figure 2. Flow chart of the proposed method.

5. Experimental validations

5.1. Vibration data description

To verify the effectiveness of the proposed method, vibration signals of rotating machinery under eight working conditions are collected on the test bench of Hunan University. Bevel gears and roller bearings are selected as the types of gears and bearings respectively. In this experiment, vibration signals of different bearing and gear faults are picked up by the accelerometer placed on the bearing bases under the condition that the motor speed is 1200rpm, the sampling frequency is 10240Hz and the load is 4Nm. The fault sizes of bearing and gear are 0.2mm and 0.6mm, respectively. The amount of samples for each working condition is set to 50, and the length of each sample is set to 1024. In each working condition, 10 samples are randomly chosen as training samples, and the remaining 40 samples are testing samples. A total of 80 training samples and 320 testing samples are collected. The detailed description about eight working conditions is listed in Table 3.

Table 3. Detailed description of eight working conditions.

Bearing condition	Gear condition	Number of training/ testing samples	Condition label
Normal	Normal	10/40	1
Inner race fault	Normal	10/40	2
Outer race fault	Normal	10/40	3
Cage fault	Normal	10/40	4
Normal	Gear crack	10/40	5
Inner race fault	Gear crack	10/40	6
Outer race fault	Gear crack	10/40	7
Cage fault	Gear crack	10/40	8

5.2. Diagnosis results and analysis

The proposed method is applied to diagnose the different bearing and gear faults. The time-domain, frequency-domain and graph-domain features are all extracted, and several of more sensitive features selected by the FS algorithm are fed into the SVM to identify the eight working conditions. To further test the advantages of the proposed method, the features extracted from any single domain and dual domains are also used to identify the eight working conditions. There are a total of seven feature extraction methods, which are called time domain (T), frequency domain (F), graph domain (G), time and frequency domains (T+F), time and graph domains (T+G), frequency and graph domains (F+G), time and frequency and graph domains (T+F+G), respectively. To avoid the interference caused by unexpected situations, each feature extraction method was tested for 10 trials. Figure 3 shows the multi-class confusion matrix of the proposed method for the first trial. Figure 4 and Table 4 display the diagnosis results of different features with FS feature selection for 10 trials.

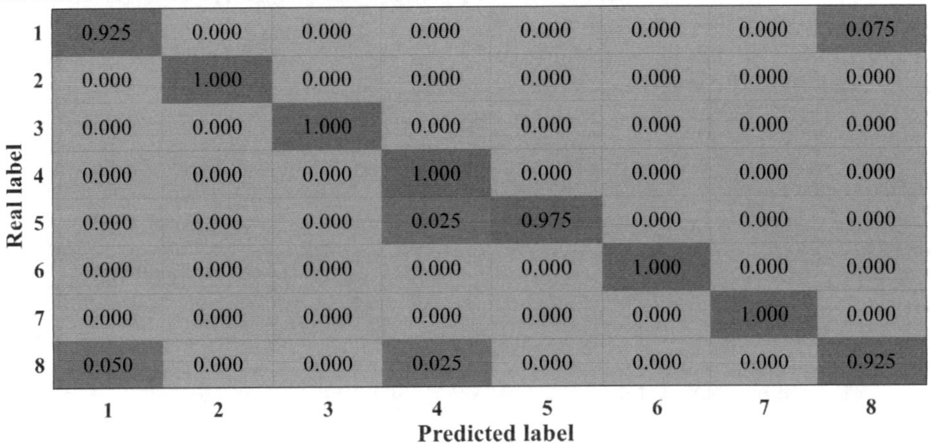

Figure 3. Multi-class confusion matrix of the proposed method for the first trial.

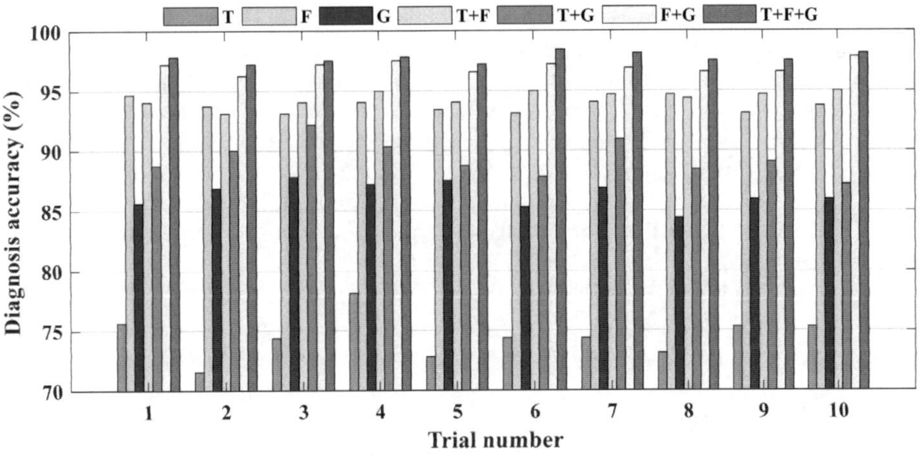

Figure 4. Diagnosis accuracy of different features with FS feature selection for each trial.

Table 4. Diagnosis results of different features with FS feature selection for 10 trials.

Different features	Diagnosis results	
	Average accuracy (%)	Standard deviation
Time domain (T)	74.50 (2384/3200)	1.80
Frequency domain (F)	93.78 (3001/3200)	0.60
Graph domain (G)	86.34 (2763/3200)	1.08

Time and Frequency domains (T+F)	94.41 (3021/3200)	0.60
Time and graph domains (T+G)	89.34 (2859/3200)	1.51
Frequency and graph domains (F+G)	96.97 (3103/3200)	0.49
Time and Frequency and graph domains(T+F+G)	**97.72 (3127/3200)**	**0.42**

It can be seen from Figure 3 that only a small amount of testing samples in conditions 1, 5 and 8 are misdiagnosed, and all testing samples in other conditions are diagnosed correctly. It can be observed from Figure 4 that features extracted from time domain, frequency domain and graph domain can be used to obtain higher diagnosis accuracy than that extracted from any single domain or dual domains in each trial. As can be seen from Table 4, the proposed method (i.e. features extracted from time domain, frequency domain and graph domain) obtains the highest average accuracy and the smallest standard deviation. These results indicate that the proposed method is effective and superior in diagnosing different faults of key rotating components.

In the absence of feature selection, the diagnosis results of each feature extraction method for 10 trials are listed in Table 5. Comparing Table 5 with Table 4, it can be concluded that for all feature extraction methods, the FS features selection is helpful for improving the average diagnosis accuracy, and reducing the standard deviation. The reason is that too many features may result in information redundancy, hence feature selection is necessary and important.

Table 5. Diagnosis results of different features without feature selection for 10 trials.

Different features	Diagnosis results	
	Average accuracy (%)	Standard deviation
Time domain (T)	70.38 (2252/3200)	2.58
Frequency domain (F)	89.38 (2860/3200)	1.59
Graph domain (G)	80.03 (2561/3200)	2.54
Time and Frequency domains (T+F)	91.31 (2922/3200)	**1.54**
Time and graph domains (T+G)	82.75 (2648/3200)	2.29
Frequency and graph domains (F+G)	90.78 (2905/3200)	3.28
Time and Frequency and graph domains(T+F+G)	**91.56 (2930/3200)**	2.74

To further research the impact of different training samples on all feature extraction methods, 2, 4, 6 and 8 samples are also chosen as training samples for each working condition, and remaining samples are testing samples. The FS feature selection and SVM classification are used for fault diagnosis in all cases. The average accuracy of each feature extraction method for 10 trials under different training and testing samples is drawn in Figure 5. As the amount of training samples increases, the average accuracy of each feature extraction method is gradually improved. Our proposed method using features extracted from time domain, frequency domain and graph domain, still has the highest average accuracy even with fewer training samples.

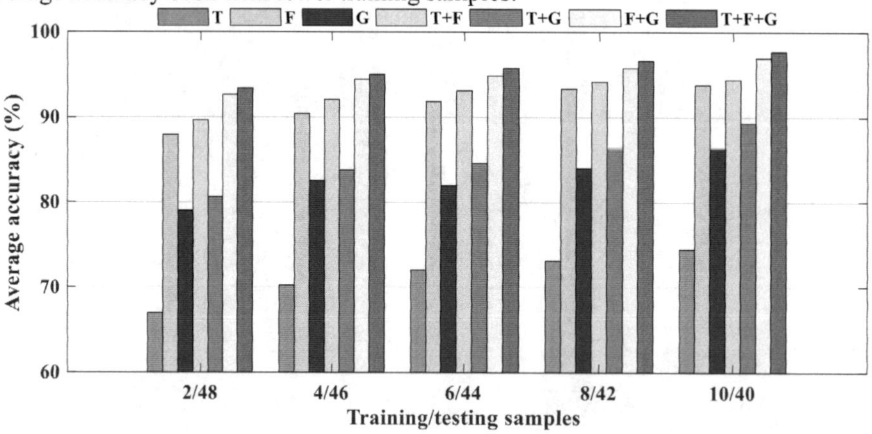

Figure 5. Average accuracy of different features under different training and testing samples.

6. Conclusions

Combining time-domain, frequency-domain and graph-domain features, a new fault diagnosis method for rotating machinery is proposed in this paper. The following conclusions are drawn by analyzing the experimental results:

(1) The graph-domain features generated from horizontal visibility graphs can extract the fault information hidden in the graph topology effectively.

(2) Features extracted from time domain, frequency domain and graph domain can be used to obtain higher diagnosis accuracy than features extracted from any single domain or dual domains.

(3) Too many features can result in information redundancy. The FS feature selection method is helpful for improving the classification performance of SVM.

Acknowledgments

This work was financially supported by the National Natural Science Foundation of China (No. 51875182) and Foundation for University Key Teacher of Henan Educational Department (No. 2018GGJS091).

References

[1] D.I. Shuman, S.K. Narang, P. Frossard, A. Ortega, P. Vandergheynst, The emerging field of signal processing on graphs: Extending high-dimensional data analysis to networks and other irregular domains, IEEE Signal Proc. Mag. 30 (2013) 83-98.

[2] L. Ou, D.J. Yu, H.J. Yang, A new rolling bearing fault diagnosis method based on GFT impulse component extraction, Mech. Syst. Signal Process. 81 (2016) 162–182.

[3] Y.Y. Gao, D.J. Yu, Total variation on horizontal visibility graph and its application to rolling bearing fault diagnosis, Mech. Mach. Theory 147 (2020) 103768.

[4] Y.Y. Gao, D.J. Yu, H.J. Wang, Fault diagnosis of rolling bearings using weighted horizontal visibility graph and graph Fourier transform, Measurement 149 (2020) 107036.

[5] L. Zhou, Z.Q. Gong, R. Zhi, G.L. Feng, An approach to research the topology of chinese temperature sequence based on complex network, Acta Physica Sinica 57 (2008) 7380-7389.

[6] Z.G. Shao, Network analysis of human heartbeat dynamics, Appl. Phys. Lett. 96 (2010) 073703.

[7] E. Estrada, Characterization of the folding degree of proteins, Bioinformatics 18 (2002) 697-704.

[8] D. Stevanovic, A. Ilic, On the Laplacian coefficients of unicyclic graphs, Linear Algebra Appl. 430 (2009) 2290-2300.

[9] L. Ou, D.J. Yu, Path graph Fourier transformation and its applications to rolling bearing fault diagnosis, J. Mech. Eng. 51 (2015) 76–83 (in Chinese).

[10] X. He, D. Cai, P. Niyogi, Laplacian score for feature selection, Advances in Neural Information Processing Systems, Vancouver, 2005, pp. 507-514.

[11] P. Gangsar, R. Tiwari, Comparative investigation of vibration and current monitoring for prediction of mechanical and electrical faults in induction motor based on multiclass-support vector machine algorithms, Mech. Syst. Signal Process. 94 (2017) 464-481.

[12] B. Luque, L. Lacasa, F. Ballesteros, J. Luque, Horizontal visibility graphs: Exact results for random time series, Phys. Rev. E 80 (2009) 046103.

[13] F. Li, J.X. Wang, B.P. Tang, D.Q. Tian, Life grade recognition method based on supervised uncorrelated orthogonal locality preserving projection and K-nearest neighbor classifier, Neurocomputing 138 (2014) 271-282.

Hydro-generators fault diagnosis with short-time-wavelet-entropy and variational auto-encoder

Ryad Zemouri[1, *], Simon Bernier[1], Olivier Kokoko[1], Arezki Merkhouf[1]

[1]Institut de Recherche d'Hydro-Québec (IREQ), Varennes, QC, J3X1S1 Canada

*Corresponding author: zemouri.ryad@hydroquebec.com

Abstract. The prognosis and health management (PHM) of hydroelectric plants are full of difficulties caused by the complexity of the hydro-generators where each machine is different and almost unique. At industrial level, several tools are used to monitor the generator condition. Among these tools, the measurement of magnetic stray flux is one which is gaining interest. This measurement is generally based on an inductive sensor and mainly mounted near the stator. The main advantages of the magnetic stray flux are the non-invasive nature and the simplicity of its implementation. In this work, the discrete wavelet transform (DWT) is used to decompose the stray flux signal. Short-Time-Wavelet-Entropy (STWE) is then applied to extract the features from the sub-bands. Finally, a variational auto-encoder (VAE) is used in an unsupervised learning process to structure the STWE signatures of more than 400 stray flux measurement collected on real hydroelectric plants. The obtained results show that the VAE has well captured the features from the wavelet entropy (WE) signatures. An analysis of the resulting latent space shows a strong correlation between a given trajectory in the reduced space and an increase of the WE.

1. Introduction

In a global reluctance to build new hydroelectric power plants and in the context of increasing demand for electricity, Hydro-Québec must maximize the production with the highest possible reliability and minimum downtime. Hydro-Quebec operates more than 350 hydro-generators, with most of them subjected to normal and slow degradation mechanisms. However, in some cases, degradation is more accelerated and can lead to premature failure and significant economic losses. To avoid this, Hydro-Québec put significant efforts on functional safety condition-based maintenance and prognosis of their equipment. A PHM platform has been developed at Hydro-Quebec's research institute (IREQ), in order to detect and prevent incipient faults in a real-time or at least at early stage of the apparition [1]. Diagnosis is based on online measurement of signals such as partial discharges [2-3], vibrations and acoustics signals [4], temperature [5], magnetic stray flux or magnetic flux in the air-gap [6]. In this work, we study the health of the hydro-generators by measuring the magnetic stray flux signal collected at the stator. This kind of signal provides useful information that can be used to monitor the health of the machine such as the static or dynamic eccentricity of the rotor, or the inter-turn short circuit [6-9]. The measurement of the magnetic stray flux is particularly interesting as it is non-intrusive.

The database processed in this paper contains more than 400 magnetic stray flux signals from 8 hydro-electric power plants operated by Hydro-Québec. The conventional approaches use Fourier transform (FT) as a signal processing tool to assess any specific changes in frequency amplitude according to the fault under study. However, it is necessary to have an a-priori knowledge of the frequencies contained in the signals under study [10]. The FT needs a high stationarity of the signal and does not provide the time evolution of the frequency patterns. This disadvantage can be partially resolved in order to follow the evolution of the frequencies by using a short time Fourier transform (STFT) where the FT is applied to a time-evolving windows. In this case, a compromise between frequency versus time resolution must be found. Narrow windows lead to poor frequency resolution, in the opposite, too wide windows lead to poor time resolution. This limitation becomes significant when the signal has a series of widely dispersed transient components. To overcome these limitations, the discrete wavelet transform (DWT) represents an interesting solution [11-12]. The DWT makes no assumptions about the stationarity of the signal; the only information needed is the time evolution. In this case, the tracking of the temporal evolution of the frequency's patterns contained in the signal becomes possible with an appropriate time-frequency resolution. The wavelet analysis is a method that relies on the introduction of an appropriate base. It allows to characterize a temporal signal given by the distribution of the amplitude in this base. If the wavelet is necessary to form an appropriate orthogonal basis, it has the advantage of being able to uniquely decompose an arbitrary time series and to invert the decomposition [11-12]. DWT is then based on the hypothesis that any signal can be approximated as the arithmetic sum of several functions called wavelets [13]. In PHM application, the wavelet theory has been widely used for stationary and non-stationary signal analysis such as power systems [14][7][8][15], electrical machines [10], rotating machinery [16], partial discharge in power transmission [17], mechanical transmission system of shearer ranging arm [18], online fuel cell remaining useful lifetime prediction [19]. Thus, the DWT as a time-frequency representation of a signal, provides an effective tool for feature extraction.

An emerging technique which allows to evaluate the degree of order/disorder of a signal is to compute the Shannon entropy to the relative energies of each wavelet decomposition [20]. Therefore, the obtained wavelet entropy (WE) provides a useful insight into the inherent dynamic process related to the signal, and has been employed in specific tasks as an indicator of faults. In fact, a highly ordered process can be thought of as a single-frequency periodic signal. For such a signal, all relative wavelet energies will be nearly zero, except for the level of wavelet resolution that includes the frequency of the representative signal. For this level, the relative wavelet energy will be almost equal to one and, therefore, the total WE will be close to zero. Conversely, a random signal can be considered to represent a very disordered behavior. This type of signal will have a wavelet mapping with all frequency bands contributing significantly. The WE will reach their maximum values since the relative energy of the wavelets will be mostly equal for all the levels of resolution.

WE has been used in specific assignments as an indicator of faults and features extraction in several PHM domains. The entropies of the wavelet decompositions have been leveraged for power transmission line fault classification and location [14][21]. In [10], a stray flux signal was first decomposed by DWT, then WE computed to feed an artificial neural network (ANN) in order to estimate the winding insulation healthiness in induction motors. WE has been used by [22] to eliminate white noise in partial discharge signals by computing a new improved threshold function. For fault diagnosis of gearbox, WE has been used by [23] to extract statistical features from wavelet coefficients of the vibration signals. The obtained features were then used to feed an ANN for fault classification. In other domains such as biomedical applications, WE has been widely used for the dynamic analysis of electroencephalogram signals [12], diagnosing congestive heart failures [24], or for brain computer interface [25].

In this study, a WE is utilized to extract the features from a stray flux signals for hydro power generators fault detection and diagnosis. As illustrated in the figure 1, the stray flux signals were first decomposed by a DWT in order to have the different time-frequency sub-bands. The energy of the Shannon WE distribution will strongly depend on the various operating and healthy conditions. In order to analyze the various WE signatures and to identify their likelihoods, we have exploited the properties of the latent space obtained by an unsupervised training of a VAE. This approach based on an unsupervised analysis is particularly adapted to our industrial context where most of the stray flux signals acquired from power plants are not labeled. This whole approach is presented in the next section. Experimental results and latent space analysis are presented in section 3. The conclusion and future works are summarized in the last section of the paper.

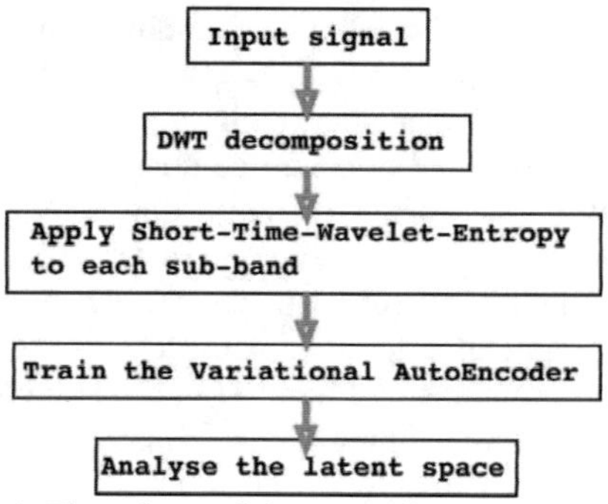

Figure 1. Flowchart of the stray flux signals analysis.

2. Signal Processing

2.1. Discrete Wavelet Transform

The DWT is used to decompose the stray flux signal with high-pass (HPF) and low-pass (LPF) filters, where the bandwidth of both filters is the same. After each decomposition step, the sampling frequency of the signal is divided by two. The output of the low-pass filter is recursively split to produce the next sub-band of the wavelet. The original signal $S(k)$ is then decomposed into details and approximations as follow:

$$S(k) = \sum_{i=1}^{j} D_i(k) + A_j(k) = \sum_{i=1}^{j+1} Sb_i(k) \quad (1)$$

Where j is the number of decomposition level, $D_i(k)$ and $A_j(k)$ are details and approximations of wavelet. The stray flux signal is then decomposed into several sub-bands $Sb_i(k)$. In this paper, as shown by the figure 2, a total of 10 sub-bands are defined and the Daubechies-8 (n = 8) is used as a mother wavelet [11]. According to Nyquist's theorem and based on sampling frequency of 50 kHz, the highest frequency contained on the signal is 25 kHz. The several frequency ranges of each sub-band are also given by the figure 2. Figure 3 illustrates an example of the different sub-bands $Sb_i(k)$ obtained by decomposing an original signal $S(k)$ with a DWT.

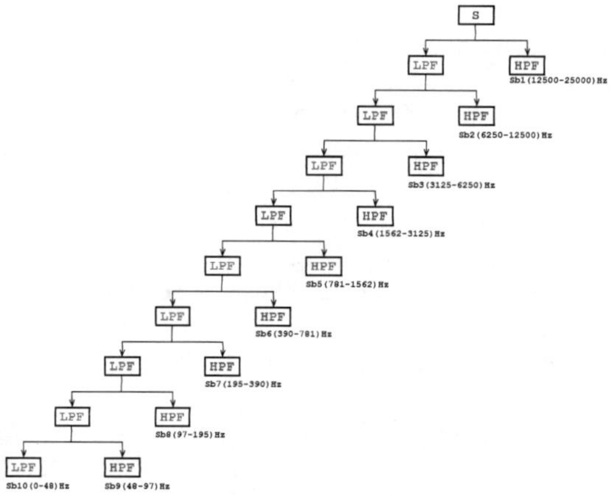

Figure 2. DWT with the corresponding frequency bands obtained from a stray flux signal with a sampling frequency of 50 kHz.

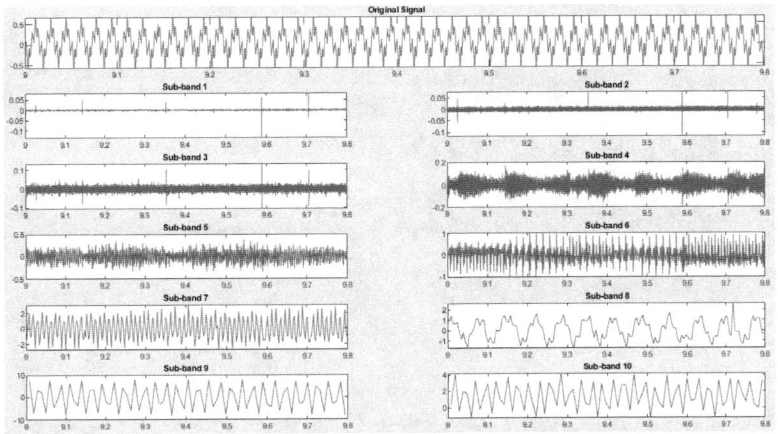

Figure 3. An example of a stray flux signal and the different sub-bands $Sb_i(k)$ obtained by a DWT decomposition.

2.2. Short-Time-Wavelet-Entropy

The Shannon entropy is a practical index for evaluating and assessing the probability distribution [7] [20]. In order to measure the uncertainty and disorder of wavelet sub-bands, we define the moving energy $E_i(k)$ of each sub-band $Sb_i(k)$ within a slide window $[k, k+\Delta k]$ as follow:

$$E_i(k) = \sum_{k}^{k+\Delta k} |Sb_i(k)|^2 \quad (2)$$

The total energy $E_{total}(k)$ of the signal $S(k)$ within the moving window $[k, k+\Delta k]$ is given by:

$$E_{total}(k) = \sum_{i=1}^{j+1} E_i(k) \quad (3)$$

The normalised energy value $P_i(k)$ of each wavelet sub-band $Sb_i(k)$, relative to the total energy $E_{total}(k)$ of the signal is given by the following equation:

$$P_i(k) = \frac{E_i(k)}{E_{total}(k)} \quad (4)$$

Where $\sum_i P_i(k) = 1$. The Short-Time-Wavelet-Entropy (STWE) of each sub-band $Sb_i(k)$ through each moving window $[k, k+\Delta k]$ is then given by:

$$W_i(k) = -P_i(k)log(P_i(k)) \quad (5)$$

The figure 4 shows some examples of different stray flux signals with their three majors corresponding STWE signatures $W_i(k)$ obtained by the equation 5. The length Δk of the sliding window used to compute the moving energy $E_i(k)$ is 1 second.

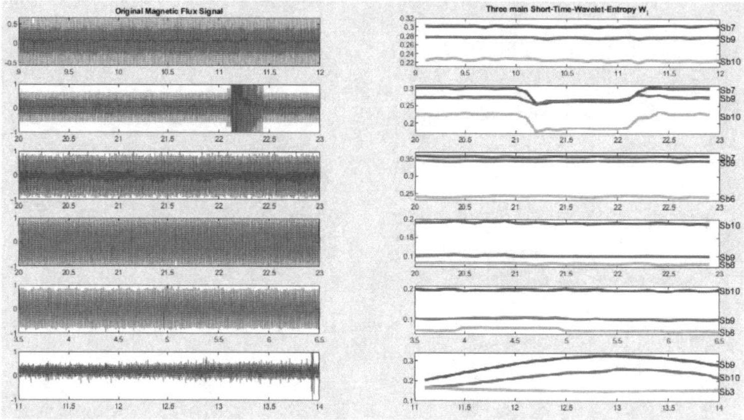

Figure 4. Some examples of various stray flux signals with their corresponding $W_i(k)$ signatures.

2.3. Variational Autoencoder

The VAE is composed by two main structures: an encoder ϕ and a decoder θ. In a first step, the input data $\mathbf{x}$ is transformed into a latent representation $\mathbf{z}$ by the encoder function $\mathbf{z} = f_{\emptyset}(x)$, whereas in the second step, the original data is reconstructed by the decoder $\hat{x} = h_{\theta}(\mathbf{z})$. The popularity of the VAE is due to the combination of the Bayesian inference and the efficiency of the ANNs to obtain a non-linear low-dimensional latent space. Recently, the VAEs are increasingly used in several PHM applications [26-39]. The Bayesian inference is obtained by a sampling additional layer $\mathbf{z}$ with a prior specified standard Gaussian distribution $p(\mathbf{z}) \equiv N(O, \boldsymbol{I})$, where $\mathbf{I}$ is the identity matrix. Each element z_i of the sampling layer is obtained as follows:

$$z_i = \mu_i + \sigma_i . \epsilon \tag{6}$$

Where μ_i and σ_i are the ith components of the mean and standard deviation vectors, ϵ is a random variable following a standard Normal distribution $\epsilon \sim N(0,1)$. Unlike the AE, the VAE generates vector of means μ_i and standard deviations σ_i which provides more continuity in the latent space than the original AE. In VAE, the loss function is composed of two terms:

$$\mathcal{L} = \mathcal{L}_{rec} + \mathcal{L}_{kl} \tag{7}$$

The first term $\mathcal{L}_{rec} = -\mathbb{E}_{q_{\phi}(\mathbf{z}|\mathbf{x})}[\log p_{\theta}(\mathbf{x}|\mathbf{z})]$ is the reconstruction loss function, which allows to minimize the difference between the input and output instances. Both the negative expected log-likelihood (e.g., the cross-entropy function) and the mean squared error (MSE) can be used. The second term $\mathcal{L}_{kl} = \mathbb{D}_{kl}(q_{\emptyset}(\mathbf{z}|\mathbf{x}) \parallel p(\mathbf{z}))$ corresponds to the Kullback–Leibler ($\mathbb{D}_{kl}$) divergence loss term that forces the generation of a latent vector with the specified Normal distribution [40]. The $\mathbb{D}_{kl}$ divergence is a theoretical measure of proximity between two densities $q(x)$ and $p(x)$ and measures how close the conditional distribution density $q_{\emptyset}(\mathbf{z}|\mathbf{x})$ of the encoded latent vectors is from the desired Normal distribution $p(\mathbf{z})$.

The figure 5 gives the detailed architecture of the VAE used to analyse the STWE signatures. A slide window of 10×10 is used as input of the VAE. Each input signal $S(k)$ will then generate a cluster of features in the reduced latent space. Two stray flux signals with two signatures having maximum likelihood will get two feature clusters projected into the same area of the latent space. On the contrary, two different signatures will have two opposite projections.

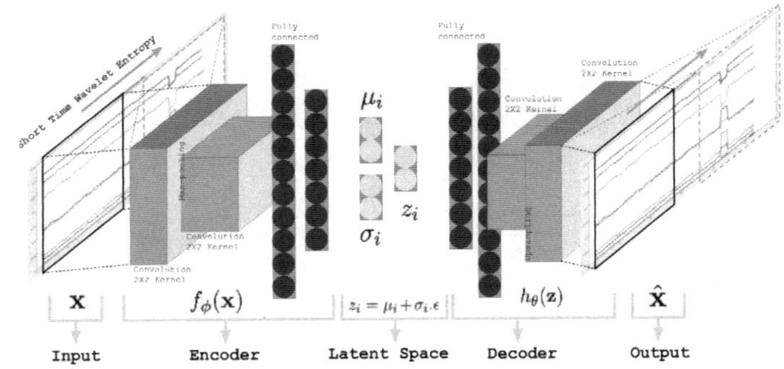

Figure 5. The detailed architecture of the VAE used to analyse the STWE signatures.

3. Experimental results

3.1. *Latent space analysis*

Figure 6 gives the projection through the encoder of the whole data base, where (z_1, z_2) are the output of the sampling layer of the VAE. The statistical distribution of the latent space is given by the figure 7. We can see that some areas of this latent space are more highly concentrated than others. These regions correspond to the most frequently encountered signatures in the database.

Once the VAE learning stage is complete, the second step is to analyze the distribution of data in the reduced space in order to have an interpretation of the health status that corresponds to each area. Since the main part of the data set is unlabeled, this identification step of the latent space distribution is crucial for the stray flux signal recognition and fault detection. We have taken some samples from this latent space and analyzed their different $W_i(k)$ signatures. As shown in Figure 6, a total of eight areas were analyzed. The shape and the value of each entropy $W_i(k)$ corresponding to each sub-band $Sb_i(k)$ of each area is provided in the Figure 8. Table 1 and 2 display respectively the mean value (in %) of the relative normalised energy $E_i(k)$ and the mean value of the $W_i(k)$ for each sub-band $Sb_i(k)$ corresponding to each area.

What we can say from this first analysis of the latent space, is that each area, among those selected, has its own STWE signature $W_i(k)$. For example, area #1 concerns the signals with a low entropy for all the sub-bands. This is due to the mono-frequency character of the original signal: the sub-band 10 dominates the total energy of the signal with $\overline{E_{10}} = 97\%$. On the other hand, the area #5 covers signals with a completely different STWE signature compared to the area #1, 90% of the signal energy is distributed among the sub-bands #7,8,9. Other types of signals are also projected in other regions, such as region #8 where 93% of the total energy of the signal is dominated by the sub-bands #1,2,3,4.

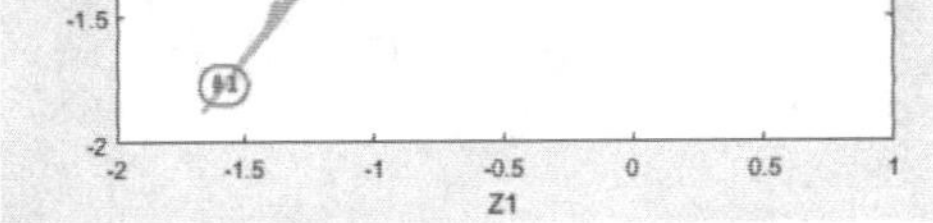

Figure 6. The obtained 2-dimensional latent space.

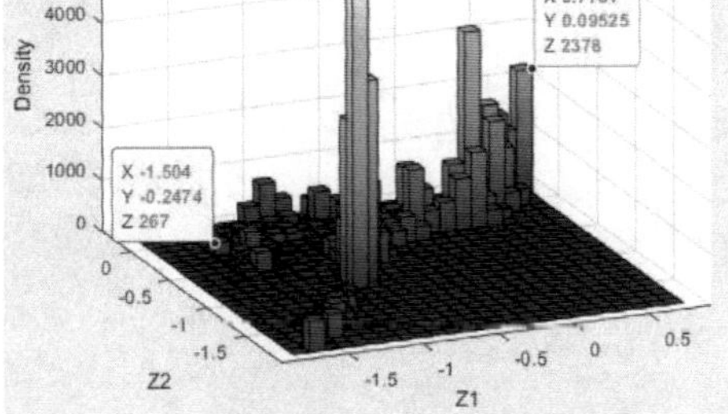

Figure 7. The density of the obtained 2-dimensional latent space

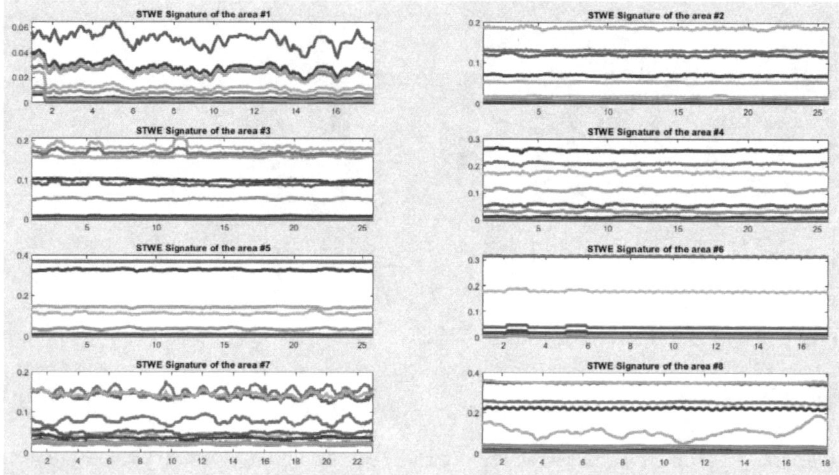

Figure 8. The $W_i(k)$ corresponding to each sub-band $Sb_i(k)$ of each area selected from the latent space.

3.2. Trend analysis

In order to investigate whether a particular trajectory in the latent space correlates with a particular trend in the evolution of the STWE $W_i(k)$, we have selected 38 samples of stray flux signals with a projection into latent space located on the trajectory shown by the figure 9. For each of these 38 signals, we have calculated the mean value $\overline{W_i}$ of the STWE $W_i(k)$ for each sub-band $Sb_i(k)$. The figure 10 illustrates the evolution of $\overline{W_i}$ according to each selected area for each sub-band $Sb_i(k)$. The total entropy is also given for each of the 38 signals by the top plot of the figure. One can see that the top graph in the figure 10 shows an overall increasing trend in WE from projected signal in area #1 to projected signal in area #38. This increasing trend is also visible for the sub-bands #7 #8 and #9. This result shows that the VAE has captured this upward trend of the WE and has organized part of its latent space according to the evolution of this criteria. This is especially important since, if we refer to some recent studies conducted on the diagnosis of hydropower generators [7-8], there is a strong correlation between the increase of the WE and some failures like static or dynamic eccentricity. The next step of our study would be to analyze the failure mechanisms according to the operating mode and/or to the machines categories. Then, a correlation with the various levels of the WE should be conducted. Therefore, the latent space of the VAE could be exploited as a visual PHM support.

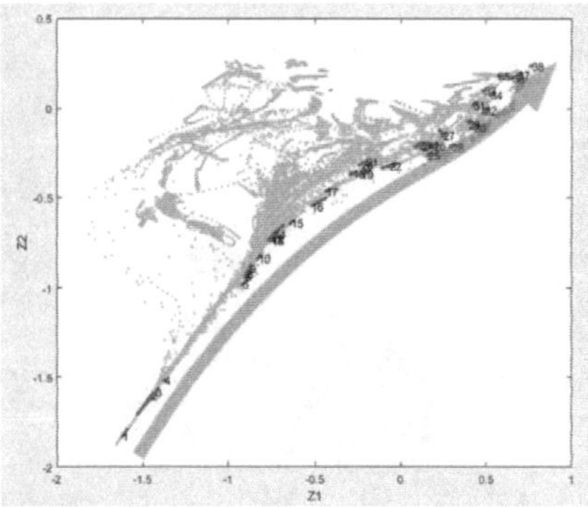

Figure 9. Selected Area for the trend analysis.

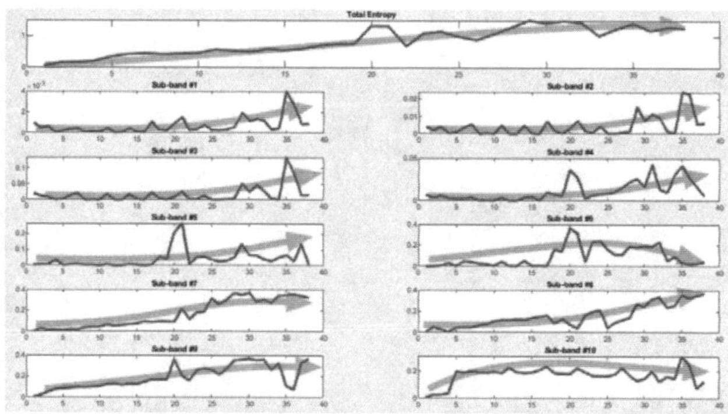

Figure 10. Trend analysis of each sub-band.

Table 1. The mean value (in %) of the relative normalised energy $E_i(k)$ of each wavelet sub-band $Sb_i(k)$ corresponding to each selected area of the figure 6.

	#1	#2	#3	#4	#5	#6	#7	#8
Sb_1	0.005643	0.004632	0.003587	0.00345	0.006461	0.00525	**4.443**	**25.15**
Sb_2	0.02645	0.06347	0.02334	0.02402	0.01906	0.006002	2.025	**11.84**
Sb_3	0.1963	0.301	0.07621	0.08365	0.06513	0.012	0.8874	**48.43**
Sb_4	0.04469	0.04693	0.1152	0.1671	0.1338	0.03352	0.4819	**9.447**
Sb_5	0.1149	0.1351	1.142	0.6156	0.7754	0.02364	0.3569	0.781
Sb_6	0.02741	1.179	**5.429**	**3.219**	**4.696**	0.03235	0.4587	0.4606
Sb_7	0.5429	1.705	2.75	**12.22**	**20.72**	0.297	0.7512	0.2809
Sb_8	1.118	**3.566**	2.37	1.239	**33.89**	0.7882	1.089	0.2247
Sb_9	0.4392	**86.01**	**81.25**	**76.06**	**36.38**	**18.77**	**5.108**	0.1418
Sb_{10}	**97.48**	**6.986**	**6.835**	**6.373**	**3.314**	**80.03**	**84.4**	**3.246**

Table 2. Mean value of the $W_i(k)$ for each sub-band $Sb_i(k)$ corresponding to each selected area of the figure 6.

	#1	#2	#3	#4	#5	#6	#7	#8
Sb_1	0.00049	0.00046	0.00037	0.00037	0.00062	0.00043	**0.14**	**0.35**
Sb_2	0.0022	0.0047	0.002	0.002	0.0016	0.00049	0.079	**0.25**
Sb_3	0.012	0.017	0.0055	0.006	0.0048	0.00092	0.042	**0.35**
Sb_4	0.0035	0.0036	0.0078	0.011	0.0089	0.0021	0.026	**0.22**
Sb_5	0.0078	0.0089	0.051	0.031	0.037	0.0017	0.02	0.038
Sb_6	0.0022	0.052	**0.16**	**0.11**	**0.14**	0.0025	0.025	0.025
Sb_7	0.028	0.069	0.099	**0.26**	**0.33**	0.017	0.037	0.017
Sb_8	0.05	**0.12**	0.089	0.054	**0.37**	0.038	0.049	0.014
Sb_9	0.024	**0.13**	**0.17**	**0.21**	**0.37**	**0.31**	**0.15**	0.0093
Sb_{10}	0.025	**0.19**	**0.18**	**0.18**	**0.11**	**0.18**	**0.14**	**0.11**

4. Conclusions

In this paper, Discrete Wavelet Transform has been used to decompose stray flux signals for hydro-generators monitoring. A Short-Time-Wavelet-Entropy was applied to extract the features from the sub-bands. Finally, a Variational Auto-Encoder has been exploited for the properties of its latent space to structure the different signatures thus obtained. The methodology has been tested on a real stray flux database signal obtained from several power plants operated by Hydro-Québec. The experimental results show that the Variational Autoencoder has well captured the features from the Wavelet Entropy signatures. A trend analysis showed a strong correlation between a given trajectory in the latent space with an upward trend in the Wavelet Entropy. This result provides us with a good basis for the next step, which will aim at a more detailed analysis of the failure mechanisms according to the machine categories operated in the various modes. The identification of the different regions of the latent space would then allow to consider a classifier at the output of the encoder with a supervised or semi-supervised learning.

References

[1] Blancke O, Tahan A, Komljenovic D, Amyot N, Lévesque M and Hudon C 2018 Reliability Engineering & System Safety 180 136 – 151 ISSN 0951-8320.

[2] Lévesque M, Amyot N, Hudon C, Bélec M and Blancke O 2017 Annual Conference of the Prognostics and Health Management Society 2017 vol 8 p 7.
[3] Zemouri R, Lévesque M, Amyot N, Hudon C, Kokoko O and Tahan S A 2020 IEEE Access 8 5438–5454 ISSN 2169-3536.
[4] Vu VH, Thomas M and Lafleur F 2014 Mechanics & Industry 15 29–37.
[5] Kahwati G, Tahan A, Merkhouf A, Hudon C and Al-Haddad K 2020 2020 International Conference on Electrical Machines (ICEM) vol 1 pp 1418–1423 ISSN 2381-4802.
[6] Kedjar B, Merkhouf A and Al-Haddad K 2020 2020 International Conference on Electrical Machines (ICEM) vol 1 pp 1384–1389 ISSN 2381-4802.
[7] Ehya H, Nysveen A, Nilssen R and Liu Y 2021 IET Electric Power Applications 15 890– 902.
[8] Ehya H and Nysveen A 2021 IEEE Transactions on Industrial Electronics 1–1 ISSN 1557-9948.
[9] Ehya H and Nysveen A 2021 IEEE Transactions on Industry Applications 57 3573–3581 ISSN 1939-9367.
[10] Zamudio-Ramirez I, Osornio-Rios R A, Romero-Troncoso R d J and Antonino-Daviu J A 2019 IECON 2019 45th Annual Conference of the IEEE Industrial Electronics Society vol 1 pp 3716–3722 ISSN 2577-1647.
[11] Daubechies I 1992 CBMS-NSF Regional Conference Series in Applied Mathematics.
[12] Rosso O A, Blanco S, Yordanova J, Kolev V, Figliola A, Schürmann M and Ba¸sar E 2001 Journal of Neuroscience Methods 105 65–75.
[13] Perez-Macias F J, Perez-Ramirez C A, Valtierra-Rodriguez M, Dominguez-Gonzalez A and AmezquitaSanchez J P 2017 IECON 2017 - 43rd Annual Conference of the IEEE Industrial Electronics Society pp 3251–3256.
[14] Zheng-you H, Xiaoqing C and Guoming L 2006 2006 International Conference on Power System Technology pp 1–6.
[15] Zamudio-Ramirez I, Antonino-Daviu J A, Osornio-Rios R A, de Jesus Romero-Troncoso R and Razik H 2020 IEEE Transactions on Industrial Electronics 67 5050–5059 ISSN 1557-9948.
[16] Liang P, Deng C, Wu J and Yang Z 2020 Measurement 159 107768 ISSN 0263-2241.
[17] Qu N, Li Z, Zuo J and Chen J 2020 IEEE Access 8 87060–87070 ISSN 2169-3536.
[18] Zheng X, Tang Y Y and Zhou J 2019 IEEE Transactions on Signal Processing 67 1696–1711 ISSN 1941-0476.
[19] Ibrahim M, Steiner N Y, Jemei S and Hissel D 2016 IEEE Transactions on Industrial Electronics 63 5057– 5068 ISSN 1557-9948.
[20] Shannon C E 1948 The Bell System Technical Journal, 27 379–423.
[21] Dasgupta A, Nath S and Das A 2012 Electric Power Components and Systems 40 1676–1689.
[22] Guojun L, Yong W, Le L, Shaofeng G, Haiqing N and Xuemei W 2017 2017 IEEE 11th International Symposium on Diagnostics for Electrical Machines, Power Electronics and Drives (SDEMPED) pp 475– 479.
[23] Heidari Bafroui H and Ohadi A 2014 Neurocomputing 133 437–445 ISSN 0925-2312.
[24] Isler˙ Y and Kuntalp M 2007 Computers in Biology and Medicine 37 1502–1510 ISSN 0010-4825 qT Variability & Heart Rate Variability.
[25] Sankar A S, Nair S S, Dharan V S and Sankaran P 2015 Procedia Computer Science 46 1476–1482 ISSN 1877-0509 proceedings of the International Conference on Information and Communication Technologies, ICICT 2014, 3-5 December 2014 at Bolgatty Palace & Island Resort, Kochi, India.
[26] Proteau A, Tahan A, Zemouri R and Thomas M 2021 Journal of Intelligent Manufacturing.
[27] Proteau A, Zemouri R, Tahan A and Thomas M 2020 The International Journal of Advanced Manufacturing Technology 111 3597–3611.
[28] Huang Y, Chen C and Huang C 2019 IEEE Access 7 139086–139096 ISSN 2169-3536.
[29] Lee S, Kwak M, Tsui K L and Kim S B 2019 Engineering Applications of Artificial Intelligence 83 13 – 27 ISSN 0952-1976.
[30] Cheng F, He Q P and Zhao J 2019 Computers & Chemical Engineering 129 106515 ISSN 0098-

1354.
[31] Martin G S, Droguett E L, Meruane V and das Chagas Moura M 2018 Structural Health Monitoring 0 1475921718788299.
[32] Wang K, Forbes M G, Gopaluni B, Chen J and Song Z 2019 IEEE Access 7 22554–22565 ISSN 2169-3536.
[33] Ren Wang Y, Jin Q, dong Sun G and fei Sun C 2019 Knowledge-Based Systems 163 438 – 449 ISSN 0950-7051.
[34] Yoon A S, Lee T, Lim Y, Jung D, Kang P, Kim D, Park K and Choi Y 2017 CoRR abs/1709.00845.
[35] Zhang Z, Jiang T, Zhan C and Yang Y 2019 Journal of Process Control 75 136 – 155 ISSN 0959-1524.
[36] Jiao R, Peng K and Dong J 2020 IEEE Transactions on Instrumentation and Measurement 69 8831–8843 ISSN 1557-9662.
[37] Zhang Z, Zhu J, liu Y and Ge Z 2020 2020 IEEE 9th Data Driven Control and Learning Systems Conference (DDCLS) pp 1370–1376.
[38] Zemouri R, Lévesque M, Hudon C and Kokoko O 2021 PHM Society European Conference, 6 494–503.
[39] Kokoko O, Hudon C, Lévesque M, Amyot N and Zemouri R 2021 2021 IEEE International Conference on Prognostics and Health Management (ICPHM) pp 1–6.
[40] Kingma D 2017 Variational inference & deep learning: A new synthesis Ph.D. thesis Faculty of Science (FNWI), Informatics Institute (IVI), University of Amsterdam.

Analysis on vibration response of bearing coupling faults in rotor-bearing system

Xiaoyun Gong*, **Chao Li, Zeheng Zhi, Wenliao Du**

School of Mechanical and Electrical Engineering, Zhengzhou University of Light Industry, Zhengzhou,450002, China

*Corresponding author: gxyun2008@sohu.com

Abstract. It is well known that the coupled fault diagnosis of rotating machinery is a challenging task, which is mainly due to the complexity of the vibration signals and the interaction of multiple fault components. In order to study the coupling vibration characteristics from unbalance and bearing fault in rotor-bearing system, A dynamic model of the coupled fault is performed to explore the phenomena of multi-fault in the rotor-bearing system in this paper. An experimental study, including seven working conditions, is also examined to further indicate the vibration response of coupling fault with unbalance and bearing fault. The influence of the rotor unbalance on the bearing vibration characteristics and the vibration rules of the coupling fault are proposed in this paper.

Keywords: Rotor-bearing system, coupling fault, fault diagnosis, vibration response.

1. Introduction

As the core of rotating machinery, the rotor system is widely used in various mechanical equipment. The vibration behavior of the rotor system plays a vital role in the performance and reliability of the equipment for fault diagnosis. When the equipment fails, multiple types of faults are usually coupled in the rotor system, especially rotor unbalance and bearing faults. Therefore, it is necessary to study the coupling vibration characteristics from unbalance and bearing fault in rotor-bearing system.

In recent years, many researches on the dynamics of the bearing-rotor system have been proposed to obtain the response of bearing coupling faults. Qin et al. [1] proposed a new dynamic model of rolling bearings to simulate the dynamic response of high-speed rolling bearings with surface defects. Yuan et al. [1] proposed a new method for modeling the vibration of the ball bearing-rotor system by analyzing the dynamic contact and path between the joint interfaces. Liang et al. [2] studied the influence mechanism of rotor offset on the nonlinear vibration response of rotating machinery under speed-increasing conditions considering the actual working conditions of the rotor system. Yang et al. [3] studied the coupling fault impulse response characteristics and the mechanism of the abrupt change of the rotor shaft motion trajectory with the finite element model. Zhao et al. [4] discussed the influence of raceway defects on the nonlinear dynamic behavior of rolling bearings by establishing a bearing dynamics model. Patel and Naik [5] studied the bearing-rotor system and proposes a mathematical analysis model. Mereles and Cavalca [6] proposed a model alternative method based on distributed parameters or continuous models, which can be applied to the mathematical modeling of complex rotor systems. Eisa et al. [7] proposed a new analysis method that uses multi-body dynamics and finite element theory to analyze and simulate lateral and torsional vibrations. Cui et al. [8] established a nonlinear vibration model for evaluating the severity of rolling bearing faults, simulating the vibration response signals of rolling bearings with different fault levels, and performing quantitative analysis on the signals. In the above research, most of the research is focused on the dynamic model and response mechanism of the bearing-rotor system, which means that there are few studies on the vibration characteristics of coupled faults.

By considering the force analysis of the rolling bearing and the model discretization theory, this paper builds a coupled motion equation of rotor-bearing system. In this equation, the Lagrange method based on the principle of virtual displacement is used to represent the bearing-rotor system. Combined with Adams dynamic analysis and experimental simulation to study the coupling vibration characteristics from unbalance and bearing fault in rotor-bearing system.

2. System modeling and simulation

In order to better describe the complexity of the rolling bearing in the coupled situation, two mechanical analyses are set up. The force analysis of the rolling bearing with the outer ring fixed and the inner ring rotating is shown in figure 1. And the force analysis of the rolling element at the defect of the outer raceway is shown in figure 2.

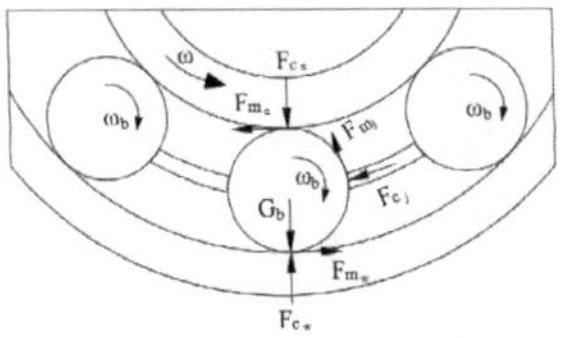

Figure 1. Bearing force analysis.

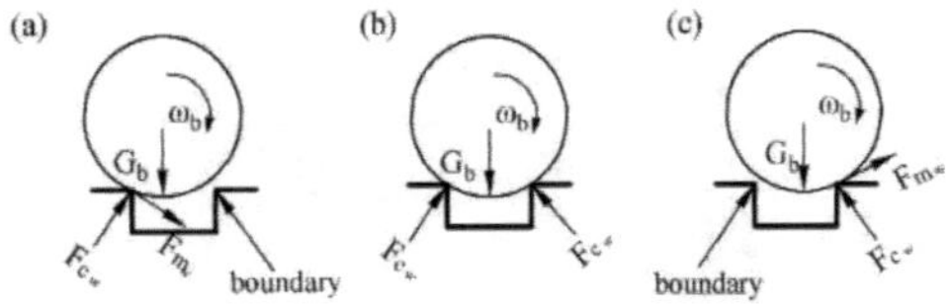

Figure 2. Force analysis of rolling elements.

As shown in figure 1 and figure 2, ω_b is the rotational angular velocity of the rolling element, G is the weight of the rolling element, F_{c_n} and F_{c_n} are the normal forces of the inner and outer rings of the bearing on the rolling element, F_{m_n} and F_{m_w} are the frictional forces of the inner and outer rings of the bearing on the rolling element, F_{c_j} and F_{m_j} are the normal force and friction force between the cage and the rolling elements.

During the rotation of the bearing, the rolling element are subjected to the frictional force of the inner ring and the radial force of the inner and outer rings, thereby exerting force on the cage and driving the cage to rotate together. The process of bearing entry and exit defects is shown in figure 2: The rolling element enters the defect will lose contact with the outer raceway, and instantaneous impact occurs when the rolling element rolls out of the defect.

2.1. Theoretical model

The equations for describing the motion model of the rotor-bearing system is shown in the following manner. In this work, equations(1) to (4) are are used as the solution method.

$$M\ddot{x}(t+\Delta t)+C\dot{x}(t+\Delta t)+Kx=F_L(t+\Delta t)+F_C(t+\Delta t)-F(t) \tag{1}$$

Where M is the element mass matrix, $\ddot{x}$ is the nodal acceleration vector, C is the element damping matrix, $\dot{x}$ is the node velocity vector, K is the element stiffness matrix, x is the displacement increment vector, t is time, F_L is the external force on the node, F_C is the equivalent nodal contact force, F is the force of the internal nodes.

Solve the equation of motion by explicit central difference method,

$$x(t+\Delta t)=x(t)+\dot{x}(t)\Delta t+\frac{1}{2}\ddot{x}(t)\Delta t^2 \tag{2}$$

$$\dot{x}(t+\Delta t)=\dot{x}(t)+\frac{1}{2}(\ddot{x}(t)+\ddot{x}(t+\Delta t))\Delta t \tag{3}$$

Substituting Eq. (2) and Eq. (3) into Eq. (1),

$$\ddot{x}(t+\Delta t)=\left[F_L(t+\Delta t)+F_C(t+\Delta t)-F(t)-K(t)x-\left(\dot{x}(t)+\frac{1}{2}\ddot{x}(t)\Delta t\right)C\right]\left(M+\frac{1}{2}\Delta tC\right)^{-1} \tag{4}$$

Then substituting Eq. (2) and Eq. (4) into Eq. (3), solving Eq. (3) to get all the parameters. (The force in the equation is obtained by the penalty function method.)

2.2. Simulation result analysis

In this part, adams software is used to simulate the dynamics response of ball bearing 6204. The simulation model of the rotor - bearing system is shown in figure 3(a), including bearings, bearing seats, shaft and unbalanced rotors. The bearing model is shown in figure 3(b), the fault degree of the inner and outer rings in the model is simulated by a pit with the diameter of 2mm and the depth of 1mm. The rolling elements fault is simulated by a cutting surface with the width of 2mm.

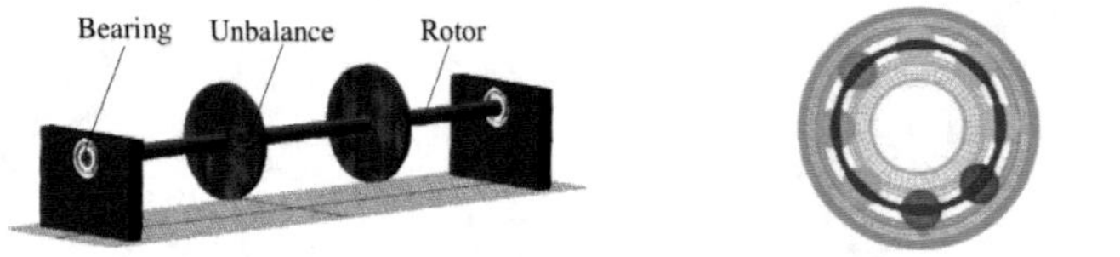

(a) Rotor-bearing system model (b) Bearing fault model

Figure 3. Schematic diagram of simulation model.

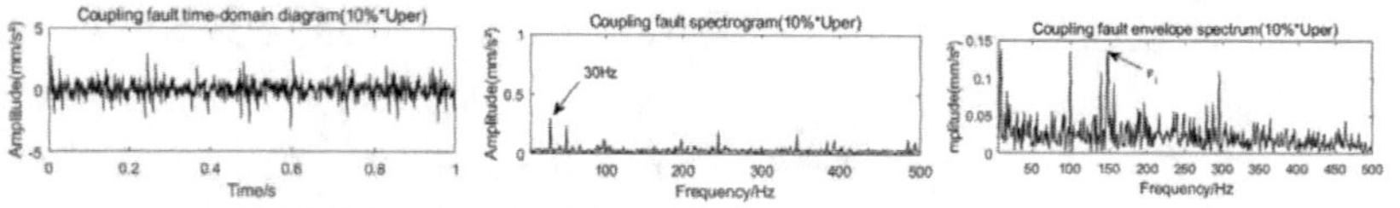

(a)10%U_{per}time domain graph; spectrogram; envelope spectrum.

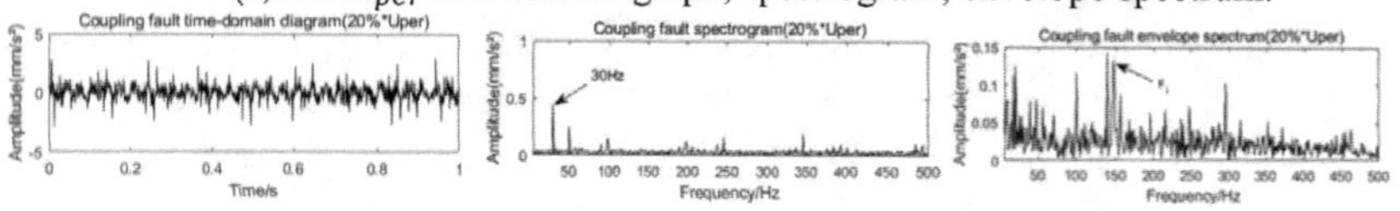

(b)20%U_{per}time domain graph; spectrogram; envelope spectrum.

Figure 4. Simulation results of different unbalance.

As shown in the figure 4, F_i is the fault frequency of bearing inner ring, U_{per} is the max allowable unbalance. Through the rolling bearing 6204 with two unbalance degrees, the influence of unbalance degree on bearing vibration response is studied. It can be seen that the amplitude of the unbalanced fault frequency increases with the increase of the unbalanced degree, and the unbalanced degree has no obvious influence on the bearing fault frequency.

3. Experimental data analysis

In this part, a deep groove ball bearing of ER-12K is used for simulation and analysis, and the bearing parameters is listed in table1. Via the fault simulation test bench of SQ Company, seven types of faults including one normal bearing, three bearing faults and three coupled faults are simulated respectively.

Table 1. Bearing parameters.

Bearing brand	Bearing type	Number of balls.	Ball diameter (in)	Pitch diameter (in)
Rexnord	ER-12K	8	0.3125	1.318

In this experiment, the rotor frequency and signal cut-off frequency are set as 40 Hz and 5000 HZ, and the number of spectral lines taken as 12800. The time-domain waveforms of three different coupled faults are shown in figure 5.

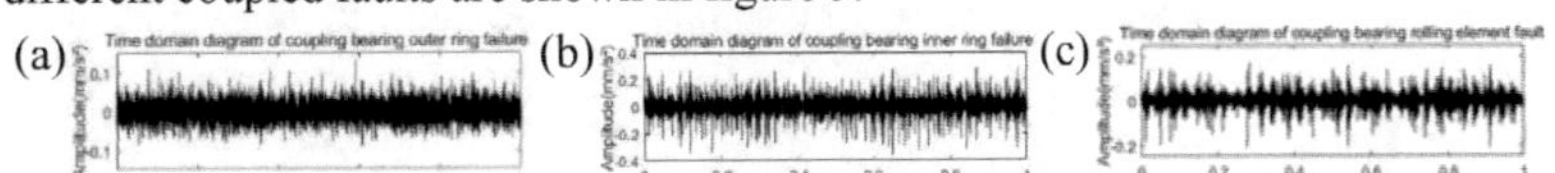

Figure 5. Coupling fault time domain waveform: (a) Unbalanced - bearing outer ring fault; (b) Unbalanced - bearing inner ring fault; (c) Unbalance-bearing rolling element fault.

Predictably, the vibration characteristic values of different fault types in the same bearing will be different. Calculate the root mean square value and kurtosis value of seven kinds experimental data, and list the point line graph and histogram are shown in figure 6.

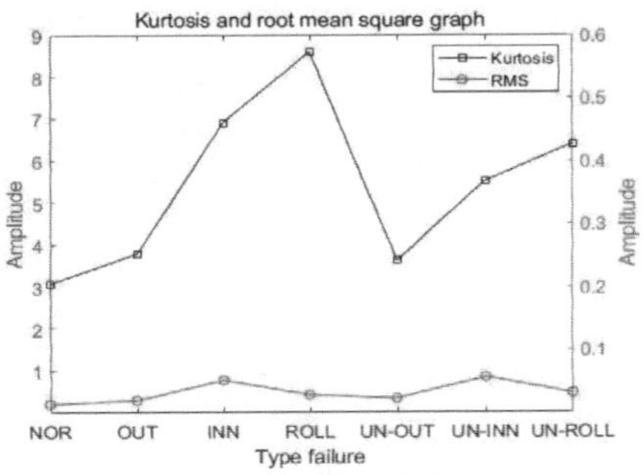

(a) Point line graph of each fault type

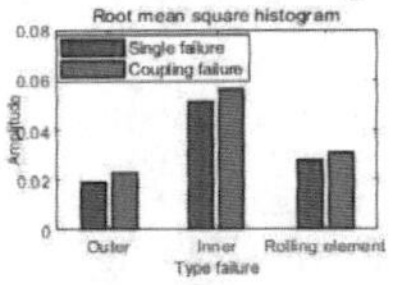

(b) RMS value histogram

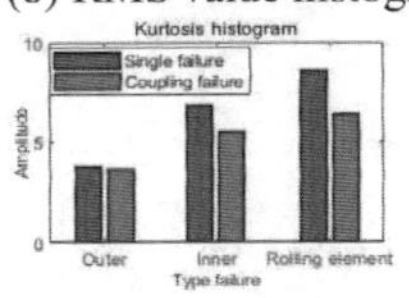

(c) Kurtosis value histogram

Figure 6. Comparison chart of characteristic values of various fault types.

Where NOR means normal, OUT means outer ring fault, INN means inner ring fault, ROLL means rolling element fault, UN-OUT means unbalance-outer ring coupling fault, UN-INN means unbalance-inner ring Coupling fault, UN-ROLL means unbalance-rolling element coupling fault.

It can be seen from figure 6(b)-(c) that with the coupling of unbalanced fault, the coupled RMS value is greater than the RMS value of bearing fault and the coupled kurtosis value is less

IOP Conf. Series: Materials Science and Engineering 1207 (2021) 012010 doi:10.1088/1757-899X/1207/1/012010

than the bearing fault kurtosis value.

3.1 Envelope analysis

Further study the mutual influence and vibration characteristics of unbalanced-bearing coupling faults, envelope analysis is used to process the experimental signals, and the single and coupled bearing fault envelope spectrum are shown in figure 7. In the test, the characteristic frequencies of bearing inner ring fault, outer ring fault and rolling element fault are calculated, and the results are as follows: 198Hz, 121.92Hz and 79.68Hz [9].

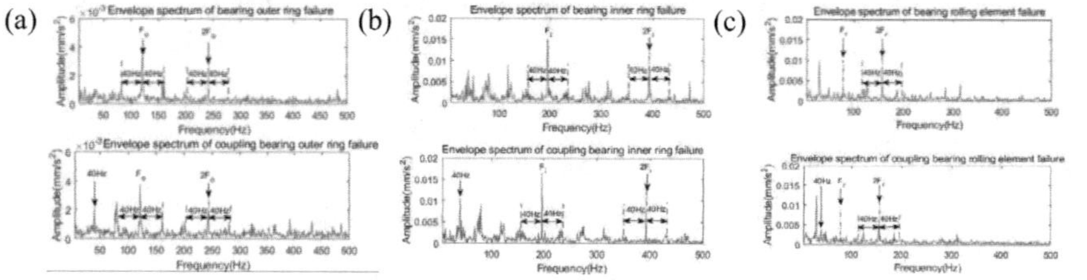

Figure 7. Single and coupled bearing fault envelope spectra: (a) Single and coupled bearing outer ring fault envelope spectra; (b) Single and coupled bearing inner ring fault envelope spectra; (c) Single and coupled bearing rolling element fault envelope spectra.

In figure 7, F_i is the fault frequency of the bearing inner ring, F_o is the fault frequency of the bearing outer ring, and F_r is the fault frequency of the bearing rolling elements. Comparing figure 7(a), figure 7(b) and figure 7(c), it can be seen that the unbalanced fault frequency (rotational frequency) is clearly visible. The coupled bearing fault frequency amplitude is higher than the amplitude in the single bearing fault envelope spectrum, and the sidebands of the unbalanced fault frequency are more obvious than the single bearing fault envelope spectrum.

3.2 Cepstrum analysis

In order to observe the periodic frequency components in the fault data more clearly, cepstrum analysis is further used to process the experimental signals, and the single and coupled bearing fault cepstrum are shown in figure 8.

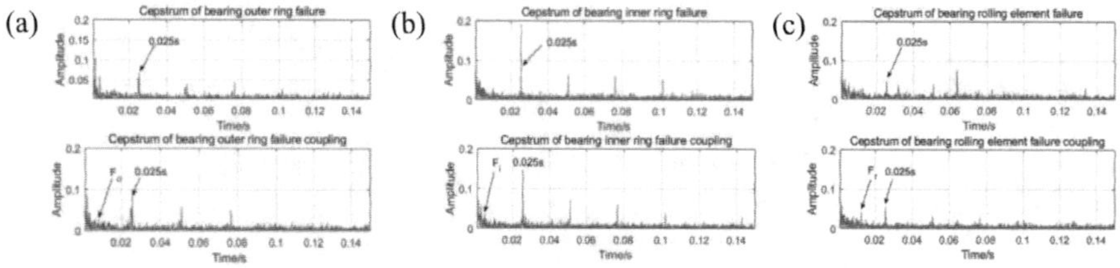

Figure 8. Single and coupled bearing fault cepstrum: (a) Single and coupled bearing outer ring fault cepstrum; (b) Single and coupled bearing inner ring cepstrum; (c) Single and coupled bearing rolling element fault cepstrum.

As shown in figure 8, it can be clearly seen that the rotation frequency conversion component is 0.025s (40 Hz). Comparing and analyzing the cepstrum transformation of the experimental data, it is found that the bearing fault frequency in the coupled fault cepstrum is more obvious.

4. Conclusions

In order to reveal the coupling vibration characteristics from unbalance and bearing fault in rotor-bearing system, A dynamic model of the coupled fault is built, and it is further study by the vibration signals from experimental simulation. The conclusions of this paper include three aspects: 1) The change of the unbalance in the rotor unbalance-bearing coupling fault has little effect on the magnitude of the bearing fault frequency; 2) With the coupling of unbalanced fault, the kurtosis value of the vibration response decreases and the root mean square value increases; 3) With the coupling of unbalanced fault, both the bearing fault frequency amplitude and the sideband amplitude in the envelope spectrum increase, and the bearing fault frequency in the cepstrum is easier to find.

Acknowledgments
This work was supported by the Science and Technology Program of Henan (Grand No. 202102210289), and the Foundation for University Key Techer of Henan Educational Department (Grand No.2018GGJS091).

References

[1] Qin Y, Li C C, Wu X G, et al. Multiple-degree-of-freedom dynamic model of rolling bearing with a localized surface defect, J. Mechanism and Machine Theory,154(2020) 104047.

[2] Yuan X, Zhu Y S and Zhang Y Y, Multi-body vibration modelling of ball bearing–rotor system considering single and compound multi-defects, J. Mechanism and Machine Theory, 154(2020) 104047.

[3] Liang M X, Yan T H, Hu J H, et al. Effect of rolling bearing parameters on the nonlinear dynamics of offset rotor, J. Mechanical Engineering Science, 2020, 234(15) 2968–2978.

[4] Yang Z Z, Zuo Y F, Shao H J , et al. Coupling analysis of rotor unbalance and bearing fault based on explicit finite element, J. Bearing, 2021.

[5] Zhao Z, Yin X and Wang W, Effect of the raceway defects on the nonlinear dynamic behavior of rolling bearing, J. Mech. Sci. Technol. 33 (2019) 2511–2525.

[6] Patel U A and Naik B S, Nonlinear vibration prediction of cylindrical roller bearing rotor system modeling for localized defect at inner race with finite element approach , J. Multi-body Dynamics, 2017, Vol. 231(4) 647–657.

[7] Mereles A and Cavalca K L, Mathematical modeling of continuous multi-stepped rotor-bearing systems, J. Applied Mathematical Modelling, 90 (2021) 327-350.

[8] Eisa A I, Li S S and Helal W M, Study on the lateral and torsional vibration of single rotor-system using an integrated multi-body dynamics and finite element analysis, J. Advances in Mechanical Engineering,2020, Vol. 12(10) 1–12.

[9] Cui L, Zhang Y, Zhang F, et al. Vibration response mechanism of faulty outer race rolling element bearings for quantitative analysis, J. Sound Vib. 364 (2016) 67–76.

[10] Sheng Z S and Yin Q L. Equipment state monitoring and fault diagnosis technology and application , M. Beijing: Chemical Industry Press, 2003: 226-228.

Investigation on failure mechanism of electrical connectors under repetitive mechanical insertion and withdrawal operations

Yuan Meng, Wanbin Ren *

School of Electrical Engineering and Automation, Harbin Institute of Technology, Harbin, China

*Corresponding author: renwanbin@hit.edu.cn

Abstract. Electrical connector is an essential accessory component for electrical and electronic interconnection circuit. In order to investigate the degradation behavior of electrical connector, a series of repetitive mechanical insertion and withdrawal operations of electrical connector have been carried out. The results indicate that there is an increasing trend in insertion/extraction force in the initial stage. Afterwards, it becomes a gradually decreasing trend attributed to the mechanical wear of the contact components. In addition, the oxidative wear of substrate copper alloy material causes the fluctuation phenomenon of contact resistance. The relevant mathematic models for insertion/extraction force and contact resistance calculation are presented to research the dynamic insertion/extraction process. Finally, the degradation behavior and associated physical mechanisms are proposed by analysing the laser confocal photographs and parameter waveforms comparison.

Keywords: Electrical connector, insertion force, contact resistance, degradation behavior.

1. Introduction

Electrical connectors are indispensable components in modern industrial and domestic application. Mechanical failure and electrical contact failure often occur in electrical connectors in field application. Field data have shown that connector degradations and failures contribute to 30~60% of the failure cases of electronic components in the system [1].

Mechanical insertion and withdrawal operations, which realizes the function of electrical connection and disconnection, is the most fundamental working mode of electrical connector. The unsuitable insertion/extraction force could result in many failure cases, such as inconvenient insertion, loose contact, instantaneous even permanent disconnection [2, 3]. The mechanical structure, shape and size of connector are considered as the key factors of the insertion/extraction force [4, 5]. Rong [6] stated that the material, dimension and assembly of contact pair could affect the insertion and extraction force and the fatigue life of connector.

Contact resistance is a significant parameter for reliability and service life of electrical connector. Reliable electrical contacts mainly demand low and stable contact resistance in a long term [7]. The gradual increases, irregular increases, periodic variations and instantaneous increases of contact resistance are the main criteria for electrical contact failure of electrical connector [8, 9]. Relevant studies have proved that repetitive insertion and withdrawal operations, fretting wear result in the destruction of surface coating material [10, 11]. It not only increases the friction coefficient and leads to the increase of insertion force, but also results in the exposure of the substrate metal and the formation of oxide film, which reduces the electrical contact performance and causes the failure of connector. The dynamic process can reflect the final running state of the connector. However, the research on the degradation law of dynamic waveforms of force and contact resistance and its influence on failure phenomenon is little known.

In this paper, a test rig specially designed for simulating the insertion/extraction operation and collecting the dynamic data of force and contact resistance in-situ in real time is described. The insertion/extraction physical model is presented to research the dynamic process of insertion/extraction operation. Under the repetitive mechanical insertion and withdrawal experiments, the parameter degradation law and associated physical mechanism are studied by analyzing the laser confocal photographs and dynamic waveforms of connector.

2. Experimental details

2.1. Description of the test rig

The test rig consists of three main modules, including mechanical modules (Actuation structure and spatial position adjusting unit, as shown in Fig.1), electrical measurement and control module and programmed PC software. The horizontal actuation of the moving plug is obtained by means of the electric slide table (RCA2-TWA4NA, IAI, Japan). And the motion velocity could be setting from 2mm/s to 100mm/s easily. The relative position of plug and socket could be realized by 2 three-dimensional linear translation stages (3D LTS). In addition, the plug is connected with the force sensor (FA404-5kg, FIBOS, China), which could measure the transient force during the insertion/extraction process. A low friction pulley, installed above a 3D LTS, is used to support the transmitting rod of the plug, and thus excludes the undesired disturbance from the transmitting rod. The electrical connector and its subsidiary mechanism are installed in the constant temperature and humidity chamber (HTC-100D-I, SANTN, China), which could control the temperature and humidity environment during the whole experiment accurately.

The contact resistance is measured by using four-wire method (as shown in Fig.2) to eliminate the connection resistance of lead and soldered joints. During the test, the force and contact resistance of the whole insertion/extraction process are collected and uploaded to the PC simultaneously with the help of data acquisition card (PCI-1706U, Advantech, China). The data acquisition and logging process are controlled by the PC with the help of LabVIEW software specifically programmed for this purpose.

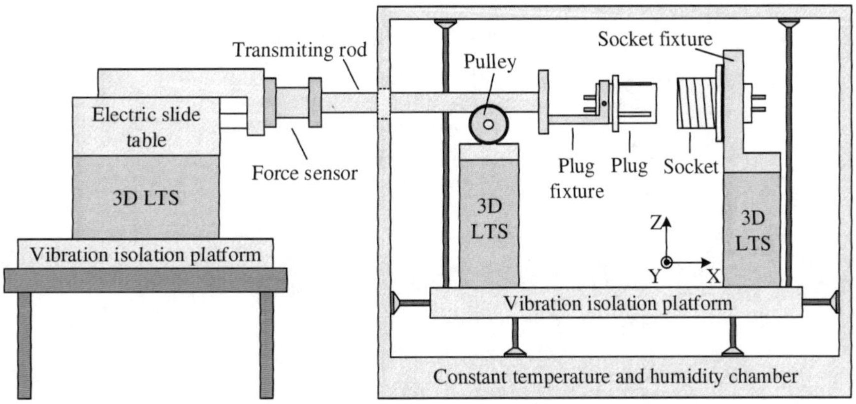

Figure 1. Schematic diagram of the test rig.

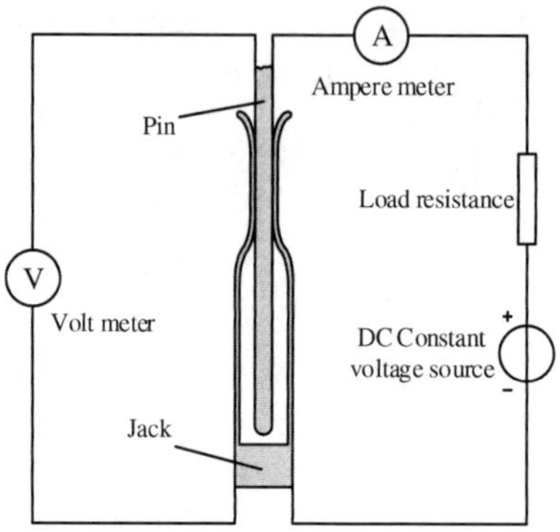

Figure 2. Circuit of four-wire method.

2.2. Experiment methods

Weipu WS16-2 type connector is selected as the experimental sample as shown in Fig.3. This connector contains 2 contact pairs and its rated current is 10A. The internal pins and jacks are made of gold-plated copper alloy. As shown in Fig.3 (d), there is a constricted structure at the front part of the jack to provide the contact force to the pin under the insertion condition. The details of experimental conditions are listed in Table 1.

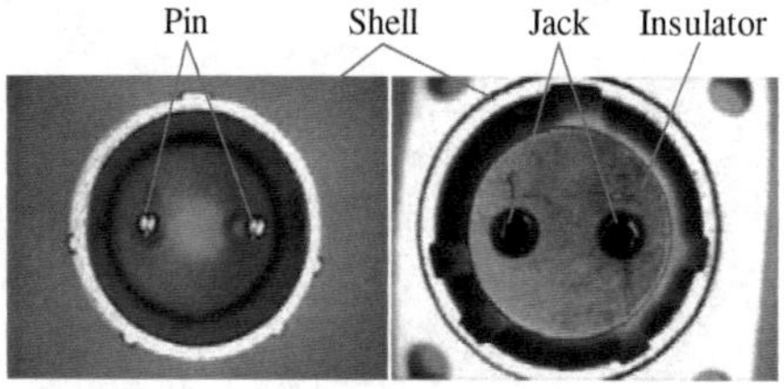

(a) Illustration of the plug **(b)** Illustration of the socket

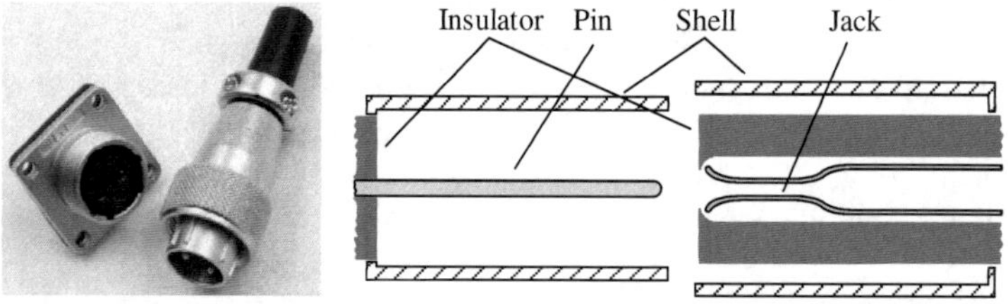

(c) Appearance of the connector **(d)** Schematic plot of the structure

Figure 3. Appearance and structural schematic diagram of the connector.

Table 1. Experimental conditions.

Parameter	Value
Temperature	25°C
Humidity	65%RH
Insertion/Extraction Velocity	100mm/s
Insertion/Extraction Cycle	12s
Supply Voltage	6V
Current Load	100mA
Load Resistance	60Ω
Sampling Rate	25kHz

3. Results and analysis

3.1. Degradation of insertion/extraction force and contact resistance

The maximum values of the force during the insertion/extraction process are defined as the insertion force F_i and the extraction force F_e, respectively. Fig.4 shows the variations of the insertion force F_i, extraction force F_e, and stable contact resistance R_1 and R_2 for two contact pairs as functions of operation cycles. It is found that the insertion force increases rapidly from 10.5N to 27.4N initially, and the extraction force increases sharply from 9.8N to 31.4N below 300 cycles. However, the contact resistance of contact pairs is low and stable during the initial stage.

There is a decrease trend after 300 cycles in insertion force and extraction force, and the reduction speed also decreases gradually. After 15000 cycles, the extraction force reduces to 4.1N, which is less than 50% of the initial value. According to the EIA standard EIA-364-13C, 5N is the lifetime limit of electrical connector with 2 contact pairs, and the extraction force could not be lower than this value. Therefore, the mechanical failure occurred in the electrical connector after 15000 cycles. In addition, there is an obvious increase trend in contact resistance after 15000 cycles, which increases to above 10mΩ. After that, a severe fluctuation in contact resistance of contact pairs is observed. According to the standard GJB 101A-97, 10mΩ is defined as the lifetime limit of electrical connector with copper alloy contact pair. So electrical contact failure occurred in the electrical connector after 15000 cycles. The electrical contact failure rate is also increasing with the increase of operation cycles.

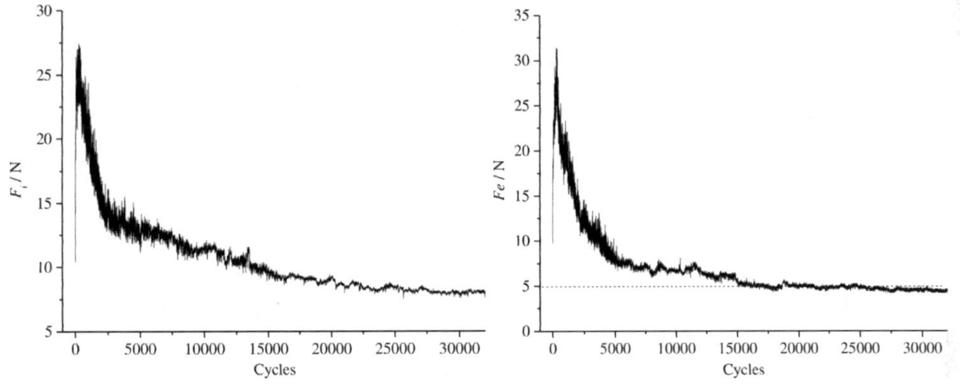

(a) Insertion force **(b)** Extraction force

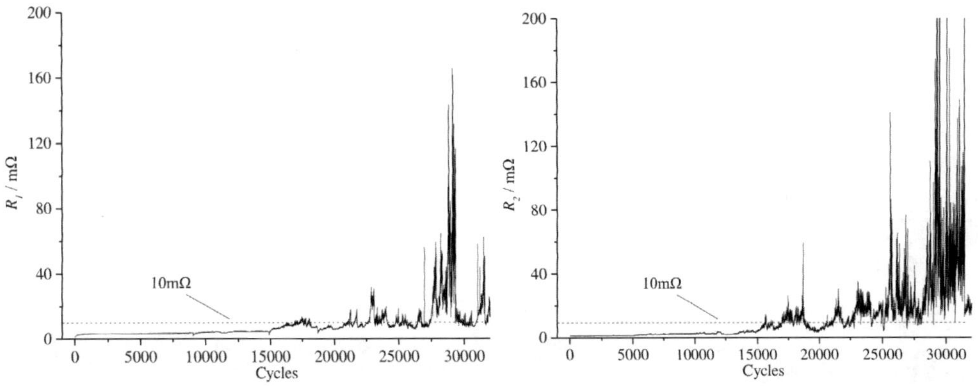

(c) Contact resistance of contact pair 1 **(d)** Contact resistance of contact pair 2

Figure 4. Variations in insertion force, extraction force and contact resistance as functions of operation cycles.

3.2. *Typical insertion and extraction process*

The typical waveforms of the insertion/extraction force and contact resistance are shown in Fig.5. At the beginning of plug insertion stage, the insertion force increases sharply to the maximum value of 12N, and the contact resistance decreases to the stable value (the contact resistance of first contact pair is 2mΩ, and the second pair is 1.3mΩ) attributed to the increase of contact force. Then, the insertion force decreases to the stable value of 8N gradually, while the contact resistance keeps relatively stable with the increase of insertion depth.

The insertion force reduces to 0 rapidly at the beginning of the withdrawal process. And the force remains at 0 in a short period of time from 475ms to 500ms due to the matching gaps between various components of the connector, such as pin/jack and insulator, insulator and shell. After that, the extraction force sharply increases to about 9.5N, while the contact resistance keeps stable attributed to the negligible change of expansion of the jack. The extraction force decreases to 0 and the contact resistance increases rapidly due to the decreasing in expansion of the jack at 535ms, when the pin separates from jack.

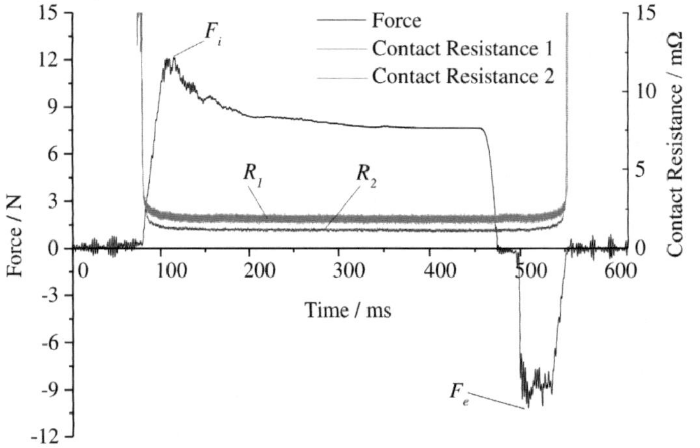

Figure 5. Typical experimental waveforms of force and contact resistance.

The insertion/extraction physical model of contact pair is shown in Fig.6. The center line of pin is taken as the horizontal axis, which is coincide with the center line of jack. The insertion process of pin could be divided into three stages: the mechanical contact between the cambered surface of pin and jack (Stage A), the expansion of jack (Stage B), and the insertion on straight part of pin (Stage C).

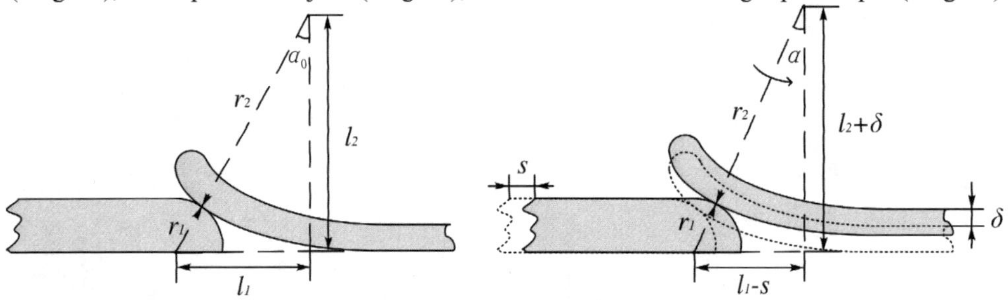

(a) Stage A **(b)** Stage B

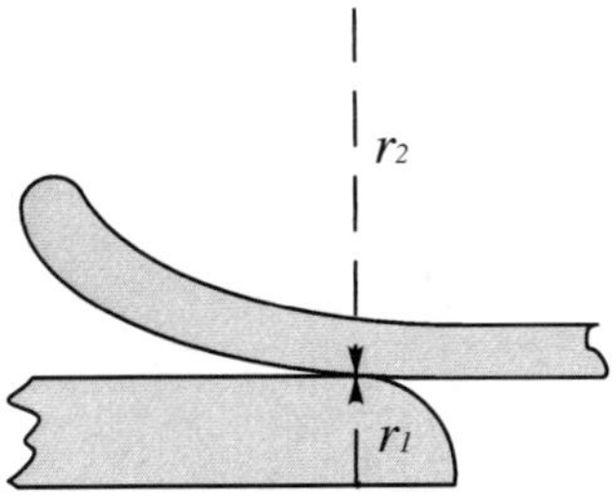

(c) Stage C

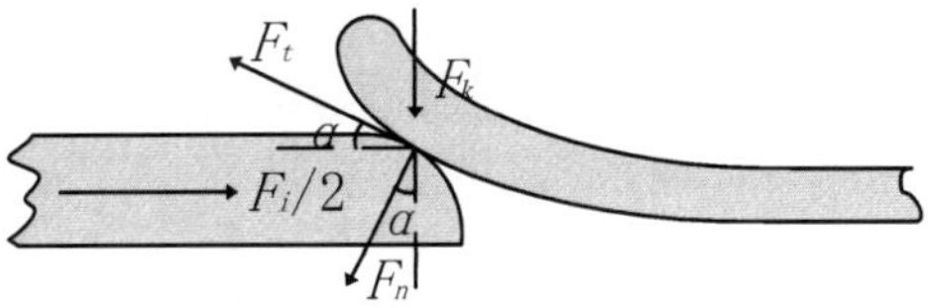

(d) Force analysis

Figure 6. Schematic plot of the insertion process.

According to Reference [5], the theoretical calculation formulas of insertion force F_i and extraction force F_e of single contact pair could be obtained as follows

$$F_i = \begin{cases} 2k\delta \dfrac{\mu\cos\alpha + \sin\alpha}{\cos\alpha - \mu\sin\alpha} & , \quad (0 < s \le l_1) \\ 2\mu k\delta_{\max} & , \quad (s > l_1) \end{cases} \tag{1}$$

$$F_e = \begin{cases} 2k\delta \dfrac{\mu\cos\alpha - \sin\alpha}{\cos\alpha + \mu\sin\alpha} & , \quad (0 < s \le l_1) \\ 2\mu k\delta_{\max} & , \quad (s > l_1) \end{cases} \tag{2}$$

And the contact force F_n could be written as

$$F_n = \begin{cases} \dfrac{k\delta}{\cos\alpha - \mu\sin\alpha} & , \quad (ds > 0) \\ \dfrac{k\delta}{\cos\alpha + \mu\sin\alpha} & , \quad (ds < 0) \end{cases} \tag{3}$$

where $ds > 0$ represents the insertion process, $ds < 0$ represents the extraction process, the angle α and the jack expansion δ could be expressed as

$$\begin{cases} \alpha = \arcsin\dfrac{l_1 - s}{r_1 + r_2} \\ \delta = \sqrt{(r_1 + r_2)^2 - (l_1 - s)^2} - l_2 \end{cases} , \quad (0 < s \le l_1) \tag{4}$$

where l_1 and l_2 could be calculated by

$$\begin{cases} l_1 = \sqrt{(r_1 + r_2)^2 - l_2^2} \\ l_2 = r_2 + \dfrac{\delta_0}{2} \end{cases} \tag{5}$$

where s is the insertion depth of pin, μ is the friction coefficient between pin and jack, k is the relevant stiffness of jack, δ_{max} is the maximum of the expansion, δ_0 is the initial width of jack, r_1 is the curvature radius of the pin and r_2 is the curvature radius of the jack.

Fig.7 shows the variations in insertion force and extraction force as functions of insertion depth of pin. It is noticed that there is a peak value of insertion force in the stage of jack expansion, which is attributed to the complex change of the angle α and the jack expansion δ (formula (4)).

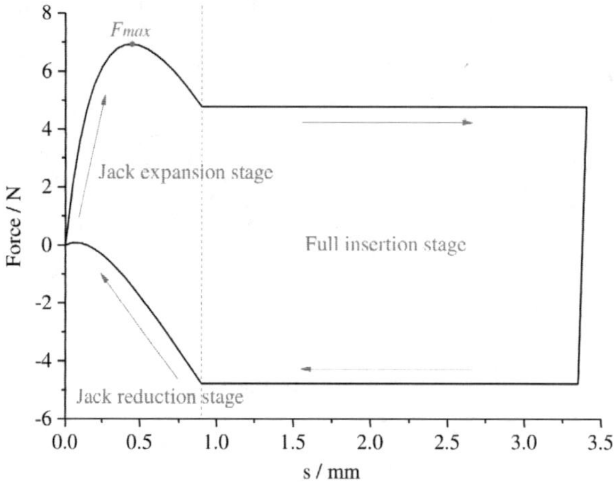

Figure 7. Theoretical calculation results of insertion and extraction force of single contact pair.

Combining formula (3) with the empirical calculation formula of contact resistance

$$R_j = K(0.102F_n)^{-m} \tag{6}$$

The theoretical calculation formula of contact resistance is revised as

$$R_j = \begin{cases} K\left(\dfrac{0.102k\delta}{\cos\alpha - \mu\sin\alpha}\right)^{-m}, & (ds > 0) \\ K\left(\dfrac{k\delta}{\cos\alpha + \mu\sin\alpha}\right)^{-m}, & (ds < 0) \end{cases} \tag{7}$$

where K and m are the experimental value related to contact material and contact form.

According to formula (7), it is easy to explain that the contact force rises sharply to the maximum value due to the jack expansion, resulting in the rapid reduction of the contact resistance to the stable value in the jack expansion stage.

4. Possible failure mechanisms

Fig.8 shows the waveforms of insertion force, extraction force and contact resistance as functions of cycles of 1, 50, 30000, respectively. It is obviously observed that the force value of insertion and extraction process in the 50th cycle increases to about 1.6 times compared with the first cycle. While the variation law of insertion/extraction force is consistent with that of the first cycle, and the contact resistance has no noticeable change.

The peak value of insertion force is disappeared after 30000 operation cycles and the insertion force increases with the increase of insertion depth. The maximum value of extraction force reduces from 11.64N to 4.4N. As shown in Fig.8 (c), the stable and reliable contact state between pin and jack could not maintain, which results in the obvious fluctuation of contact resistance during the insertion process. The stable value of contact resistance is related to the fluctuation state during the insertion process, so the contact resistance R_1 and R_2 show a strong fluctuation phenomenon after 30000 operation cycles.

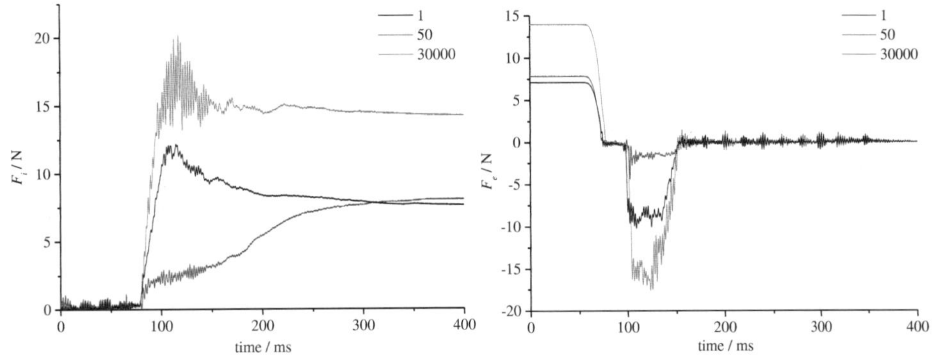

(a) Waveform comparison of insertion force **(b)** Waveform comparison of extraction force

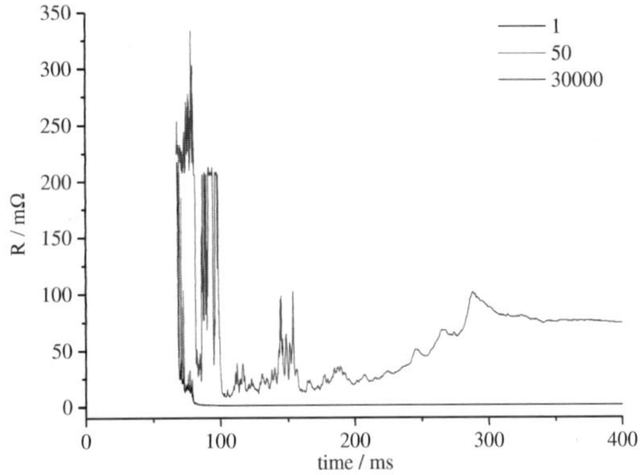

(c) Waveform comparison of contact resistance

Figure 8. Waveform comparison of insertion force, extraction force and contact resistance at 1, 50, 30000 cycles.

In order to better explain the degradation law of above parameters and associated physical mechanism of connector, the surface topologies of the pin before and after experiments are captured by the laser confocal microscopy. As shown in Fig.9, there is a serious wear on the surface resulting in an obvious inclined surface on the front of the pin due to the repetitive mechanical insertion and withdrawal wear. Combined with the insertion/extraction physical model and the comparison of dynamic waveforms at different stages mentioned above, it is speculated that the surface morphology of the contact pair changes as shown in Fig.10.

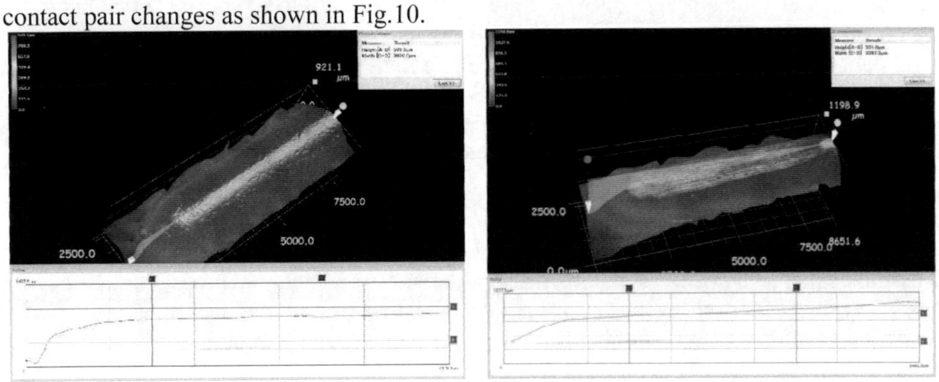

(a) Before experiment **(b)** After experiment

Figure 9. Surface topology of the pin.

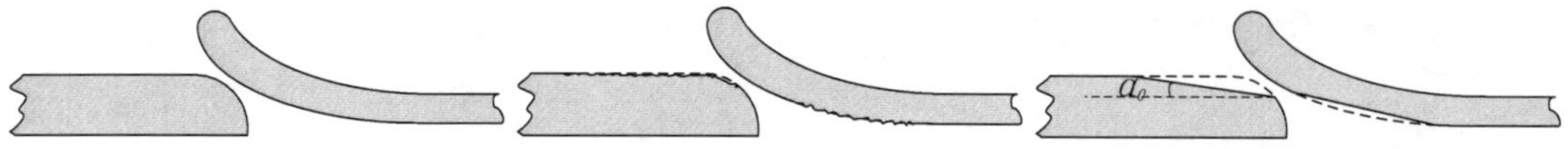

(**a**) Before experiment (**b**) Initial stage (**c**) Last stage

Figure 10. Schematic plot of the change of the surface morphology of contact pair.

At the beginning of the experiment, the wear of contact position leads to the increase of friction coefficient (as shown in Fig.10 (b)), resulting in a sharp increase trend on insertion force and extraction force (as shown in Fig.4 (a), (b) and the 50th waveforms in Fig.8 (a), (b)). However, there is no obvious change in contact resistance due to the slight wear on the surface of gold coating (as shown in the 50th waveform in Fig.8 (c)).

With the increase of operation cycles, deformation occurs on the contact surface of pin and jack due to the repetitive mechanical insertion and withdrawal wear (as shown in Fig.9 (b)). And the accumulation of wear results in an obvious inclined surface on the front of the pin (as shown in Fig.10 (c)). Therefore, it is necessary to correct the insertion/extraction mathematical model established above. The theoretical calculation formulas of insertion and extraction force are still formula (1) and (2). However, the angle α changes from a function of the insertion depth to a constant value α_0, and the expansion δ of jack changes from a complex function about the insertion depth s to a positive proportion function

$$\delta = s \tan \alpha_0 \tag{8}$$

The theoretical calculation formula of insertion and extraction force could be corrected as

$$F_i = 2ks \tan \alpha_0 \frac{\mu \cos \alpha_0 + \sin \alpha_0}{\cos \alpha_0 - \mu \sin \alpha_0} \tag{9}$$

$$F_e = 2ks \tan \alpha_0 \frac{\mu \cos \alpha_0 - \sin \alpha_0}{\cos \alpha_0 + \mu \sin \alpha_0} \tag{10}$$

Therefore, the insertion force is a positive proportion function of insertion depth s in the last stage of the experiment theoretically, resulting in the peak value disappearance of insertion force and the increase of the insertion force with the increase of insertion depth, which reflects in the 30000th waveform of insertion force in Fig.8 (a). In addition, the fatigue and stress relaxation phenomenon, which is caused by the repetitive mechanical insertion and withdrawal operations, occurs on the jack resulting in the decrease of stiffness k of jack. Combined with the formulas (9) and (10), this phenomenon aggravates the decrease of the insertion and extraction force of the connector.

The contact resistance of the connector keeps relatively stable in the initial stage of the experiment due to the inactivity chemical properties of the gold coating on the surface of the contact pair, which is difficult to form oxide film. The following four reasons leads to the increase and strong fluctuation of contact resistance in the last stage of the experiment. Firstly, the expansion δ of the jack decreases when the pin inserted due to the repetitive mechanical insertion and withdrawal wear, resulting in the decrease of the contact force F_n between jack and pin. Furthermore, the fatigue and stress relaxation phenomenon, which is caused by the repetitive mechanical insertion and withdrawal behaviors, occurs on the jack resulting in the decrease of contact force F_n by decreasing the stiffness k of jack. These two reasons mentioned above induce an obvious reduction of the contact force F_n provided by jack, which is unable to maintain the reliable contact state between the pin and the jack. And it would make the contact resistance increase and be unstable. Thirdly, the substrate metal is exposed to the environment and oxidized, which is caused by the destroyed gold plating due to the accumulation of wear. The oxide film adheres to the contact surface, resulting in the increase of contact resistance. The insertion and extraction behavior could destroy the formed oxide film continuously, and the substrate metal exposed to the environment and oxidized again. This physical process causes the increase and fluctuation phenomenon of the contact resistance in the last stage of the experiment. Finally, the insulation debris produced by insulator wear, oxide particles and metal particles adhere to the contact surface, reducing the real areas for current conduction and increasing the contact resistance. And these debris would displace with the repetitive insertion and withdrawal behaviors, resulting in the strong fluctuation of contact resistance value. These four physical processes mentioned above interact with each other and they would become more and more intense with the increase of operation cycles, resulting in the fluctuation phenomenon of contact resistance more and more frequent in the last stage of the experiment.

5. Conclusions

A test rig which allows for simulating the insertion and withdrawal behavior and collecting the dynamic data of force and contact resistance of electrical connector in-situ in real time under a controllable temperature and humidity environment is deigned. The degradation behavior and associated physical mechanism of repetitive mechanical insertion and withdrawal experiments are investigated under 25℃ and 65%RH ambient stress. This study reveals that there is an obvious inclined surface on the front of the pin due to the accumulation of wear from repetitive insertion and withdrawal behaviors. Coupled with fatigue and stress relaxation phenomenon, the insertion force and extraction force are reduced and the mechanical failure occurs on connector. The reduction of contact force caused by wear and fatigue, the oxide film formed by oxidation of substrate metal and the debris produced by wear attached to the contact surface are the main factors for the increase and strong fluctuation of the contact resistance in the last stage of the experiment, which results in the electrical contact failure.

References

[1] J. Swingler, J.W. McBride, C. Maul, Degradation of road tested automotive connectors, IEEE Trans. Compon. Packag. Technol., 2000, 23(1) 157–64.
[2] L. Xu, S.Q. Ling, D.W. Li, "Insertion force quality consistency investigation of electrical connector," in Proc. 11th Conf. Ind. Electron. Appl., Hefei, China, 2016, pp. 2395-2399.
[3] R. El Abdi, N. Benjemaa, Experimental and analytical studies of the connector insertion phase, IEEE Trans. Compon. Packag. Technol., 2008, 31(4) 751-758.
[4] S.Q. Ling, L. Xu, G.F. Zhai, "Insertion Force Investigation of Electrical Connector with Convergent Angle and Concentricity," in Proc. 6th Int. Conf. Instrum. Meas. Comput. Commun. Control, Harbin, China, 2016, pp. 270-273.
[5] W.B. Ren, Y. Meng, T.Y. Wang, "Preliminary experimental study on the insertion and withdrawal characteristics of blade and receptacle pair," in Proc. 65th Holm Conf. Elect. Contacts, Milwaukee, WI, USA, 2019, pp. 184-191.
[6] R.Yuan, H.Q. Li, Q.Y. Wang, Simulation-based design and optimization and fatigue characteristics for high-speed backplane connector, Adv. Mech. Eng., 2019, 11(6) 1-10.
[7] W.B. Ren, D.Y. Du, Y. Du, Electrical contact resistance (ECR) of connector response to mechanical vibration environment, IEEE Trans. Compon. Pack. Manuf. Technol., 2020, 10(2) 212-219.
[8] M. Andrusca, M. Adam, R. Burlica, A. Munteanu, A. Dragomir, "Considerations regarding the influence of contact resistance on the contacts of low voltage electrical equipment," in Proc. Int. Conf. Expo. Elect. Power Eng., Iasi, Romania, 2016, pp. 123-128.
[9] W.B. Ren, P. Wang, Y.H. Fu, C.F. Pan, J. Song, Effects of temperature on fretting corrosion behaviors of gold-plated copper alloy electrical contacts, Tribol. Int., 2015, 83 1-11.
[10] J. Song, H.M. Yuan, C. Koch, "Accelerated Testing of Electromechanical Connectors Considering Thermal and Mechanical Loads," in Proc. 64th Holm Conf. Elect. Contacts, Albuquerque, NM, USA, 2018, pp. 467-474.
[11] H. Essone-Obame, L. Cretinon, B. Cousin, N. Ben Jemaa, E. Carvou, R. El Abdi, "Investigation on friction coefficient evolution for thin-gold layer contacts," in Proc. 26th Int. Conf. Elect. Contacts, Beijing, China, 2012, pp. 411-416.

An integrated condition monitoring method for rolling element bearings based on perceptual vibration hashing and SOM

Shiwei Yan[1], Haining Liu[1*], Fajia Li[1], Huanyong Cui[1],

[1]School of Mechanical Engineering, University of Jinan, Jinan, China

*Corresponding author: me_liuhn@ujn.edu.cn

Abstract. Rolling element bearings are widely used in rotating machinery. Bearing faults will result in damage to property. So, the condition monitoring of bearings is of great significance, but few methods can achieve both degradation assessment and fault diagnosis. In this paper, an integrated condition monitoring method for rolling element bearings based on perceptual vibration hashing (PVH) and self-organizing maps (SOM) is proposed. Distance matric based on PVH is used as a health indicator for degradation assessment, in which the baseline of healthy state is selected based on the clustering centre of SOM instead of experience. When the value of health indicator exceeds the pre-set threshold, visualized fault diagnosis can also be achieved by training the SOM network. The effectiveness of the developed method is verified with the vibration data from accelerated degradation tests of rolling element bearings.

Keywords: Bearing fault diagnosis; bearing degradation assessment; Perceptual Vibration Hashing; Self-organizing maps.

1. Introduction

The rolling element bearings are one of the most widely used parts in rotating machinery. According to statistics, most failures of the rotating machinery are caused by bearing faults, which leads to loss of property. So, the degradation assessment and accurate identification of the bearing fault types are of great significance to ensure the safe operation of mechanical equipment and avoid catastrophic accidents.

A feature is a compact representation containing the information of the machine condition. Statistical parameters in the time and frequency domain such as root mean square (RMS), kurtosis, and crest factor are widely used for condition monitoring. Time-frequency joint analysis methods such as empirical mode decomposition (EMD) [1] and wavelet-based methods [2] are widely studied and used. However, one feature can only represent one aspect of the bearing health state. It might result in the loss of some important information so that the effectiveness of the feature extraction is not satisfactory. Perceptual vibration hashing by sub-band coding is proposed [3]. The extracted sub-band features can effectively reveal the frequency distribution pattern in the vibration signals, which are further compacted and symbolically represented by condition hashes. It can represent the machine condition comprehensively and improve the computational speed. The distance matric against healthy machine condition hash (MCH) is obtained which can be used for degradation assessment.

In recent years, with the gradual maturity of the artificial intelligence (AI) technology, using the AI to fault diagnosis of bearings has become a research hotspot. BP neural network with its improved algorithm is an extensively researched method [4,5], but it is a supervised learning method which need advance knowledge. Its learning speed slows with the increasement of the data dimension. Self-organizing maps (SOM) is a kind of neural network with simple structure and good classification effect, which is an unsupervised learning method with high learning speed. SOM does not need too much experience knowledge to achieve a good classification effect. It also has the advantages of efficient data visualization.

Based on above analysis, it can be seen that the degradation assessment and fault diagnosis of rolling element bearings is of great significance, but few methods can achieve both two process simultaneously. In this paper, an integrated condition monitoring method for rolling element bearings based on PVH and SOM is proposed. Distance matric based on PVH is obtained for degradation assessment, in which the healthy state is selected by the clustering centre of SOM network instead of experience. When the health indicator exceeds the pre-set threshold, a visualized bearing fault diagnosis method based on PVH and SOM is introduced. The effectiveness of proposed methods is validated through the experiment.

2. Theoretical Foundation

2.1. Perceptual Vibration Hashing by Sub-band Coding

The purpose of the perceptual vibration hashing is to transmit the vibration signal into MCHs, which can represent the equipment condition in a compact form. The process is shown below:

1) Sub-band division by wavelet packet transform (WPT) [6].
2) Blocking: The sub-band signal is truncated into several blocks so that a sub-band matrix can be made for two-dimensional discrete cosine transform (2D-DCT) [7].
3) 2D-DCT: 2D-DCT is implemented on each frequency band for decorrelation and then a coefficient matrix is obtained.
4) Feature extraction: A sub-band feature vector for raw signal is obtained by cascading feature vectors extracted from each sub-band coefficient together.
5) Hashing with Symbolic Aggregate Approximation (SAX) [8]: The MCH which is the symbolic representation of feature vectors is obtained through SAX.

Finally, a piece of a 1024-point vibration signal can be transformed into a compact MCH with a few numbers. A distance function can be used for results of SAX to quantize the similarity of MCHs. A more detailed description of it can be found in [3].

2.2. Self-organizing Map Network

SOM proposed by Kohonen [9] is a kind of artificial neural network (ANN) which is suitable for data clustering. It is trained using unsupervised competitive learning, which can produce a low-dimensional space to represent the high-dimensional input data. The training process of it is show below.

1) Initialize all weights w_{ij} with small random values. Set the initial topological neighborhood, initial learning rate, total number of iterations, and the number of the nodes in competitive layer.
2) Choose an input vector $\mathbf{x}$ randomly from the training data.
3) Calculate the distance between each neuron and the input vector so that the neuron *j* called the winner neuron or best matching unit (BMU) is determined, of which weight vector $\mathbf{w}_j$ is closest to the input vector. The closest distance $D_{\min}$ is given:

$$D_{\min} = \{D_i\} = \min\{(\mathbf{x} - \mathbf{w}_i)^2\} \tag{1}$$

4) Update the weight vectors of the BMU and its neighbors to make the nodes more like the input vector. The closer the node and BMU are, the more the weight is adjusted.
5) Repeat from the step 2 until the specified number of iterations.

3. An Integrated Condition Monitoring Method Based on PVH and SOM

3.1. Selection of Optimum Healthy State Based on SOM

For a mechanical system, the beginning of the operation indicates that the machinery is new but the machine condition at the beginning cannot represent the optimum healthy state. Due to the advantages of visualization and unsupervised clustering, SOM can present the data distribution clearly in the U-Matrix and weight space. In order to select an optimum healthy state as the baseline in distance metric, a new method is proposed below:

1) Raw vibration signals are processed by PVH to get MCHs.
2) MCHs are used to train SOM neural network and the weight coefficient of neurons can be obtained from the trained SOM neural network. The mean of distance sum between one neuron and others can be calculated and then the neuron located in the cluster centre with minimum mean distance can be obtained.
3) Calculate the distance between samples and the central neuron. Then, the sample with minimum distance can be obtained, which represent the optimum healthy state.
4) When new vibration signals are obtained, repeat from step 1 to achieve the update of optimum healthy state selection.

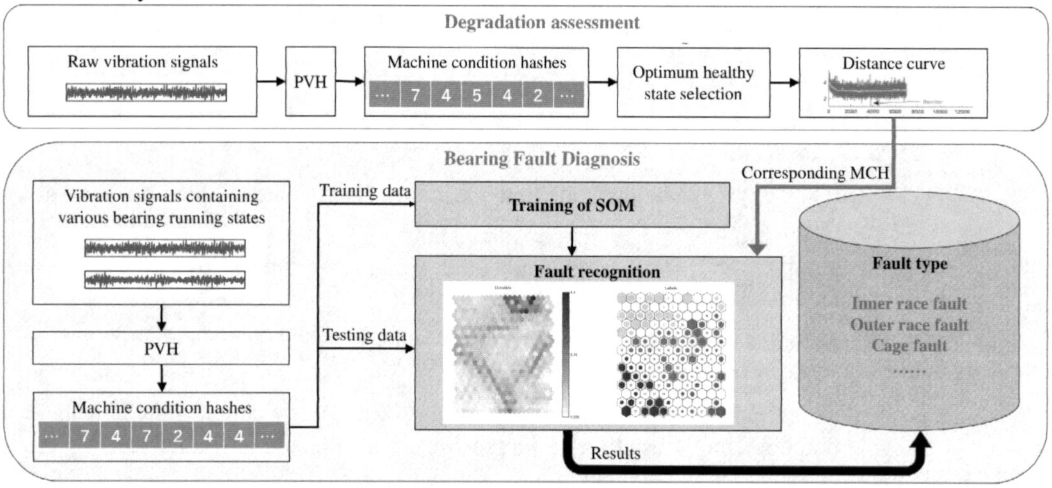

Figure 1. Flowchart of the proposed integrated condition monitoring method.

3.2. Degradation Assessment

For the degradation assessment, as is shown in figure 1, the distance matric based on PVH is set as the health indicator. Firstly, raw vibration signals are transmitted into compact MCHs through PVH. Secondly, select an optimum healthy state using the method in section 3.1 as the baseline in distance metric. Finally, a distance matric against a condition hash in optimum healthy state is obtained which can provide a quantitative assessment of degradation. When the value of this health indicator exceeds the pre-set threshold, a bearing fault diagnosis method based on PVH and SOM is used for diagnosis.

3.3. Fault Diagnosis

The proposed diagnosis method is shown in figure 1, firstly, the raw vibration signals containing various bearing running states are initially selected. Secondly, the selected raw signals are processed by PVH to get MCHs. Thirdly, the MCHs are divided into training data and testing data randomly to be input feature vector of the SOM. After the input feature vectors are normalized, the SOM is trained iteratively. The fault type of the bearings can be described by the matched regions in the SOM. Finally, when testing data is input the trained SOM, the fault types of bearings can be recognized automatically, which can be visualized clearly though the hit units on the map. When early degradation occurs in distance curve, corresponding MCH is put into the trained SOM and then the fault diagnosis can be achieved.

4. Experimental Results

4.1. Run-To-Failure Bearing Vibration Data

In order to validate the effectiveness of the proposed approach, the vibration signal data is obtained from the XJTU-SY bearing datasets [10] which provides the run-to-failure data of 15 rolling element bearings. Table 1 shows the information of the dataset used in this paper. The type of the bearings is LDK UER204. In the process of the experiment, the sampling frequency is 25600 Hz and the rotating speed is 2250 r/min under the load of 11 kN. The dataset contains four types of the bearing health condition including the normal, inner race fault, outer race fault and cage fault.

Table 1. Description of the bearing dataset.

Bearing type	Fault type	Dataset	Rotating speed (r/min)	Working condition (kN)
LDK UER204	Normal	Bearing 2_1	2250	11
LDK UER204	Inner race fault	Bearing 2_1	2250	11
LDK UER204	Outer race fault	Bearing 2_5	2250	11
LDK UER204	Cage fault	Bearing 2_3	2250	11

4.2. Degradation Assessment with Distance Curves

To show the effectiveness of the proposed method in degradation assessment, 15712 pieces of MCHs were obtained from Bearing 2_1. Then the distance curves calculated based on different MCHs as baselines were obtained. The comparison of different distance curves is shown in figure 2. Figure 2(a) shows the distance curve of the whole life calculated based on the first MCH as the baseline of the healthy bearing and figure 2(b) presents the distance curve calculated based on the MCH selected by the proposed method in section 3.1.

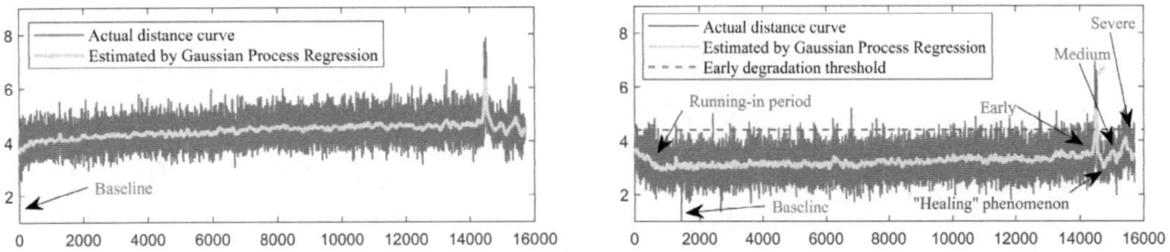

(a) (b)

Figure 2. Comparison of distance curves

Through the comparison of the distance curves, it can be seen that the decreasing tendency at the beginning in figure 2(b) can indicate the running-in period of the new machinery. Then, the distance curve remains at a relatively stable value, which indicates that the bearing is in a healthy state before the degradation. After the early degradation, the "healing" phenomenon is presented clearly and the increasing tendency is stronger compared with figure 2(a). When the value of distance exceeds a pre-set threshold, corresponding MCHs can be used as input of the trained SOM to achieve fault diagnosis.

4.3. Bearing Fault Diagnosis

4.3.1. Data Processing based on PVH. 300 rows of 1024-point vibration signal of each health condition were obtained and 1200 rows were got in total. In the process of PVH, the bearing dataset of each health condition is transformed respectively into 300 samples with 128-diamensional sub-band feature vectors and the sub-band features were hashed with SAX. The hashing procedure is shown in figure 3, in which 16 segments are used for PAA and a number size of 8 is used for symbolic representation. For example, a piece of vibration signal under normal condition in figure 3 is transformed into the MCH {7874333447454222}. Finally, 300 pieces of MCH under each health condition were obtained.

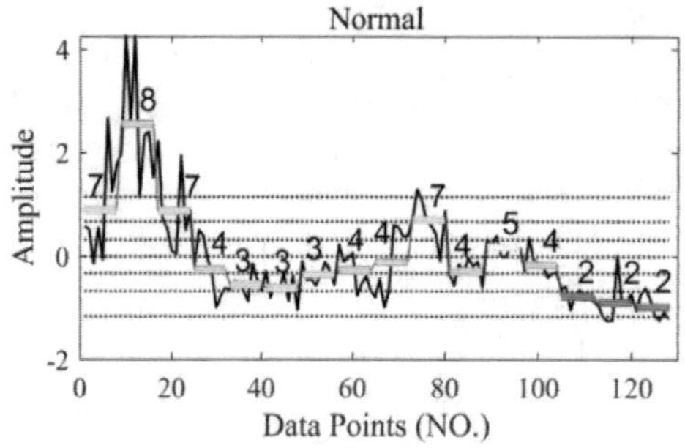

Figure 3. Illustration of SAX hashing

4.3.2. Fault diagnosis using SOM. 5-fold cross-validation was used to generate training data and testing data. The processed dataset which was arranged at random was divided into 5 folds. For each training process, 4 folds which included 960 samples containing 4 types of health states were selected as training data and 1 fold including 240 samples was left as testing data.

Figure 4. Training and testing results.

The size of the competitive layer was 14×11. After repeated training, as is shown in figure 4, the neuron labels of clustering results on training data are presented by various letters in the hexagons where the "N, I, O, C" respectively represent the normal, inner race fault, outer race fault and cage fault.

Table 2. Diagnosis accuracy of testing data.

Fault type	Accuracy of model 1 (%)	Accuracy of model 2 (%)	Accuracy of model 3 (%)	Accuracy of model 4 (%)	Accuracy of model 5 (%)
Normal	96.67	93.33	93.33	95	93.33
Inner race fault	98.33	91.67	96.67	95	93.33
Outer race fault	100	100	100	96.67	100
Cage fault	96.67	100	100	100	98.33
Average	97.92	96.25	97.50	96.67	96.25

After the training of the network was finished, to diagnose the bearing faults and validate the accuracy of the model, testing data was put into the network. Bearing health states could be classified automatically. Diagnosis results are shown in figure 4 where red, green, blue and yellow respectively represent the normal, inner race fault, outer race fault and cage fault. Most of the faults were clustering in the true area. The accuracy of the SOM models under 5-fold cross-validation is calculated which is shown in table 2.

The SOM accuracy of different bearing faults is listed and an overall diagnosis accuracy can be calculated which is 96.92%. The results show that the SOM neural network can achieve a good performance on classification and visualization.

5. Conclusions

This paper presents an integrated condition monitoring method for rolling element bearings based on PVH and SOM, which can achieve both degradation assessment and fault diagnosis. Dynamic degradation assessment can be tracked with the distance curve of MCHs, which is calculated based on an optimum healthy state as a baseline. After early degradation occurs, fault diagnosis starts directly by putting corresponding MCHs into the trained SOM, which can achieve an excellent effectiveness on fault classification and visualization. Convenience of the whole process of degradation assessment and fault diagnosis is improved by the proposed integrated condition monitoring method.

Acknowledgments

This work was supported in part by the National Natural Science Foundation of China under Grant 51605191, in part by the Shandong Provincial Key Research and Development Program under Grant 2019GGX105009.

References

[1] Lei Y, Lin J, He Z, et al. A review on empirical mode decomposition in fault diagnosis of rotating machinery[J]. Mechanical systems and signal processing, 2013, 35(1-2): 108-126.
[2] Yan R, Gao R X, Chen X. Wavelets for fault diagnosis of rotary machines: A review with applications[J]. Signal processing, 2014, 96: 1-15.
[3] Liu H, Wang Y, Li F, et al. Perceptual vibration hashing by sub-band coding: an edge computing method for condition monitoring[J]. IEEE Access, 2019, 7: 129644-129658.
[4] Wu J, Yang X L. Fuzzy BP neural network and its applications in fault diagnosis[J]. Systems Engineering and Electronics, 2001, 10.
[5] Tang H, Wu Y, Hua G, et al. Internal leakage fault diagnosis of hydraulic cylinder using PCA and BP network[J]. Journal of Central South University of Science and Technology, 2011, 42(12): 3709-3714.
[6] Mallat S G . A Wavelet Tour of Signal Processing: The Sparse Way. China Machine Press, 2009.
[7] Ahmed N, Natarajan T, Rao K R. Discrete cosine transform[J]. IEEE transactions on Computers, 1974, 100(1): 90-93.
[8] Lin J, Keogh E, Lonardi S, et al. A symbolic representation of time series, with implications for streaming algorithms[C]//Proceedings of the 8th ACM SIGMOD workshop on Research issues in data mining and knowledge discovery. 2003: 2-11.
[9] Kohonen T . Self-Organizing Maps[J]. series in information science, 2001.
[10] Wang B , Lei Y , Li N , et al. A Hybrid Prognostics Approach for Estimating Remaining Useful Life of Rolling Element Bearings[J]. IEEE Transactions on Reliability, 2018:1-12.

Comparative analysis and forecasting of isentropic efficiency of gas turbine compressor with ARIMA, VAR, NARNN and ANFIS approaches

Jiachi Yao[1,2], Chao Liu[1,2,*], Yunfeng Jin[1,2], Gaofeng Deng[3], Yunlong Guan[3], Jiangang Hao[4], Haizhou Huang[4], Dongxiang Jiang[1,2]

[1]Department of Energy and Power Engineering, Tsinghua University, Beijing 100084, China
[2]State Key Laboratory of Control and Simulation of Power System and Generation Equipment, Tsinghua University, Beijing 100084, China
[3]State Key Laboratory of Building Safety and Environment, China Academy of Building Research, Beijing 100013, China
[4]HuaDian Electric Power Research Institute Co., Ltd., HangZhou, China

*Corresponding author: cliu5@tsinghua.edu.cn

Abstract. It is extremely important to monitor the status of gas turbine to ensure its safe and reliable operation. In this work, the variation trend of isentropic efficiency of compressor is analysed based on the measured data of F-class heavy-duty gas turbine in practical industrial application. The actual measured data of F-class heavy-duty gas turbine includes the data under start-stop and unstable working conditions, which cannot be directly used for calculation and analysis. To solve this problem, the data selection rules are designed and determined according to the operating conditions of gas turbine to select the data under effective working state. The isentropic efficiency of compressor is calculated based on the selected data. Then the forecasting effects of four forecasting methods on the variation trend of isentropic efficiency of compressor are studied. Four indexes, namely, symmetric mean absolute percentage error (SMAPE), mean absolute percentage error (MAPE), root mean square error (RMSE), and similarity (SIM) values are utilized to evaluate the forecasting accuracy. The research results indicate that the Adaptive Neuro-Fuzzy Inference System (ANFIS) method has better forecasting effect than Autoregressive Integrated Moving Average (ARIMA), Vector Autoregression (VAR) and Nonlinear Autoregression Neural Network (NARNN) for this F-class heavy-duty gas turbine. Through the ANFIS method, the SIM up to 96.77%, the SMAPE and MAPE are less than 0.1, and the RMSE is only 0.1157. Therefore, the ANFIS method is suitable for forecasting the isentropic efficiency of this F-class heavy-duty gas turbine compressor.

Keywords: Gas turbine, compressor, isentropic efficiency, forecasting, ANFIS.

1. Introduction

Gas turbine is a high-end equipment of power generation equipment, which plays a very important role in national economy and energy and power industry. Gas turbines can generate lots of electricity with very little pollution. Thus power plants adopt gas turbines. The CO_2 emissions of natural gas-fired power plants are about half of that of supercritical coal-fired power plants. Gas turbines have become the most important mechanical equipment in carbon maximum and carbon neutral environment because of its very low pollution emission and excellent performance [1].

Nowadays, some gas turbines have been operating for many years, resulting in degradation of their performance. In order to ensure the safe and reliable operation of gas turbines, it is important to carry out condition monitoring and predictive maintenance of gas turbines [2-6]. The commonly used methods are model-based methods, data-driven methods and hybrid methods. The model-based methods need to establish the mathematical model of gas turbine. The structural parameters and actual data of gas turbine in actual industrial applications are difficult to obtain, and it is difficult to accurately verify the reliability of the established model to the gas turbine in actual industry. Thus it is difficult to accurately establish the gas turbine model. The data-driven methods mainly use the information and characteristics in the historical data of gas turbine to predict the future operation state, so as to evaluate the gas turbine performance. The hybrid methods combine the model and data-driven methods to evaluate the gas turbine performance [7-10].

In this study, the F-class heavy-duty gas turbine in actual industrial operation is taken as the research object. Due to commercial reasons, it is difficult to obtain the structural parameters of gas turbine to establish the gas turbine model, thus this work mainly studies the variation trend of gas turbine compressor performance based on data-driven method. For this F-class heavy-duty gas turbine, the measured data related to the compressor mainly include the ambient temperature, ambient pressure, total pressure difference of the filter, compressor outlet pressure and compressor outlet temperature. Based on these data, the isentropic efficiency of compressor can be calculated. The isentropic efficiency can reflect the operating status and performance of the gas turbine compressor. By predicting and analyzing the change trend of isentropic efficiency of gas turbine compressor, it can provide reference for early warning and predictive maintenance of gas turbine.

In order to predict and analyze the isentropic efficiency of gas turbine compressor, the classical methods include ARIMA and VAR are studied. In recent years, more and more forecasting methods based on neural networks have emerged with the rapid development of artificial intelligence, such as NARNN, CNN, RNN, LSTM, GRU, ANFIS, etc. The neural network-based method can effectively extract the nonlinear and non-stationary features. Moreover, some neural network-based methods can adaptively optimize the network structure, which can improve the forecasting performance [11-15]. Thus the neural network-based methods are adopted to forecast and analyze the isentropic efficiency of the gas turbine compressor in this work.

The main goal of this study is to establish a reliable and accurate forecasting method for predicting the variation trend of compressor isentropic efficiency, which can provide reference for predictive maintenance of gas turbine. The main work includes: (I) According to the operating conditions of this F-class heavy-duty gas turbine, the data selection rules for the effective working state are determined. Then the isentropic efficiency of compressor can be accurately and effectively calculated based on the selected data. (II) Four forecasting methods are studied to forecast the isentropic efficiency of compressor, and the forecasting effects are quantitatively evaluated based on quantitative evaluation index.

The paper is organized as follows. In section II, the fundamentals of four forecasting methods are introduced. In section III, the actual measured F-class heavy-duty gas turbine data are analysed and processed. In Section IV, the calculation results of four forecasting methods are compared and analysed. Finally, the conclusions are summarized in section V.

2. Methodology

2.1. *Autoregressive Integrated Moving Average (ARIMA)*

The ARIMA model is a widely used time series forecasting model. The calculation formula of ARIMA (p, D, q) is as follows [16].

$$(1-\sum_{i=1}^{p}\varphi_i L^i)(1-L)^D x_t = (1-\sum_{i=1}^{q}\theta_i L^i)\varepsilon_t \tag{1}$$

where L is represented the lag operator, φ_i are the parameters of the autoregressive part, θ_i are the parameters of the moving average part, and ε_t is represented the error terms.

In this study, there are three steps to predict the isentropic efficiency of compressor based on the ARIMA model. The first step is to construct the stationary time series and select the model type according to the autocorrelation function and partial autocorrelation function. The second step is to determine the order q and p of the ARIMA model based on the Akaikes Information Criterion (AIC) and Bayesian Information Criterion (BIC). The last step is to use the established ARIMA model to predict the isentropic efficiency and analyze the prediction results.

2.2. *Vector Autoregression (VAR)*

VAR is a model based on the statistical properties of data, which can be used to predict time series. A VAR(p) model can be written as the following formula [17].

$$Y_t = c + A_1 Y_{t-1} + A_2 Y_{t-2} + \cdots A_p Y_{t-p} + \varepsilon_t \tag{2}$$

where c is an n by 1 constant vector, A is an n by n matrix, and ε_t is an n by 1 error vector.

In this study, The VAR (4) model is constructed for the prediction analysis of isentropic efficiency of F-class heavy-duty gas turbine sompressor.

2.3. *Nonlinear Autoregression Neural Network (NARNN)*

The NARNN method is a widely used artificial neural network-based prediction method. The basic model calculation formula is as follows [18].

$$Y_t = f(Y_{t-1}.Y_{t-2},\cdots,Y_{t-d}) \tag{3}$$

where Y_t is the value of Y in time t, Y_t is the function of past d number.

In this study, the number of hidden neurons and the number of delays are respectively set 10 and 2 , and the training algorithm is adopted the scaled conjugate gradient algorithm.

2.4. *Adaptive Neuro-Fuzzy Inference System (ANFIS)*

The ANFIS has the learning ability and self-adaptation. It mainly includes fuzzification layer, rule reasoning layer, normalization layer, defuzzification layer and output layer [19]. The fuzzification layer is responsible for fuzzing the input variable. The rule reasoning layer is responsible for multiplying the input signals to obtain the excitation intensity of the fuzzy rules. The normalization layer normalizes the above-mentioned excitation intensity. The defuzzification layer is responsible for defuzzifying the normalized result. The output layer is responsible for calculating the total output result of the entire network.

3. Analysis and processing of measurement data of F-class heavy-duty gas turbine

The isentropic efficiency of gas turbine compressor is calculated as follows [20].

$$\eta_c = (T_1 \times (P_2 / P_1)^{k-1/k} - T_1)/(T_2 - T_1) \tag{4}$$

where η_c is the isentropic efficiency, T_1 and T_2 are the inlet and outlet temperature of compressor, P_1 and P_2 are the inlet and outlet pressure of compressor, the ideal gas adiabatic index k is 1.4.

According to the calculation formula of the isentropic efficiency of the gas turbine compressor, the isentropic efficiency is related to the inlet and outlet pressure and temperature of the compressor. For the F-class heavy-duty gas turbine in actual industrial applications, the compressor outlet pressure (P_2) and compressor outlet temperature (T_2) are measured by sensors. The compressor inlet temperature

(T_1) can be regarded as the ambient temperature (T_A). The compressor inlet pressure (P_1) is the ambient pressure (P_A) minus the total pressure difference of the filter (P_F).

3.1. *Analysis of gas turbine measurement data*

The F-class heavy-duty gas turbine has been used for many years in practical industrial applications. Here, the measured data from January 2019 to December 2019 are selected for analysis. In the process of measuring data, the unit of compressor outlet temperature (T_2) and ambient temperature (T_A) is degree centigrade. The unit of ambient pressure (P_A), total pressure difference of the filter (P_F) and the compressor outlet pressure (P_2) are the mmHg(ABS), mmH$_2$O(ABS) and bar, respectively. In the following analysis and calculation, the units of temperature and pressure are uniformly converted into thermodynamic temperature (K) and Pa. The measured data of F-class heavy-duty gas turbine are shown in Fig. 1.

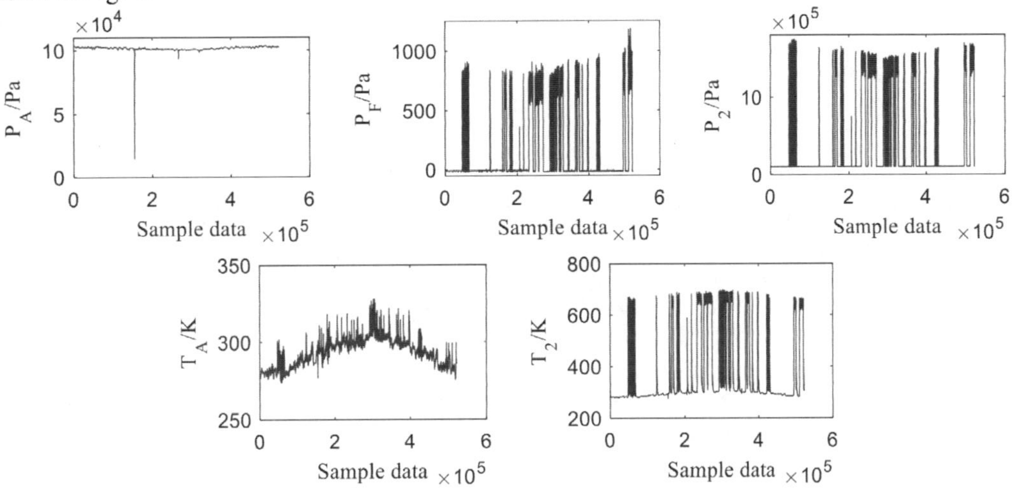

Figure 1. The measured data of F-class heavy-duty gas turbine.

For this F-class heavy-duty gas turbine, the data collection system collects data every 1 minute. This work focuses on analyzing the collected data from January 2019 to December 2019, which can be seen in Fig. 1. Since the data is collected continuously and uninterrupted, the collected data includes the data of gas turbine startup and unstable working state. It is obvious that the collected data cannot be directly used to calculate and analyze the isentropic efficiency of the compressor. Therefore, it is necessary to select the data of the gas turbine under effective working state for analysis and processing.

3.2. *Determine and select the data of gas turbine under effective working state*

In order to select the data of gas turbine under effective working state, this work adopts the following data selection rules:

(1) Since the stable working speed of this F-class heavy-duty gas turbine is 3000 rpm, the data with speed fluctuations within 0.1% is selected first.

(2) When the speed of the gas turbine reaches 3000±3 rpm for the first time from the start up, the gas turbines may not operate completely stably. In order to ensure the stable operation of gas turbine and consider the influence of acquisition system, the data of the gas turbine is selected after the speed reaches 3000±3 rpm for the first time and runs steadily for 10 minutes.

(3) In order to select stable and effective data, the data that runs stably for less than 10 minutes in the 3000±3 rpm range is discarded.

(4) In order to select stable and effective data, the data with the difference between two consecutive data points of compressor outlet pressure less than 0.5 MPa is selected.

(5) The data with natural gas flow greater than 10.5 kg/s is selected.

Based on the data selection rules, the selected data of gas turbine are shown in Fig. 2.

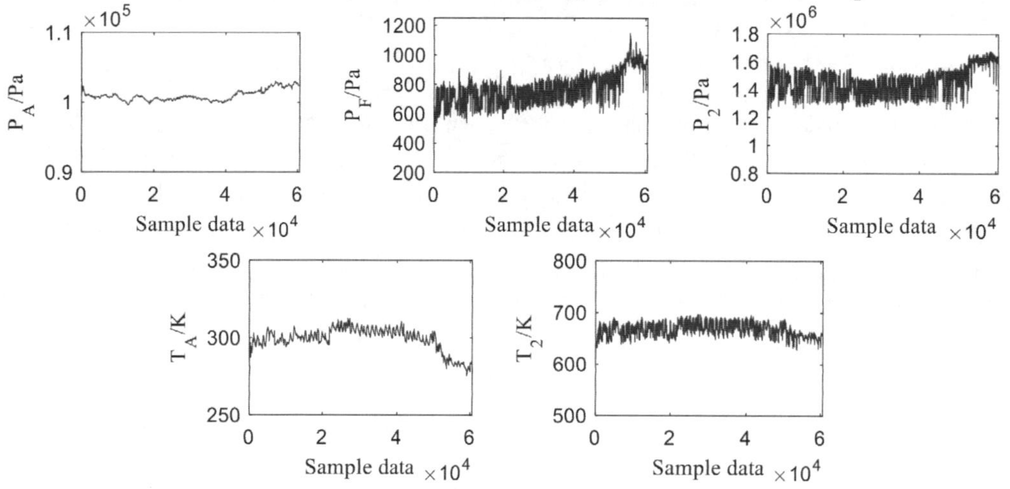

Figure 2. The selected data of F-class heavy-duty gas turbine.

Compared the Fig. 2 with Fig. 1, it can be seen that the selected data of F-class heavy-duty gas turbine are stable, without wildly fluctuating data. Moreover, the fluctuation of ambient pressure and temperature is consistent with the actual situation within a year. Therefore, the selected data used to analyze the isentropic efficiency of the compressor is effective and reasonable.

3.3. *Calculation results of isentropic efficiency of gas turbine compressor*

When stable and effective data of gas turbine are selected, then the isentropic efficiency of compressor can be accurately calculated, and the calculation results are shown in Fig. 3.

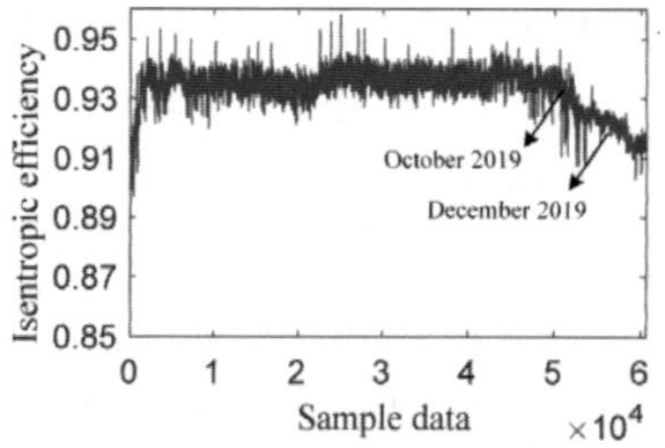

Figure 3. The calculation results of isentropic efficiency of compressor.

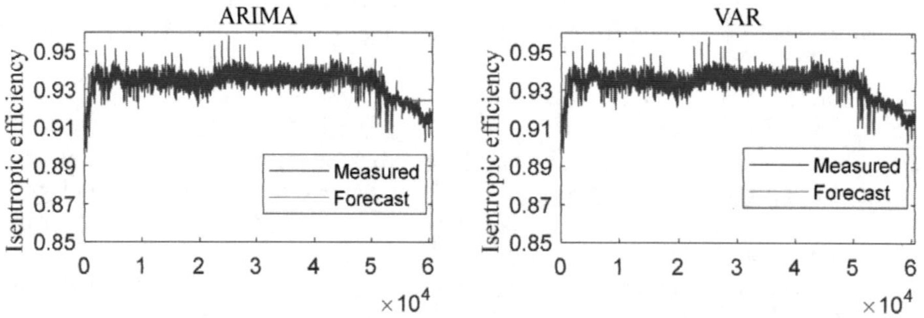

Figure 4. The calculation results by ARIMA and VAR.

From Fig. 3, the isentropic efficiency of the compressor under normal and steady operation is about 93.5%. However, the isentropic efficiency of the compressor began to gradually decrease from October 2019. The decrease of isentropic efficiency may be due to the degradation of gas turbine performance, or it may be influenced by the environment. This requires further analysis.

4. Comparative analysis of forecasting effects of different forecasting methods

In order to forecast the changing trend of isentropic efficiency of gas turbine compressor, the classical ARIMA and VAR method are firstly adopted. The ARIMA(1,1,1) model which has the lowest AIC (-588582.49) and BIC (-588575.45) is adopted to forecast the isentropic efficiency. The VAR(4) model is established utilized to forecast the isentropic efficiency. The calculation results are shown in Fig. 4.

From Fig. 4, it can be seen that the forecasting effect is not good. The isentropic efficiency of gas turbine compressor is non-stationary and nonlinear, but the ARIMA and VAR are usually used to forecast the linear and stationary time series. Moreover, due to the large volatility and burr of gas turbine measurement data, the calculated isentropic efficiency fluctuates sharply, which increases the difficulty of accurate prediction. Thus the forecasting effect by ARIMA and VAR is not good.

In order to extract the nonlinear and non-stationary features of isentropic efficiency, the NARNN model is adopted. In NARNN model, the number of hidden neurons is 10 and the delay number is 2. The training algorithm is the scaled conjugate gradient algorithm. The forecasting results by NARNN are shown in Fig. 5.

From Fig. 5, it can be seen that the forecasting effect of NARNN method is better than that of VAR and ARIMA method. This is because the NARNN method adopts a nonlinear model, which is suitable for predicting nonlinear and non-stationary isentropic efficiency. However, the forecast result based on NARNN method is larger than the actual result. This may be because the NARNN model is not optimized enough, and each parameter cannot be adjusted adaptively. To overcome this problem, the ANFIS method is employed to forecast the isentropic efficiency. The ANFIS method can adaptively adjust the structure and parameters of the neural network. The forecasting results by ANFIS are shown in Fig. 6.

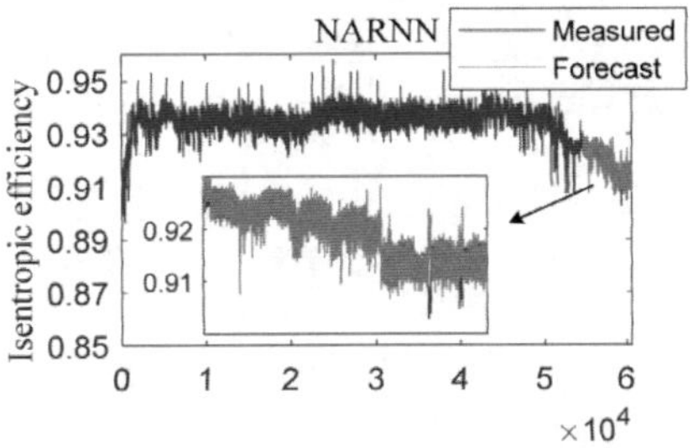

Figure 5. The calculation results by NARNN.

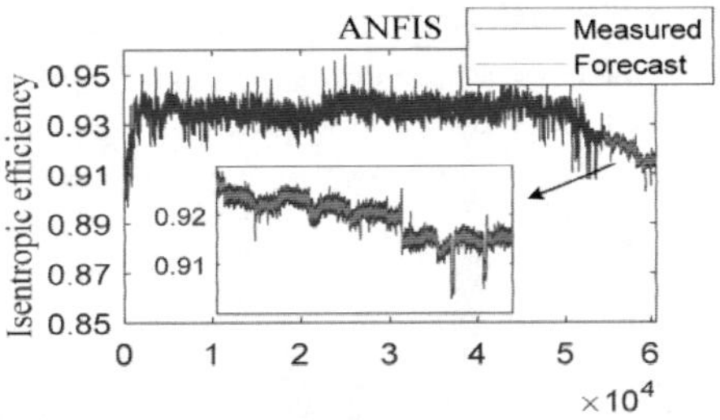

Figure 6. The calculation results by ANFIS.

From Fig. 6, the ANFIS method can accurately forecast the change trend of isentropic efficiency. This is because the ANFIS method can adaptively adjust the network structure and parameters according to the nonlinear and non-stationary characteristics of isentropic efficiency, so as to improve the forecasting accuracy.

In order to quantitatively evaluate the forecasting effects, the four quantitative evaluation indexes, namely symmetric mean absolute percentage error (SMAPE), mean absolute percentage error (MAPE), root mean square error (RMSE), and similarity (SIM) are employed. For SMAPE, MAPE and RMSE, the smaller the value is, the better the forecasting effect is. For SIM, the larger the value is, the better the forecasting effect is. The calculated results are shown in Table 1.

Table 1. Comparison of calculation results of different forecasting methods.

Forecasting methods	SMAPE	MAPE	RMSE	SIM
ARIMA	0.5873	0.5900	0.6737	84.21%
VAR	0.3810	0.3815	0.4216	88.76%
NARNN	0.2296	0.2296	0.2851	92.94%
ANFIS	0.0975	0.0975	0.1157	96.77%

From Table 1, the SMAPE, MAPE and RMSE are respectively the 0.0975, 0.0975 and 0.1157 by ANFIS, which is the lowest value among these four forecasting methods. For SMAPE, MAPE and RMSE, the smaller the value is, the better the forecasting effect is. Thus the ANFIS has the best forecasting effect on the isentropic efficiency of compressor. In addition, the SIM value by ANFIS is 96.77%. The higher the similarity (SIM) between the forecast result and the actual result is, the better the forecasting effect is. Therefore, the ANFIS method is suitable for the forecasting of isentropic efficiency of gas turbine compressor.

5. Conclusions

The F-class heavy-duty gas turbines are widely used in power plants. At present, some F-class heavy-duty gas turbines have been running for many years and are approaching the maintenance deadline. In order to ensure the safe and stable operation of the F-class heavy-duty gas turbine, it is necessary to monitor and analyze the operating status of the gas turbine. This paper adopts the F-class heavy-duty gas turbine measurement data in actual industrial operation to analyze and research the change trend of isentropic efficiency of compressor, which can provide reference for the predictive maintenance of this F-class heavy-duty gas turbine.

For the measured gas turbine data for one year in 2019, the data collected by the data acquisition system includes the gas turbine start-stop and unstable working state data. In order to select the data of gas turbine under effective working conditions, the data selection rules are determined according to the actual operation of the gas turbine. By determining the data selection rules, it is beneficial to calculate and analyze the isentropic efficiency of gas turbine compressor.

In order to accurately forecast the isentropic efficiency, this paper studies four forecasting methods. Firstly, the ARIMA and VAR are respectively adopted to forecast the isentropic efficiency, but the forecasting effect is not good. This is because the nonlinear and non-stationary characteristics of the isentropic efficiency are not well extracted by ARIMA and VAR. In order to overcome this problem, the NARNN is employed to forecast the isentropic efficiency, and the forecasting effect is good. In order to further improve the forecasting accuracy, the ANFIS is utilized to forecast the isentropic efficiency. The ANFIS can adaptively adjust the structure and parameters of neural network. Based on the ANFIS, the similarity between the forecasting results and the actual results is 96.77%. Moreover, both the SMAPE and MAPE are less than 0.1, and the RMSE is only 0.1157. The research results indicate that the forecasting effect of ANFIS is the best in these four forecasting methods for the forecasting of isentropic efficiency of gas turbine compressor.

In this work, the actual F-class heavy-duty gas turbine data is utilized to calculate the isentropic efficiency of the compressor, and the ANFIS method is suitable to forecast the change trend of isentropic efficiency. In the future, the transferability and applicability of ANFIS method to different gas turbines can be studied.

Acknowledgments
This work was financially supported by the National Key R&D Program of China (Grant No. 2019YFF0216104, 2019YFF0216103, 2019YFF0216101).

References

[1] B Swa, Z A Lu, L A Chao, et al. Techno-economic-environmental evaluation of a combined cooling heating and power system for gas turbine waste heat recovery, ENERGY. 231 (2021) 120956.

[2] R J Seeking, A S Verma, Y Li, et al. Offshore wind turbine operations and maintenance: A state-of-the-art review, RENEW SUST ENERG REV. 144 (2021) 110886.

[3] H Hanachi, C Mechefske, J Liu, et al. Performance-based gas turbine health monitoring, diagnostics, and prognostics: A survey, IEEE T RELIAB, 67 (2018) 1340-1363.

[4] M Tahan, E Tsoutsanis, M Muhammad, et al. Performance-based health monitoring, diagnostics and prognostics for condition-based maintenance of gas turbines: A review, APPL ENERG. 198 (2017) 122-144.

[5] H D Mo, G Sansavini, M Xie. Performance-based maintenance of gas turbines for reliable control of degraded power systems, MECH SYST SIGNAL PR. 103 (2018) 398-412.

[6] M Zagorowska, F S Spüntrup, A M Ditlefsen, et al. Adaptive detection and prediction of performance degradation in off-shore turbomachinery, APPL ENERG. 268 (2020) 114934.

[7] Y Z Chen, X D Zhao, H C Xiang, et al. A sequential model-based approach for gas turbine performance diagnostics, ENERGY. 220 (2021) 119657.

[8] S Amirkhani, A Tootchi, Chaibakhsh A . Fault detection and isolation of gas turbine using series-parallel NARX model, ISA T. 115 (2021) 1-17.

[9] T Olsson, E Ramentol, M Rahman, et al. A data-driven approach for predicting long-term degradation of a fleet of micro gas turbines, ENERGY AI. 4 (2021) 100064.

[10] A Alblawi. Fault diagnosis of an industrial gas turbine based on the thermodynamic model coupled with a multi feedforward artificial neural networks, ENERGY REP. 6 (2020) 1083-1096.

[11] Z M Liu, I A Karimi. Gas turbine performance prediction via machine learning, ENERGY. 192 (2020) 116627.

[12] M B Rahmoune, A Hafaifa, A Kouzou, et al. Gas turbine monitoring using neural network dynamic nonlinear autoregressive with external exogenous input modelling, MATH COMPUT SIMULAT. 179 (2021) 23-47.

[13] Y Y Shen, K Khorasani. Hybrid multi-mode machine learning-based fault diagnosis strategies with application to aircraft gas turbine engines, NEURAL NETWORKS. 130 (2020) 126-142.

[14] D J Zhou, Q B Yao, H Wu, et al. Fault diagnosis of gas turbine based on partly interpretable convolutional neural networks, ENERGY. 200 (2020) 117467.

[15] X S Yang, M L Bai, J F Liu, et al. Gas path fault diagnosis for gas turbine group based on deep transfer learning, MEASUREMENT. 5 (2021) 109631.

[16] O Celestino, S L Fernando, R P Javier, et al. A hybrid ARIMA–SVM model for the study of the remaining useful life of aircraft engines, J COMPUT APPL MATH. 346 (2019) 184-191.

[17] W B Nicholson, I Wilms, J Bien, et al. High dimensional forecasting via interpretable vector autoregression, J MACH LEARN RES. 21 (2020) 1-16.

[18] W Xing, M Razi, R M Kirby, et al. Greedy nonlinear autoregression for multifidelity computer models at different scales, ENERGY AI. 1 (2020) 100012.

[19] H Hanachi, J Liu, C Mechefske. Multi-mode diagnosis of a gas turbine engine using an adaptive neuro-fuzzy system, CHINESE J AERONAUT. 31 (2018) 1-9.

[20] H Aygün. Effects of isentropic efficiency of turbomachinery components on entropy production for small turbojet engine, AIRCR ENG AEROSP TEC. 93 (2021) 711-722.

IOP Conf. Series: Materials Science and Engineering 1207 (2021) 012014 doi:10.1088/1757-899X/1207/1/012014

A gas turbine thermal performance prediction method based on dynamic neural network

Yansong Hao[1, 2], Yunfeng Jin[1], Chao Liu[1, *], Jiangang Hao[3], Haizhou Huang[3], Dongxiang Jiang[1, 2]

[1]Department of Energy and Power Engineering, Tsinghua University, Beijing, China
[2]StateKey Laboratory of Control and Simulation of Power System and Generation Equipment, Beijing, China
[3]HuaDian Electric Power Research Institute Co., Ltd., HangZhou, China

*Corresponding author: cliu5@tsinghua.edu.cn

Abstract. In order to ensure safety and reliability of energy transportation, it is necessary to understand and predict the performance of the gas turbine components. A prediction frame of the gas turbine compressor isentropic efficiency is established using the neural time series theory based on the Dynamic Neural Network. In order to obtain appropriate parameters for the network, a validation set is introduced to generalize the model. The compressor isentropic efficiency can be predicted based on the suggested model which provides an effective technical mean for the early warning of gas turbine performance. The experiment verified that the performance calculation model and the isentropic entropy efficiency prediction model based on the neural time series are effective.

Keywords: Gas turbine, State prediction, Dynamic Neural Network, Neural time series.

1. Introduction

The core components of gas turbine work continuously in complex conditions like extreme temperature, high pressure and high speed, and are easy to fail in such working condition [1, 2]. On the one hand, the main systems will deteriorate in long-term operation, and the untimely maintenance will lead to the increase of inlet pressure loss or partial seal leakage. On the other hand, the frequent shutdown caused by excessive maintenance will also reduce unit operation benefit. Therefore, it is of great significance for the safe and economic performance of gas turbine unit to study condition evaluation methods for inlet filtration system performance of the gas turbine [3, 4].

Prediction is becoming an indispensable part of the field of modern industrial condition monitoring in recent years. As an important construction machinery in the industry, gas turbines has been studied for years based on predictions. For example, Wang [5] et al. applied Markov nonlinear estimation to the tracking of aero-engine performance degradation, taking into account the progressive degradation and sudden failure of the engine. The results show that the performance tracking and degradation prediction errors are within 1%. Since there are many parameters that affect the operation of gas turbines, the use of a single time parameter to predict the performance trend of gas turbines in the process of gas turbine performance prediction will have large errors. In contrast, the use of multi-parameter prediction methods will obtain more good result. Rahman [6] et al. took micro gas turbines as the research object and selected 7 types of degradation parameters (compressor isentropic efficiency, compressor flow, turbo-entropy efficiency, turbine flow, heat exchanger efficiency, compressor exhaust leakage, Shaft efficiency) to discuss the effect of one or more degradation parameters on performance. However, an overly complex model will bring computational pressure, and too many parameters will cause the algorithm adjustment process to be more time-consuming.

In recent years, many neural networks have been used for mechanical state degradation and prediction [7, 8]. However, most popular deep networks have the same static reasoning paradigm. After the training is completed, the network structure and parameters remain unchanged during the testing phase, which limits the representation ability of model, reasoning efficiency and interpretability to a certain extent. Dynamic neural networks, which include tapped delay lines are used for nonlinear filtering and prediction [9]. The dynamic network can adaptively adjust its own structure or parameters according to the input in the inference stage, so that it has good characteristics that the static network cannot have in many aspects. In this paper, a prediction model of the gas turbine compressor isentropic efficiency is established using the neural time series theory based on the Dynamic Neural Network.

2. Basic theory

The prediction model of the compressor and the basic theory of the dynamic neural network is described in this section.

2.1. prediction model of the compressor

In the thermal system, the main performance parameters of the compressor include pressure ratio, speed, flow, efficiency and so on. It is worth noting that the performance parameters in the compressor are not completely independent. Under the condition that the internal structure and working fluid of the compressor remain unchanged, when the compressor inlet pressure, inlet temperature, flow rate, and speed are constant, the corresponding outlet pressure and outlet temperature are also determined.

The pressure ratio of the compressor PR_c and the isentropic efficiency η_c can be expressed function as follows

$$PR_c = f_{c1}\left(G_{cor}, n_{cor}\right), G_{cor} = G\sqrt{T}/P \tag{1}$$

$$\eta_c = f_{c2}\left(G_{cor}, n_{cor}\right), n_{cor} = n/\sqrt{T} \tag{2}$$

The functional relationship f_{c1} and f_{c2} is determined by the design structure of the compressor. In theory, f_{c1} and f_{c2} remain unchanged in the same types. Where, G is the real mass flow rate of the compressor inlet, T and P are the inlet temperature and inlet pressure.

Therefore, a prediction model for the isentropic efficiency of the compressor is established (the training sample of the model is $\left(In_{(i)}, out_{(i+n)}\right)$. $In_{(i)}$ is the input vector, include the model sample compressor isentropic efficiency η_c, compressor inlet temperature t, compressor inlet pressure p and rotating speed r. Where n is the time step, that is, the current data is used to predict the future n time.

$$\begin{Bmatrix} In_{(i)} = [\eta_i, t_i, p_i, r_i] \\ Out_{(i+n)} = \eta_{i+n} \end{Bmatrix} \tag{3}$$

2.2. Time series model based on dynamic neural network

The structure and parameters of the network remain unchanged after the training is completed, which is the same static similarity that most popular deep networks have. This static property limits the representation ability, reasoning efficiency and interpretability of a model to a certain extent.

Considering the different importance of inputs at different moments, the dynamic network is designed to adaptively decide whether to allocate calculations to the inputs at each moment. The dynamic network can also skip a certain amount of input at each moment and decide to read the input from the certain area of data.

The dynamic network can adaptively adjust its own structure or parameters according to the input in the inference stage. Therefore, it has good characteristics in terms of computing efficiency, adaptability and expressive ability by comparing with static networks.

The defining equation of the DNN model is shown as equation (4) and the basic construct is shown as Figure 1.

$$y(t) = F(y(t-1), y(t-2),\ldots, y(t-my), u(t-1), u(t-2),\ldots, u(t-mu)) \tag{4}$$

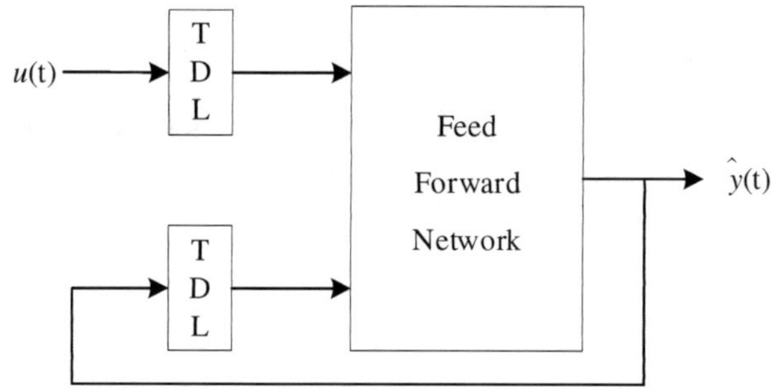

Figure 1. Basic construct of time series model, where the output y(t) is estimated based on a previous independent input and the corresponding output signal.

3. Experimental verification

The actual operating gas turbine data is collected for experiment to verify the effectiveness of the method. The actual degradation data was collected from M701F4 gas turbines operating in a certain area of East China. The collection time starts in January 2019 and ends in June 2020.

In order to improve the practicability of the model, the time step n is selected as 1 day. The gas turbine thermal parameters of the day are used to predict the isentropic efficiency of the compressor at 1d in the future. Since the historical data of gas turbine thermal parameters are tested once every 1 minute, a large number of historical data samples will affect the calculation time of the model. It is necessary to process the sample data before establishing the prediction model due to the large temperature change within a day. Take average of daily sample data of the M701F4 gas turbine to obtain the average compressor inlet temperature, compressor inlet pressure, speed and other parameters within a day. 98 data samples (each sample has a time interval of 1d) are obtained.

The results of time series response and relative errors of the isentropic efficiency prediction model of the compressor are shown in Figure 2. In addition, regression prediction results are shown in Figure 3. The results show that the whole state prediction can be achieved based on the suggested method. An obvious sample point deviation appeared because the gas turbine has been shut down at the corresponding time.

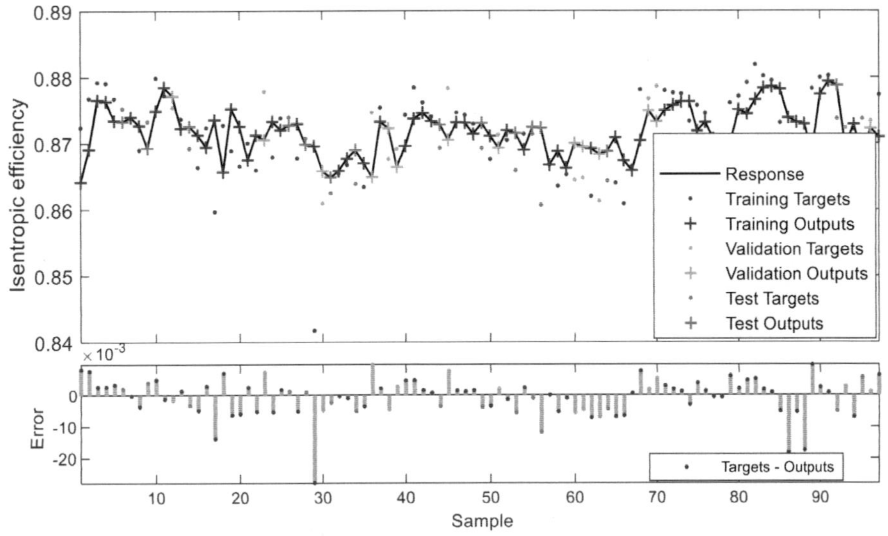

Figure 2. Time series response and relative errors.

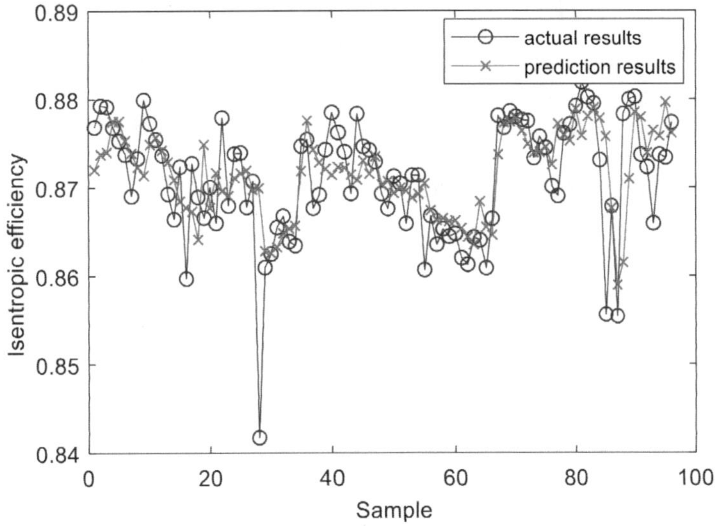

Figure 3. Regression prediction results.

4. Conclusions

A prediction model of the gas turbine compressor isentropic efficiency is established using the neural time series theory based on the Dynamic Neural Network. The model for the isentropic efficiency of the compressor is established based on isentropic efficiency, compressor inlet temperature, compressor inlet pressure and rotating speed. Data samples (each sample has a time interval of 1d) are obtained by taking average of daily sample data, which are utilized to predict the gas turbine during a long time.

The compressor isentropic efficiency can be predicted based on the suggested model which provides an effective technical mean for the early warning of gas turbine performance. The experiment verified that the performance calculation model and the isentropic entropy efficiency prediction model based on the neural time series are effective.

Acknowledgments
This work was financially supported by the National Key R&D Program of China (Grant No. 2019YFF0216104, 2019YFF0216101).

References
[1] Zhang B, Jiaqiang E, Gong J, et al. Influence of structural and operating factors on performance degradation of the diesel particulate filter based on composite regeneration. Applied Thermal Engineering, 2017, 121, pp. 838-852.
[2] Hanachi H, Mechefske C, Liu J, et al. Performance-based gas turbine health monitoring, diagnostics, and prognostics: A survey. IEEE Transactions on Reliability, 2018, 67(3), pp. 1340-1363.
[3] Joly R, Ogaji S, Singh R, et al. Gas-turbine diagnostics using artificial neural-networks for a high bypass ratio military turbofan engine. Applied energy, 2004, 78(4), pp. 397-418.
[4] Hanachi H, Jie L, Mechefske C. Multi-mode diagnosis of a gas turbine engine using an adaptive neuro-fuzzy system. Chinese Journal of Aeronautics, 2018, 31(1), pp. 1-9.
[5] Wang P, Gao R X. Markov nonlinear system estimation for engine performance tracking. Journal of Engineering for Gas Turbines and Power, 2016, 138(9).
[6] Rahman M, Zaccaria V, Zhao X, et al. Diagnostics-oriented modelling of micro gas turbines for fleet monitoring and maintenance optimization. Processes, 2018, 6(11), pp. 216.
[7] Han T, Liu C, et al. Deep transfer network with joint distribution adaptation: A new intelligent fault diagnosis framework for industry application. ISA Transactions, 2020, 97, pp. 269-281.
[8] Han T, Liu C, et al. A novel adversarial learning framework in deep convolutional neural network for intelligent diagnosis of mechanical faults. Knowledge-Based Systems, 2019, 165, pp. 474-487.
[9] Han Y , Huang G , Song S , et al. Dynamic Neural Networks: A Survey. 2021.

Development of sliding contact resistance testing equipment for carbon brush confluence ring

Daozhuo Feng[1], Bin Liang[2], Fangyao Zhong[1], and Yilin Zhou[1, 3*]

[1] School of Artificial Intelligence, Beijing University of Posts and Telecommunications
[2] Automation School, Beijing University of Posts and Telecommunications
[3] Corresponding Author: ylzhou@bupt.edu.cn

Abstract. A carbon brush and a confluence ring form a pair of sliding electrical contacts. To reduce the contact resistance between the carbon brush and the confluence ring, the time for the running-in wear of the carbon brush increased during the machining process to increase the actual contact area. As a result, the welding and ablation failure between the carbon brush and the confluence ring happened under the impact of high current during service. To explore the effect of carbon brush wear on sliding electrical contact performance, it is necessary to monitor the dynamic contact resistance of the carbon brush and the confluence ring. In this paper, a set of 8-channel contact resistance testing equipment is developed. The device uses a relay array to achieve 1-8 channel gating and uses a data acquisition card to detect the contact resistance of each carbon brush and confluence ring pair combining with a four-point method. Based on the Labview software, it realizes the selection of acquisition channels, data acquisition card driving, real-time data processing, displaying, and storage. The range of contact resistance that this set of equipment can detect is 10mΩ ~2Ω, and the accuracy can reach 1% of the full scale. Finally, the simulation experiments show that the testing equipment meets the technical requirements, and the continuous test runs stably and reliably. It can be used for the testing and research of the sliding contact resistance between the carbon brush and the confluence ring during the running-in wear process.

Keywords: Carbon brush confluence ring; running-in wear; sliding contact resistance; testing equipment

1. Introduction

The carbon brush and the confluence ring are used in a radar power supply system to realize the signal and power transmission between the rotating part and the fixed part [1]. One of the factors that affect the reliability of electrical contact in the carbon brush and the confluence ring structure is the increase in contact resistance caused by the oxidation accumulation of carbon brush wear debris, the other is the insulation failure between adjacent confluence rings. To make the carbon brush and the confluence ring form a consistent contact interface curvature and maximize the contact area as much as possible, the manufacturer adjusted the carbon brush running-in time from 2 hours to 60 hours in the carbon brush pre-running in process. But after that, the carbon brush and the confluence ring experienced carbon brush ablation soon in the field application. It is speculated that the prolongation of the pre-running-in process time leads to an increase in the wear debris between the carbon brush and the confluence ring, which causes the interface welding, even ablation. To dynamically monitor the contact resistance of the carbon brush and the confluence ring during the pre-running in process, it is necessary to develop a set of multi-channel sliding contact resistance testing equipment.

For multi-channel contact resistance data acquisition equipment, Zhu Meng et al. [2] developed a 4-channel sliding electrical contact test equipment based on the NI M series PCI-6281 data acquisition card. The sliding contact resistance is dynamically measured, the normal force of the contacts can be adjusted from 0-300g, the sliding frequency (1-5Hz) and the sliding distance (1cm) of the sliding mechanism can be set, and 40 contact resistance data can be collected per sliding cycle. Liang Zhichao [3] developed a 4-channel synchronous voltage data acquisition based on FPGA, the highest sampling rate is up to 105MHz, using off-chip high-speed large-capacity FIFO to buffer the collected sine wave voltage data, the system has data playback function, and uses Labview as the upper computer software. Sun Wen [4] took C8051F060 and STM32F207 as the core to design a portable, general-purpose, high-precision 4-channel data acquisition system, mainly to achieve the acquisition and processing of analog and digital voltage signals, and through the Ethernet interface, the data is uploaded to the host computer to realize functions such as signal analysis and display. Pan Xiaorong [5] used the M series NI PCI-6221 data acquisition card to develop a multi-channel sensor signal data acquisition system based on Labview based on this hardware. The system has 4 input channels and the maximum acquisition range is between -10V and 10V, the sampling rate is 250kB/s.

Reference to the principles of the existed data acquisition schemes, and combining the dynamic contact resistance measurement range and the accuracy, as well as the number of channels, a multi-channel contact resistance testing equipment is designed with a high sampling rate and a high precision for the carbon brush and confluence ring based on Labview. This testing equipment can dynamically detect, display and store 1~8 channels of contact resistance at the same time to monitor the electrical contact performance of the carbon brush and the confluence ring contact pair in real-time, and assist in the study of carbon brush wear in the pre-running in process.

2. The requirements of sliding contact resistance testing equipment

The object to be tested is a contact pair between a carbon brush and a confluence ring. As shown in Figure 1, each ring is connected in parallel with 4 carbon brushes. The design requires a contact pair of a single carbon brush and a confluence ring to have a resistance value of less than 50mΩ. According to the GJB1217 of the electrical connector contact resistance test method, the contact resistance measurement adopts a four-point method, that is, a constant current of 100mA is applied to both ends of the contact pair, then the voltage across the contact is detected to calculate the contact resistance.

Figure 1. The contacts between the carbon brush and the confluence ring

To avoid the test cable entanglement during the rotation of the confluence ring, it is necessary to connect two adjacent confluence rings in series, so the detected contact resistance is the series value of two contact pairs between two carbon brushes and two confluence rings. The four-point method of wiring is shown in Figure 2. The contact resistance between the carbon brush and the confluence ring that meets the design standards should be in the order of tens of milliohms, and the failed contact resistance that causes the ablation is also below the ohm level. Therefore, the detection range of the contact resistance is set between 10mΩ and 2Ω. The detection accuracy is 1% of the full-scale. Since the minimum full-scale of the data acquisition card is 0.2V, and the four-point method needs to pass a current of 100mA to measure the contact resistance, the maximum allowable full-scale error of the device is 0.2V/0.1A × 1%=20mΩ. The accuracy of the device can reach 1 milliohm while testing dozens of the milliohm contact resistance, which meet the requirement of the contact resistance between the carbon brush and the confluence ring at the level of 10~60mΩ.

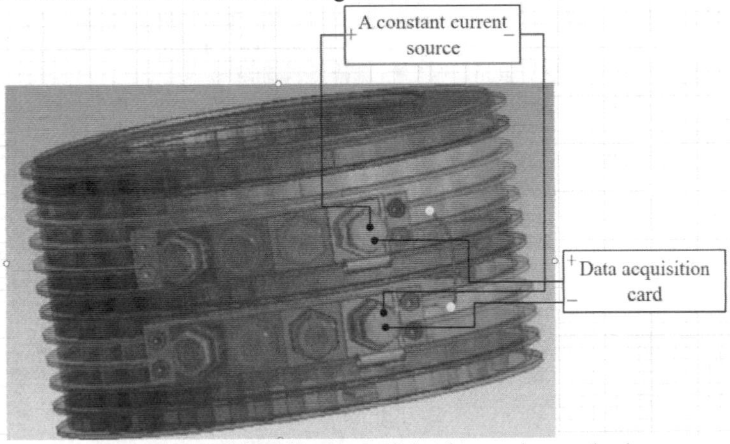

Figure 2. Wiring diagram of the four-point method between the carbon brushes and the confluence rings

Since the carbon brush-confluence ring system is running multiple groups at the same time, the testing equipment needs to be able to detect the sliding contact resistance of 1-8 channels. In the multi-channel contact resistance detection, the constant current source only connects to one pair of contacts switching by the relays, and then the multi-channel contact resistance is measured in turn.

The circumference of the confluence ring is 650mm, and the length of the contact surface of a carbon brush is about 7.6mm. Therefore, when each carbon brush slides around the confluence ring, 85 complete contact interfaces between the carbon brush and the confluence ring will be formed, that is, 650mm/7.6mm=85. Since the contact between a carbon brush and a confluence ring is surface contact type, the contact resistance will not change much in a short period. Therefore, it is recommended to take 85 contact resistance test positions during the process of one revolution of the confluence ring. Take one reading for each position and save it. After saving the data, the experiment can be reprocessed and graphed. According to the rotation speed of the confluence ring 20r/min, the interval time for each test point is about 60s/20r/85=0.035s.

The user can set the detection channel through the software interactive interface according to the detection requirements, and select the real-time display of the contact resistance curve of different channels over time. Timed storage of test data is available for subsequent research for data analysis.

According to research requirements, the design functions and technical indicators can be listed as follows:

1. Measuring range of contact resistance: between 20mΩ and 2Ω;
2. The measurement accuracy of contact resistance is 1% of the full-scale;
3. Measuring 85 contact resistance data per rotation cycle of the confluence ring;
4. Realize 1-8 channels contact resistance measurement options;
5. Man-machine interaction interface for inputting of experimental conditions;
6. Realize contact resistance data storage and real-time curve display of contact resistance changing over time.

3. Design of sliding contact resistance testing equipment

The design of the sliding contact resistance testing equipment for the carbon brush and the confluence ring is divided into two parts: hardware and software. The hardware part includes a constant current source, an 8-way relay array, a data acquisition card, an upper computer, as shown in Figure 3. The constant current source is mainly used to provide current to the contact to measure the voltage at both ends of the contact point; the 8-way relay array is mainly used for gating the measured channel according to the user's setting, and the constant current source power supply can be input. The appropriate precision data acquisition card is selected to measure the contact resistance of the selected channel; the upper computer implements control instructions and data acquisition, storage, and real-time display of the sliding contact resistance curve. The software is based on the Labview program to design the relay array driver, data acquisition program, data processing, resistance curve display program, data storage program, man-machine interaction interface.

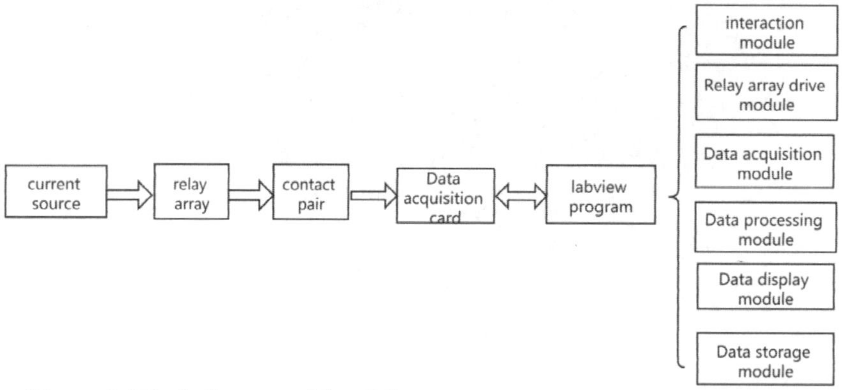

Figure 3. Block diagram of the sliding contact resistance testing equipment for the carbon brush and the confluence ring.

4. Hardware Design

4.1 Relay array

A relay array is used to switch the constant current source to a selected channel to detect the contact resistance, as shown in Figure 4.

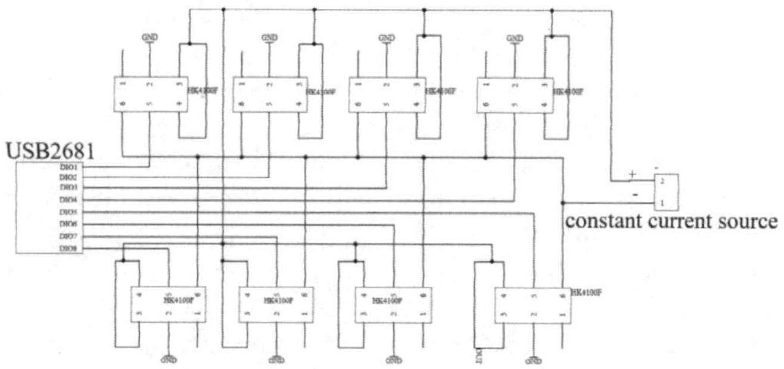

Figure 4. Schematic diagram of a relay array to realize gating the tested channel

4.2 Data acquisition card selection and technical indicators

Due to the need to monitor 8 channels of contact resistance between the carbon brush and the confluence ring, at least 16 analog input and output channels and 8 digital output channels are required, so that an Altai data acquisition card, model USB2861, is selected. The data acquisition card has 16 differential input channels and 24 digital I/O channels. The highest sampling rate can reach 500Ks/s and the accuracy can reach 1mV. The pins of the data acquisition card are shown in Figure 5.

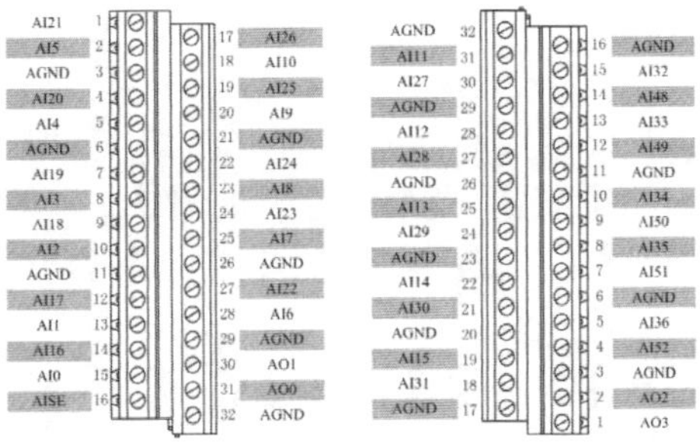

Figure 5. Pins of the data acquisition card

5. Software Design

The software design is divided into three parts: a user interface, a system control software, and a data acquisition software, as shown in Figure 6.

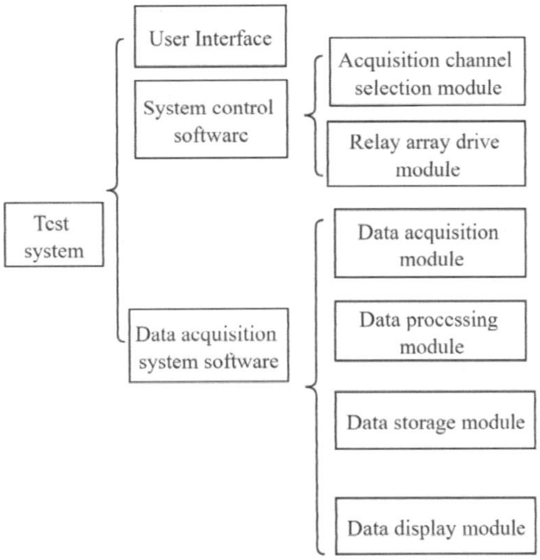

Figure 6. System software architecture

5.1 Design of Man-Machine Interaction Interface of Test System

The entire man-machine interaction interface is divided into a data display panel 1, a data acquisition channel selection panel 2, a parameter setting panel 3, as shown in Figure 7. The data display panel 1 can display the data collected by each channel in real-time, and you can choose to display the data of the corresponding acquisition channel in the display menu bar, and the general map display of each acquisition channel. The user can toggle the switch on the data acquisition channel selection panel 2 to the channel that needs data collection. The sampling rate, the number of readings sampled per channel, and the input of test system parameters such as file names can be set up on the parameter setting panel 3.

Before starting the data collection, the user must first select the physical channel to be collected in the data collection channel selection panel, enter the data collection parameters such as the sampling rate, the number of samples reads per channel, and the file name in the parameter setting page, and then click "Start" button for data collection. If you want to observe the data display in real-time, you can click to switch to the data display layout menu bar and select the data display situation of the physical channel collection that needs to be observed.

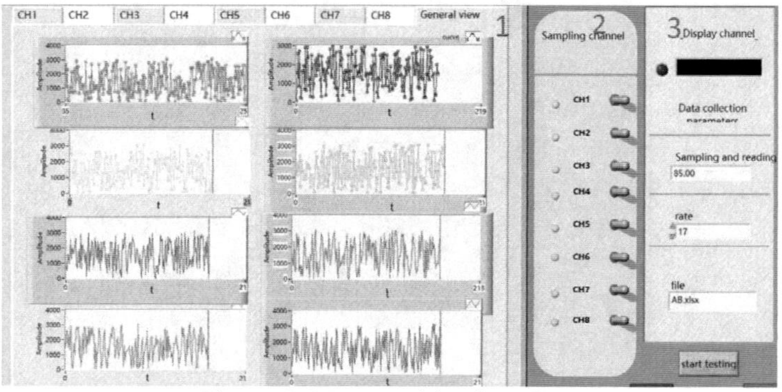

Figure 7. Man-machine interface

5.2 Channel selection control module

This system does 1-8 channels of data collection, at the same time it can do at most 8 channels and at least 1 channel of data collection. Use Labview to write a control program that allows users to choose which physical channels for data collection. The flow chart and software implementation are shown in Figure 8.

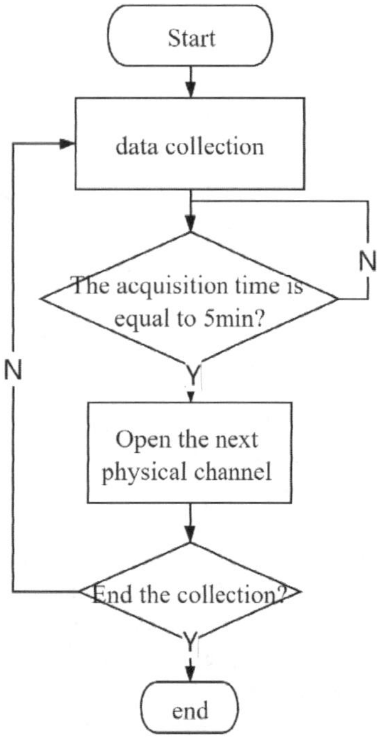

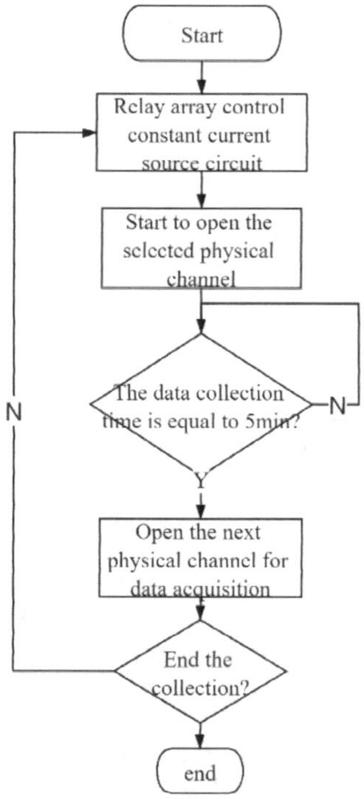

Figure 8. Flow chart of the physical channel gating Figure 9. Flow chart of the relay array

5.3 Relay Array Driver

Since the four-point method of measuring contact resistance needs to provide a constant current to the contact pair, the existing constant current source only provides one current output, if you need to collect the contact resistance of multiple contact pairs at the same time, you need to design the relay array circuit to supply the constant current source in turn. The relay array is controlled by the instructions issued by the data acquisition card. To make the relay array work as required, the corresponding relay driver program needs to be written. The gate control flow chart of the relay array is shown in Figure 9.

5.4 Design method of data acquisition software

The data acquisition software part includes four modules: data acquisition module, data storage module, data processing module, and data display module. The software flow chart and software implementation of data collection are shown in Figure 10. The software flow chart of the data display is shown in Figure 11.

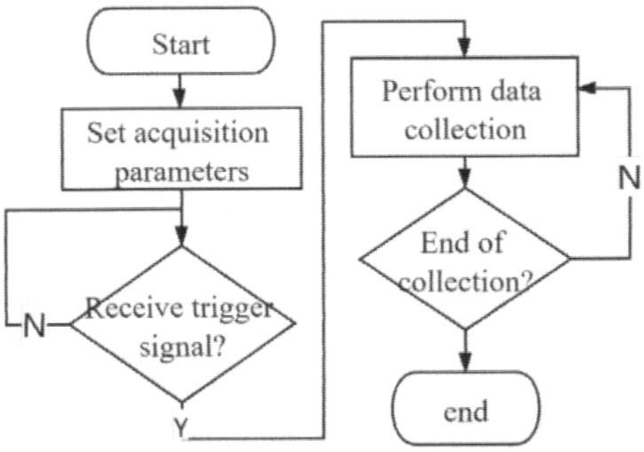

Figure 10. Flow chart for data collection

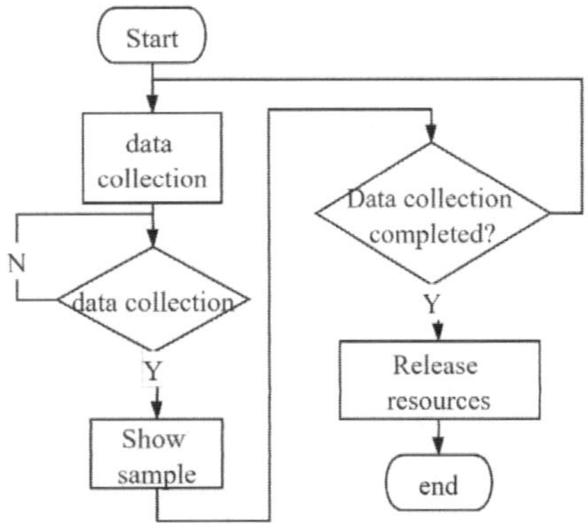

Figure 11. Flow chart for data display

Save the collected data separately to the corresponding worksheet according to the channel number. Each physical channel gated during each rotation of the confluence ring will collect 85 points of data, so set the program to save these 85 data to a specific physical channel. Each row in the worksheet corresponds to one cycle in one channel, which is convenient for subsequent data processing and data analysis. The software flow chart of the data storage is shown in Figure 12.

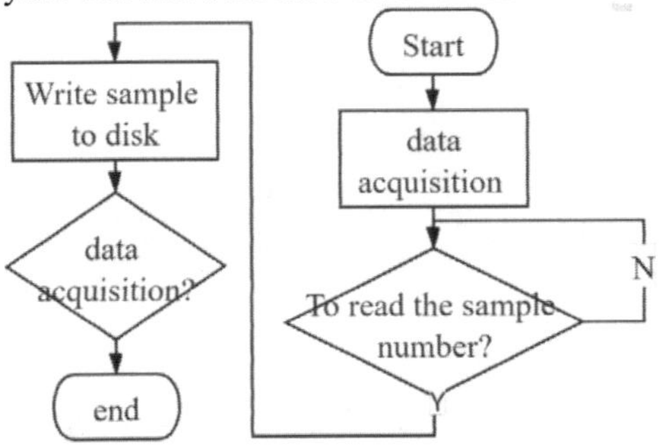

Figure 12. Flow chart of the data storage

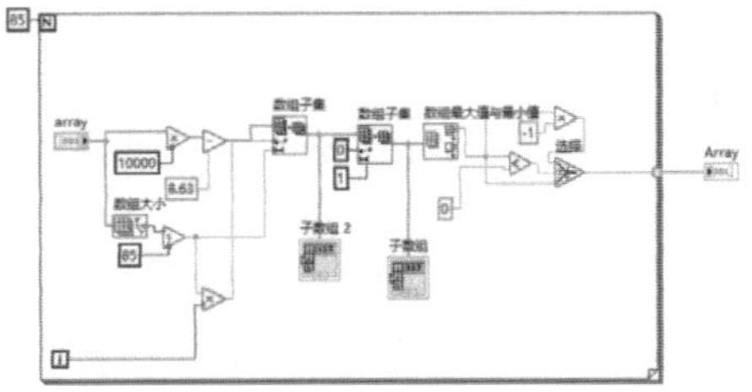

Figure 13. Elimination of the wire resistance

Since the wire is used to short the adjacent confluence ring when measuring the carbon brush-confluence ring contact resistance, it is necessary to remove the wire resistance from the collected data. The Labview program to remove the wire resistance is shown in Figure 13.

6. Verification of the equipment function and application

6.1 Verification of functional indicators

The sliding contact resistance testing equipment for the carbon brush and the confluence ring can measure 85 contact resistance data per rotation cycle of the confluence ring, as shown in Figure 14, and realize real-time display of 8-channel contact resistance measurement, as shown in Figure 15.

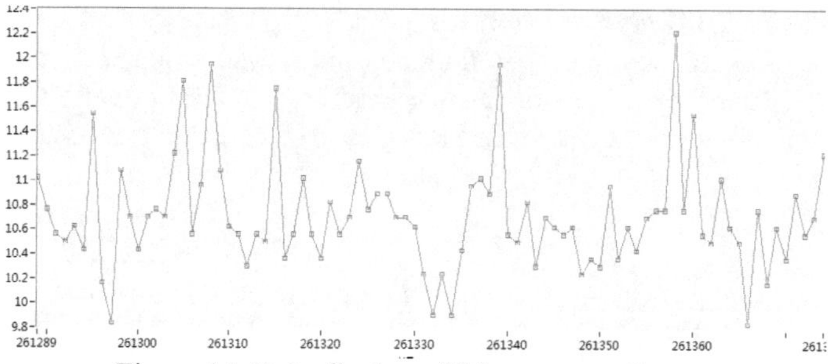

Figure 14. Data display of 85 contact resistance

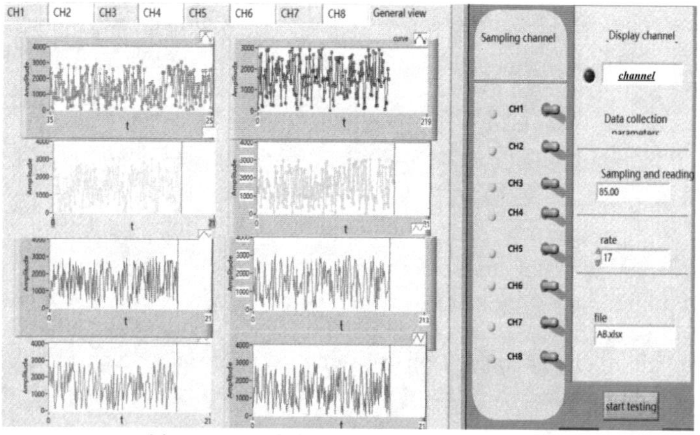

Figure 15. Eight-channel data display

To verify the measuring range of contact resistance is between 10mΩ and 2Ω, this device is used to collect data of 20 sliding cycles for standard contact resistances of 10mΩ and 2Ω respectively, which mean 1700 contact resistance values are collected. The average contact resistance and its standard error are calculated and listed in Table 1. It shows that the design accuracy is satisfied in the range of 10-2000mΩ.

Table 1. Resistance measurement accuracy

Standard value (mΩ)	Measurement value (average value) (mΩ)	Standard error (mΩ)	1% of full-scale(mΩ)	Up to standard
10	10.75	0.74	20	YES
2000	2002.64	2.66	20	YES

And then take the average contact resistance of 85 data per cycle to draw curves as shown in Figure 16 and Figure 17 respectively. Experiments have directly verified that the device can collect the contact resistance data in the range of 10mΩ~ 2Ω, the error of the measured two boundary resistance data is also less than 1% of the full scale.

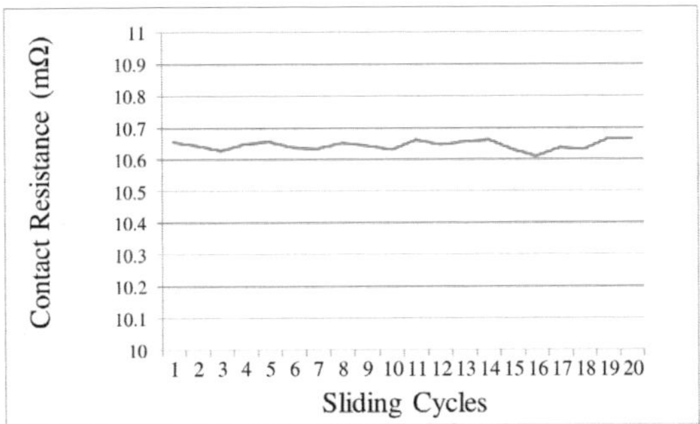

Figure 16 Average contact resistance per cycle for the 10mΩ standard resistance

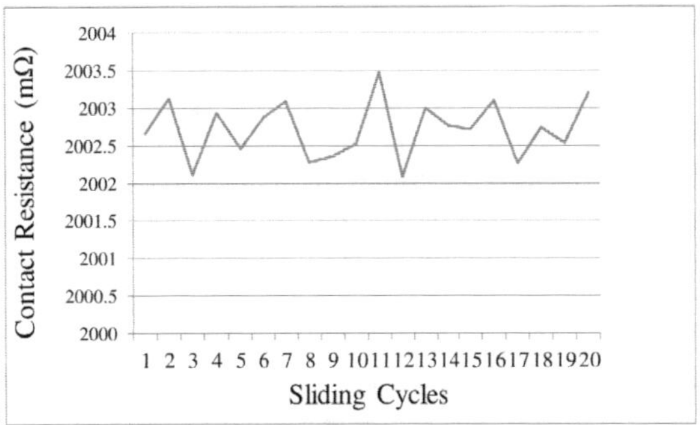

Figure 17 Average contact resistance per cycle for the 2Ω standard resistance

6.2 Application of equipment

The four-point method for measuring carbon brush-confluence ring contact resistance is shown in Figure 18. Among them, the two carbon brushes of the upper and lower ring are connected in series, so the actual measured sliding contact resistance is the contact resistance of the two contact pairs.

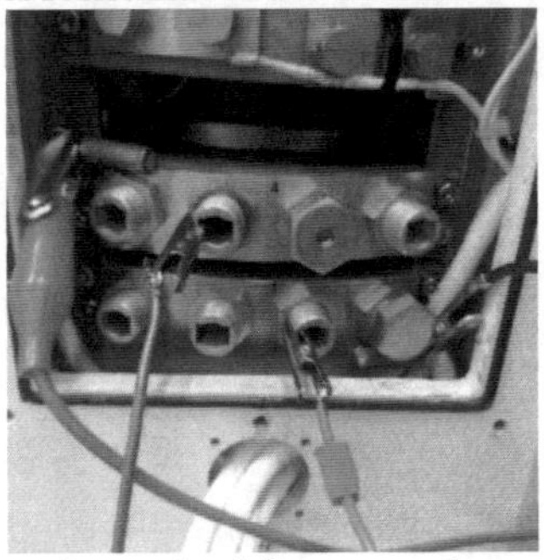

Figure 18. Four-point method for measuring contact resistance

The curves of the collected sliding contact resistance data every other 500 sliding cycles are drawn in Figure 19, which shows that the contact resistance of carbon brush-the confluence ring is unstable, and instantaneous high contact resistance appears during the sliding process. It may be caused by the oxidation accumulation of wear debris. Take the average, minimum, and maximum values of the contact resistance data every 120 cycles in 3600cycles to draw curves as shown in Figure 20, which also show the fluctuation of the contact resistance during sliding wear.

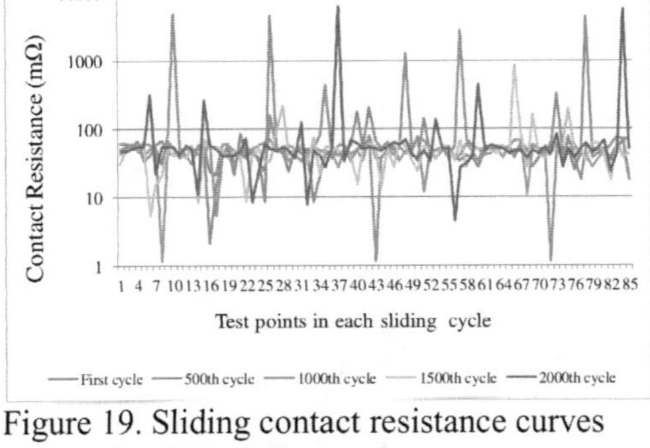

Figure 19. Sliding contact resistance curves every other 500 sliding cycles

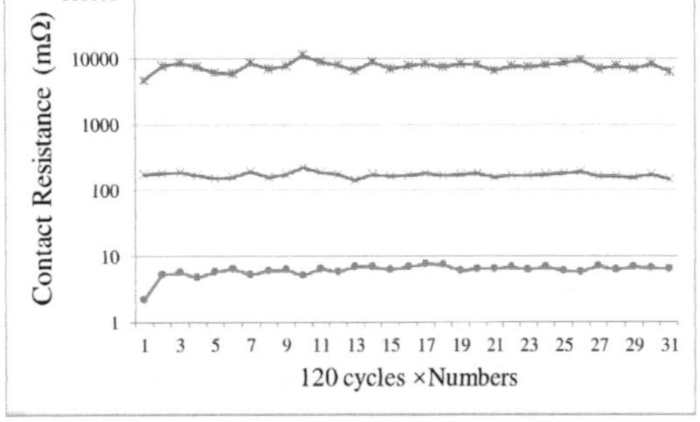

Figure 20. The average, minimum and maximum contact resistance of every 120 cycles in 3600 sliding cycles

7. Discussion

7.1 The function of sliding contact resistance testing equipment

The designed multi-channel sliding contact resistance detection equipment realizes the data acquisition, display, and storage functions of the contact voltage of 1-8 channels. Combined with the four-point method, it satisfied the contact resistance measurement range between 20mΩ and 2Ω and was verified by standard resistance. The experiment verified that the measurement accuracy of the contact resistance satisfied less than 1% of the full-scale. The confluence ring measures 85 contact resistance data per cycle and can update and display the curve of contact resistance with time in cycles, which provides a guarantee for the monitoring of electrical contact characteristics during the running-in wear of the carbon brush and the confluence ring.

In the design of the relay array driver, the Labview driver interface is highly integrated, which makes the upper layer Labview program logic more complicated when the host computer realizes more complex timing control tasks. Since the Altai data acquisition card provides control Digital I/O port driver, C++ can be used to write lighter and flexible driver programs, and it is easier to implement complex control tasks.

7.2 Application of sliding contact resistance testing equipment

The developed sliding contact resistance detection equipment was applied to the dynamic monitoring of the contact resistance of the carbon brush- confluence ring contacts. The contact resistance data can be collected in real-time, and the fluctuation of the contact resistance can be observed directly through the display curve, which is conducive to the study of contact resistance and carbon brush-confluence ring wear. The equipment operates stably for tens of hours at a sampling frequency of 10KS/s. Subsequently, a gross error processing module will be added to the data processing module to better research the running-in wear and electrical contact characteristics of the carbon brush confluence ring.

8. Conclusion

In this paper, a set of test equipment that can dynamically monitor 1-8 channels of sliding contact resistance data is developed by using data acquisition cards, relay arrays, and cooperating with Labview and data acquisition card drivers. This device can realize data acquisition, data display, and data storage. The design can meet the requirements of the functional indicators and technical indicators. When the device is applied to the dynamic test of sliding contact resistance for the carbon brush and the confluence ring, the performance is stable and the accuracy can meet the 1% of full-scale.

The detection of contact resistance during the sliding process of the carbon brush and the confluence ring shows that the oxidation accumulation of interface wear debris during the wear process will cause the contact resistance to increase instantaneously, which increases the possibility of welding of the contact interface under the high current impact.

References

[1] Tang X. 2015, Fault processing and maintenance of CINRAD/CC weather radar confluence ring, *China, J. Agriculture and Technology.* **35**, 177.
[2] Zhu M. 2011, Sliding electrical contact test system, *Dessertation*. Beijing University of Posts and Telecommunications, China.
[3] Liang Z. 2010, Design and implementation of universal multi-channel data acquisition system *Dessertation*, Xidian University, China.
[4] Sun W. 2013, The design and implementation of multi-channel data acquisition system. *Dessertation.* Hunan University, China.
[5] Pan X. 2018, Design of multi-sensor channel data acquisition system based on virtual instrument, *China, J. Mechanical Design and Manufacturing Engineering*, **47**, 79-81.

Remaining Useful Life Prediction of Aero-engines by Appropriate Utilization of Multi-sensor Signals

Bin Zhang*, **Ning Huang, Jiufei Luo, Yi Zhang**

School of Advanced Manufacturing Engineering, Chongqing University of Posts and Telecommunications, Chongqing, China

*Corresponding author: zhangbin@cqupt.edu.cn

Abstract. This paper presents the prediction of remaining useful life (RUL) with appropriate fusion of multi-sensor signals for the aero-engine, which is the heart of an aircraft. With the rapid development of information technology, health condition of one aero-engine is usually monitored with multiple sensors. To properly utilize these multi-sensor condition information for degradation modeling and RUL prediction is one of the key challenges for condition-based maintenance of the whole aircraft. Thus this paper proposes one statistical method based on health indicator (HI) construction and empirical parametric model for aero-engines RUL prediction. The method is validated with run-to-failure data sets of an aircraft gas turbine engine test-bed developed by NASA. Results show that the proposed method can effectively fuse multi-sensor signals to describe the degradation and predict RUL of the aero-engine.

Keywords: Remaining useful life prediction, multi-sensor fusion, aero-engine, health indicator, statistical modeling.

1. Introduction

Health condition of an aero-engine directly affects the operational safety, reliability, and efficiency of the whole aircraft. To monitor operational health of the aircraft engine, multiple sensors of varying types are usually embedded or equipped and thus brings a multiple sensor environment for health analytics of aero-engines. How to utilize the multi-sensor signals for condition-based maintenance is one of the major challenges for aircraft applications [1]. Recently, remaining useful life (RUL) prediction, for its potential in enhancing reliability and maintenance efficiency as well as reducing operational costs, has been attracting attention from both academics and industry.

Practically, RUL prediction aims to prognosticate the left time of a component or system from the current time to the time of functional failure. Then with the estimated RUL, maintenance actions can be prepared beforehand and be done as required. According to the literature, the approaches for the RUL prediction of aero-engines can be classified into model-based and data-driven methods [2]. In model-based methods, a model based on the physics-of-failure knowledge is normally constructed to describe the engine health degradation and do future prediction. Markov model [3] and particle filtering [4] are typical model-based methods for the aero-engine health condition and RUL prediction. However, certain assumptions or simplifications of model-based methods pose limitations on their practical deployment. With the development of big data technique, data-driven methods may be the more appealing solution for aero-engines that are with limited knowledge of physics-of-failure but multi-sensor signals. For the data-driven methods, machine learning and statistical approaches are the two popular branches. The machine learning methods based on traditional shallow and emerging deep learning architectures are suitable for prognosis of complex systems as aircraft engines, however the predicted results are hard to explain due to the lack of transparency [5]. On the contrary, the aero-engine monitoring data can be fitted by some parametric or non-parametric statistical models for RUL prediction [6]. Thanks to the good mathematical properties and easy interpretations, RUL prediction of aero-engines using statistical methods have been recently gaining popularity.

In statistically data-driven RUL predictions of one aircraft engine, a health indicator (HI) that describes its condition degradation is constructed by appropriate utilization of the monitoring sensor data. Traditionally, one dominant sensor that reflects the engine health condition is directly chosen as the HI [7]. However, this may be invalid in many cases as single sensor only contains partial or limited information, which is far from enough to comprehensively capture the physical degradation process of the aero-engine. Therefore, combination of multi-sensor signals into a synthesized and virtual HI may be more reasonable. In general, utilization of more sensor signals can be helpful. Unfortunately, it is possible that some sensors are irrelevant to the aero-engine health evolution tracking, duo to low sensitivity to the degradation, large interferences or inconsistency among the engine population. To address these issues, sensors that are more degradation-related should be selected. Currently, sensors are mainly selected based on some degradation relevance metrics. Then with the optimally selected sensors, HI can be constructed with one information fusion method, such as PCA [8] and linear data transformation [7]. To secure accurate and efficient RUL predictions, it is highly desired that the constructed HI enables the degradation of one aero-engine be characterized and also the degradation trajectory of each aero-engine from the whole population be consistent. For this purpose, a generic HI construction method to fuse the optimally selected sensors is presented in this work.

After the fusion of multi-sensor signals of the aero-engine into HI, statistical models are applied to fit the HI for degradation modeling and RUL prediction. Random coefficient models and stochastic process models are the two frequently used groups of statistical approaches to model the available data for prognostic inference. Lasheras et al [9] proposed a hybrid multivariate adaptive regression splines method for the aircraft engine RUL prediction. In [10], an exponential model was proposed for prognosis and validated on degradation data set of an aircraft gas turbine engine. The parametric models are effective when the HI can be well modeled with a specific model, however there are also numerous cases where the aero-engine HI may not be properly fitted by parametric ones. Thus, more flexible degradation modeling and RUL prediction methods are needed for practical aero-engine applications. To partially make up this gap, one functional data analysis method was proposed for health prognostics of aircraft engines in [11]. For its effectiveness and easy of application, the empirical parametric approach is studied to fit the constructed HI for RUL prediction of aero-engines in this paper.

2. The proposed method

In this section, the procedures for fusion of multi-sensor signals into a generic HI and the studied parametric models for RUL prediction of aero-engines are detailed.

2.1. A generic HI construction method

To secure the effectiveness and efficiency of RUL prediction of aero-engines, some pre-processings should be carried out before integration of the available multi-sensor signals into a HI. As not all the original monitoring sensors are relevant to the true health degradation, informative sensors need to be selected. In view that effectiveness of the four metrics of Corr, Mon, Rob and Pre has been validated for relevant sensor selection of aircraft engines prognostics in [11], they are also studied here for informative sensor selection. Nevertheless, these metrics were proposed for the scenario of one operational condition with one failure mode. In this regard, one two-step strategy is presented for multiple operation conditions and multiple failure modes. In the first step, the multi-sensor signals of each aero-engine under multi-operation conditions are divided into each specific operational condition with clustering or normalization. In the second step, the failure of individual aero-engine is attributed to one mode by clustering the sensor measurements upon failure instant or tracking the varying trend of the multi-sensor signals. Through the above pre-processings, the original multi-sensor signals can be selected using the four metrics. Specifically, denote s^{mk} as the thus selected kth sensor for the mth failure mode ($m = 1, 2, ..., M$; $k = 1, 2, ..., K_m$).

In addition, certain kind of reference baseline for construction of a HI should be established. Most often, the initial stage when one aero-engine is deemed to operate at its normal state is chosen as the baseline. Then later aero-engine operational conditions are compared to this baseline, and a HI can be constructed using some kind of metrics, such as Euclidean distance or probability density matching. Actually, the thus constructed HIs measure the deviations from the aero-engine normal operating state. Nevertheless, due to material impurities or manufacturing errors etc., there are large variations in the initial normal stage, which impacts its effectiveness as a baseline. Also, aero-engines that are with similar normal operating status can fail of distinct modes. Thus it is more reasonable to take the sensor measurements upon failure as the baseline.

Based on the above analysis, a generic one-dimensional HI is constructed by fusion of the selected muti-sensor signals with consideration of multi-failure modes as follows.

$$y_i^m = \frac{\left(\sum_{k=1}^{K_m}\left(s_i^{mk} - s_F^{mk}\right)^2\right)^{1/2}}{\left(\sum_{n=1}^{N}\sum_{k=1}^{K_m}\left(s_{nF}^{mk} - s_F^{mk}\right)^2 \Big/ (N-1)\right)^{1/2}} \tag{1}$$

Where y_i^m and s_i^{mk} are respectively the HI and the kth sensor measurement at the time t_i for the aero-engine under the mth failure mode, s_{nF}^{mk} is the measurement upon failure instant of the kth sensor under the mth failure mode for the nth aero-engine, and s_F^{mk} is the mean of the measurements upon failure instants of N historically failed aero-engines.

2.2. *The empirical parametric models*

With the accumulation of damages (such as wear and fatigue), the health condition of an aero-engine gradually degrades. Therefore, properly tracking the degradation and accurately predicting RUL is of great significance for condition-based maintenance of aero-engines. That is the constructed HI should be modeled and be extrapolated for future prediction.

For a statistical method, the HI trajectory is usually fitted with one kind of empirical model for degradation modeling and prediction. The exponential (Exp) and power law (Pow) are two empirical parametric models that have been frequently researched in statistical modeling. They are utilized to fit the aero-engine HI with the functional form as Eq. (2) in this work. Additionally, the aero-engine HI trajectory is also fitted by the general pth polynomials (Poly).

$$\begin{aligned} y_i &= x(t_i) + \varepsilon \\ Exp: \ x(t) &= a \times \exp(bt) + c \\ Pow: \ x(t) &= a \times t^b + c \end{aligned} \tag{2}$$

where y_i and $x(t_i)$ is the observed and latent true HI at the time t_i of one aero-engine, ε is the noise term that is assumed to be normally distributed as $N(0, \sigma^2)$, and a, b, and c are fitting coefficients.

To fairly compare performance of the above empirical parametric models for the HI modeling, the three commonly used evaluation criteria as in Eq. (3), i.e. sum of square error (SSE), root mean square error (RMSE) and the coefficient of determination (R-square), are employed in the following analysis.

$$\begin{aligned} &\text{SSE} = \sum_{i=1}^{I}\left(y_i - \hat{y}_i\right)^2, \ \text{RMSE} = \left(\sum_{i=1}^{I}\left(y_i - \hat{y}_i\right)^2 \Big/ I\right)^{1/2} \\ &\text{R-square} = 1 - \frac{\text{SSE}}{\text{SST}}, \ \text{SST} = \sum_{i=1}^{I}\left(y_i - mean(y_i)\right)^2 \end{aligned} \tag{3}$$

Where y_i and $\hat{y}_i$ are respectively the true and estimation at the time t_i of one aero-engine, and I is the total number of monitoring time instants.

3. Case studies

To validate the proposed method, run-to-failure data sets of turbofan engine was simulated using the dynamic tool of C-MAPSS that was developed by NASA [12]. In the output data sets, each individual run-to-failure instance of the same aero-engine unit was represented by multi-sensor signals. To be specific, 21 sensor measurements were recorded for every cycle of each aero-engine run-to-failure instance. The multi-sensor signals were contaminated with measurement noises and each engine started with different initial health conditions and manufacturing variations which were unknown. In this paper, FD003 data sets containing 100 training and 100 testing data of aero-engines is studied to illustrate and demonstrate the effectiveness and performance of the presented method. These engines were simulated under one operational condition with two failure modes (i.e. HPC and fan). More details about the simulation and data set can be found in [12]. Also, note that deterioration of the aero-engine with multiple usage conditions can also be analyzed by the method with the pre-processing as discussed in Subsection 2.1.

3.1. Results and analysis of the constructed HI

As there are two failure modes in the aero-engine data set, clustering of the failures is conducted. The results indicate that there are respectively 56 and 44 training engines for failure modes of HPC and fan. Then the run-to failure multi-sensor signals of the training aero-engines from each failure mode are evaluated by the four metrics of Corr, Mon, Rob and Pre. Since the sensor selection for the HPC failure mode has already been studied in [11], only the results for the fan failure mode are presented here in Table 1. For the metrics of Corr, Mon and Rob that are defined on sensors from one engine, statistics of the their means are given based on results of the 44 engines.

Table 1. Evaluation results for the fan failure mode.

Sensor	Corr	Mon	Rob	Pre	Sensor	Corr	Mon	Rob	Pre
T2	*NaN*	*1.000*	*1.000*	*NaN*	phi	0.8509	0.2956	0.9995	0.9500
T24	0.7699	0.0517	0.9997	O.8086	NRf	0.8458	0.2233	0.9996	0.9310
T30	0.8066	0.0548	0.9982	0.8219	NRc	0.8524	0.1742	0.9997	0.9547
T50	0.7932	0.0755	0.9979	0.8729	BPR	0.8518	0.0680	0.9982	0.8833
P2	*NaN*	*1.000*	*1.000*	*NaN*	farB	*NaN*	*1.000*	*1.000*	*NaN*
P15	0.5039	0.6569	0.9999	*NaN*	htBleed	0.8144	0.2527	*NaN*	0.8459
P30	0.8519	0.2605	0.9994	0.9502	Nf_dmd	*NaN*	*1.000*	*1.000*	*NaN*
Nf	0.8443	0.2251	0.9998	0.9352	PCNfR_dmd	*NaN*	*1.000*	*1.000*	*NaN*
Nc	0.8508	0.1660	0.9997	0.9418	W31	0.8226	0.0767	0.9981	0.8413
epr	0.7254	*1.000*	*1.000*	0.8300	W32	0.8235	0.0610	0.9982	0.8190
Ps30	0.8076	0.1190	0.9984	0.9072					

From the results, it can be observed that there are default values (i.e. NaN) or 1 for eight sensors (i.e. T2, P2, P15, epr, farB, htBleed, Nf_dmd and PCNfR_dmd). A further analysis shows that the above sensor measurements are constant values or having a step-wise trend with the engine operating cycle. These sensors are not helpful for the RUL prediction, so they are excluded. For the other 13 sensors, there seems no significant difference. The Nf and NRf are the same kind of sensors, and Nf is shown in Fig. 1. It fluctuates largely at the early operating stage and is with a very small varying range. Thus the two sensors should be excluded for the aero-engine HI construction. Compared to the failure mode of HPC researched in [11], the two sensors Nc and NRc demonstrate a consistently increasing tendency. This may be because that the Nc and NRc can well capture the engine degeneration under the fan failure mode while these two sensors are poor for sensing the HPC failure mode. Considering that it is difficult to identify the true failure mode of an in-service aero-engine at its early operational

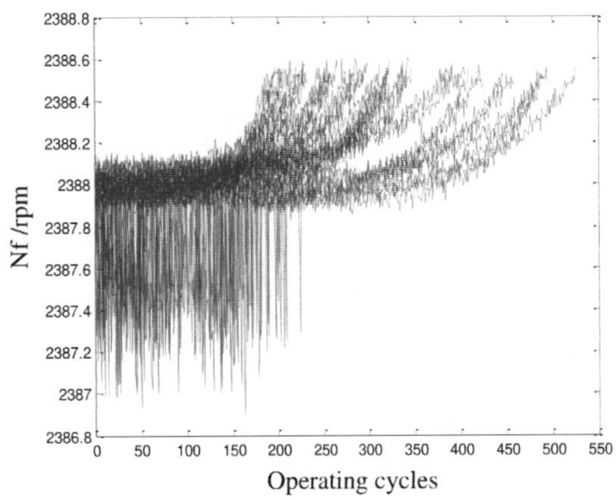

Figure 1. Nf signal for the training engines with fan failure mode.

stage, the Nc and NRc are not selected to have a balance between the two failure modes. That is the remaining nine sensors are finally selected for HI construction of the fan failure mode.

With the selected nine-dimensional sensors and clustering of the sensor measurements upon failure for each failure mode, the aero-engine HI is constructed following Eq. (1). Results of the 100 training engines under the two failure modes are shown in Fig. 2, which depicts a general decreasing tendency. Fig. 2 (a) shows that under the HPC failure mode, the trend of the constructed HI for the 56 engines is relatively consistent and the degrading rate increases gradually during the whole life. While for the fan failure mode, the health of the 44 engines degrades slowly at the beginning and exacerbates at the later stage. It can also be observed that there is large dispersion for the engine lives even under the same failure mode.

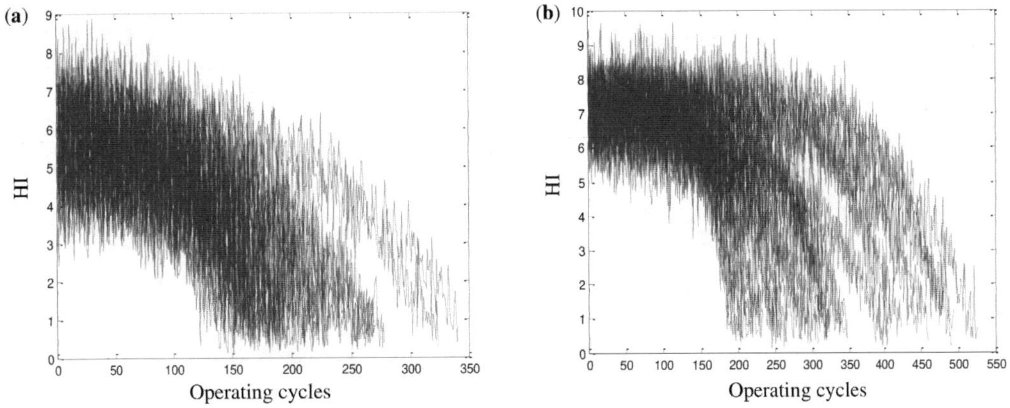

Figure 2. HI for the the training engines with failure mode of: (**a**) HPC; (**b**) fan.

3.2. Results and analysis of the RUL prediction

As stated in Subsection 2.2, the constructed HI is fitted using the empirical parametric models of Exp, Pow and Poly. Specifically, 3rd polynomials (Poly3) from the Poly type is considered. The HI modeling results for the two failure modes of FD003 training engines are summarized in Table 2. Statistically, the means for the three criteria of SSE, RMSE and R-square are given. To have a view of the curve fitting, the modeling for one engine HI is plotted in Fig. 3.

Referring to the results in Table 2, it can be concluded that the three empirical parametric models fit the trend of the noisy HI accurately. When it comes to the two kinds of failure modes, the two

Table 2. HI fitting results of the 100 training engines.

Failure mode	Method	SSE	RMSE	R-square	Failure mode	Method	SSE	RMSE	R-square
	Exp	88.83	0.6650	0.7964		Exp	102.72	0.5814	0.8864
HPC	Pow	88.87	0.6669	0.7955	fan	Pow	103.43	0.5834	0.8855
	Poly3	87.86	0.6648	0.7976		Poly3	103.61	0.5857	0.8852

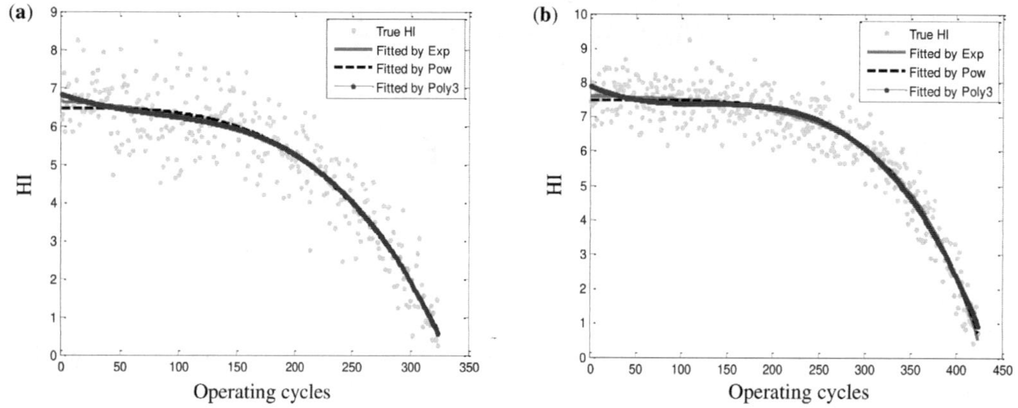

Figure 3. HI fitting results for one training engine from the failure mode of: (**a**) HPC; (**b**) fan.

criteria of RMSE and R-square show that better degradation modeling results are obtained for the fan failure mode, which may be explained by the fact that more number of monitoring HIs are generally available to fit the three models for aero-engines have much longer operating cycles (lifetimes) under this failure mode. This is also in accordance with the bigger SSE values.

To further study effectiveness of the proposed method for RUL prediction, original monitoring sensor data sets of the testing aero-engines from FD003 are processed following the same procedure as for the training units. With pre-processing the multi-sensor signals of the 100 testing engines, it is found that 39 engines fail of the HPC mode and 42 engines fail of the fan mode, while the remaining 19 ones can not be classified to either of the two failure modes. Then HI that describes the health degeneration of each testing engine is constructed by fusing the selected nine-sensor signals.

In statistical prognostics, the HI is extrapolated to the future for RUL prediction, thus a failure threshold requires to be predefined. Although there are some ISO standards that may be helpful for

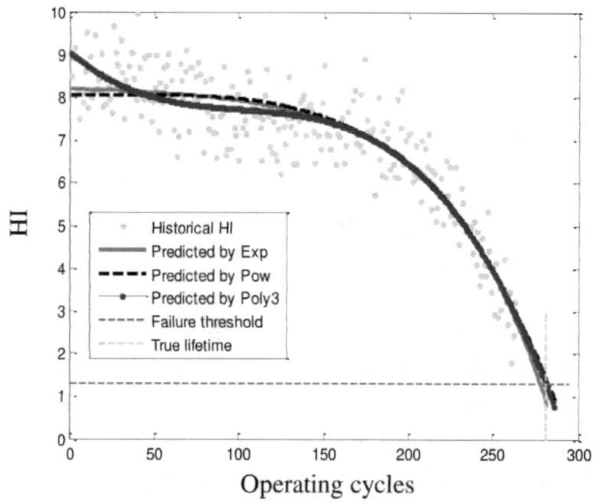

Figure 4. Illustration for RUL prediction of one testing engine.

Table 3. Comparisons of RUL prediction results for the 100 testing engines.

Failure mode		Method	RMSE	Rang of prediction errors
Known	HPC	Exp	36.83	[-96, 110]
		Pow	41.78	[-111, 113]
		Poly3	29.74	[-111, 43]
	fan	Exp	40.20	[-91, 113]
		Pow	47.93	[-68, 113]
		Poly3	49.31	[-178, 150]
Unknown	Treated as HPC	Exp	60.20	[-126, 65]
		Pow	57.90	[-91, 113]
		Poly3	53.97	[-97, 120]
	Treated as fan	Exp	128.60	[-100, 287]
		Pow	125.62	[-45, 200]
		Poly3	95.58	[-83, 168]

determination of the failure threshold, they are documented for general applications and may not be directly applicable for specific situations, especially for complex aircraft engine systems whose health deterioration is tracked with the constructed virtual HI as in this work. Thus the threshold for the two failure modes of HPC and fan are respectively determined through a statistical analysis of the constructed HI from the training engines. In Fig. 4, the RUL prediction for one testing engine from the HPC failure mode is illustrated.

With the constructed HI and the failure threshold, RULs of the 100 testing engines are predicted with the three empirical parametric models. The results are listed in Table 3, where RMSE and Rang of prediction errors (i.e. predicted RUL−true RUL) are calculated to demonstrate the performance of the methods. For the 19 engines with unknown failure mode, they are artificially treated as one of the two failure modes to predict RULs.

The statistical results in Table 3 indicate that RUL can be predicted more accurately when the failure mode information is known. Further, the prediction RMSE of the HPC failure mode is smaller than that of the fan failure mode. This may partially because the different varying patterns of the HI for the two failure modes, as shown in Fig.2 and discussed in Subsection 3.1. Among the three empirical models, the Poly3 model generally gives the best RMSE performance for the testing engines. However, all the three models are not with satisfactory performance in view of the long range of prediction errors. To make a sense of that, the RUL prediction result for each testing engine is shown in Fig. 5 and Fig. 6.

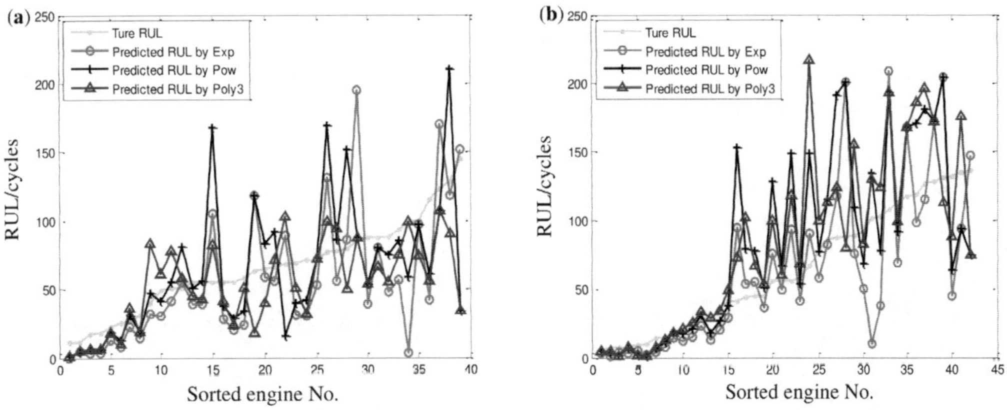

Figure 5. RUL prediction results for testing engines of the known failure modes: (**a**) HPC; (**b**) fan.

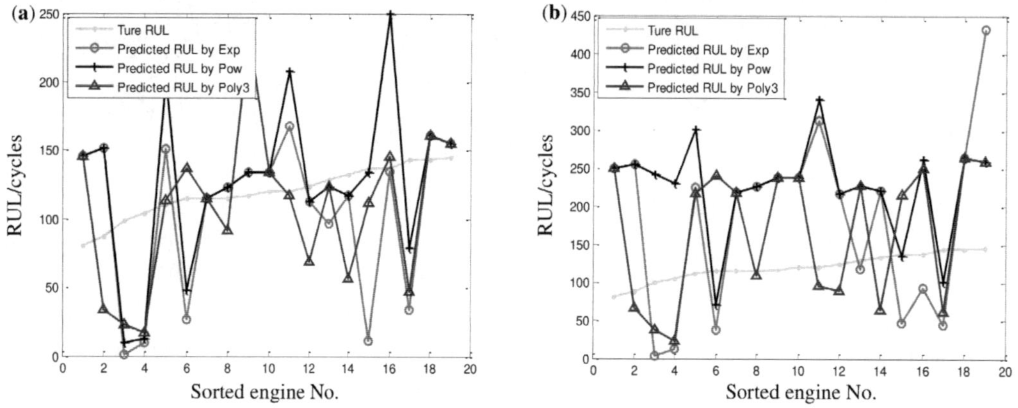

Figure 6. RUL prediction results for testing engines with unknown failure modes: (**a**) treated as HPC; (**b**) treated as fan.

Note that the testing engine No. is sorted with the increasing RULs in both Fig. 5 and Fig. 6. From these two figures, similar conclusions as those of Table 3 can be firstly drawn. Furthermore, it can be seen that more accurate RUL predictions can be achieved with the approaching of failure (i.e. small RULs), especially with the case of known failure mode as in Fig. 5. To artificially assign engines with certain failure modes may obtain both relatively accurate and poor predictions (Fig. 6). Thus the failure mode information is helpful to secure the accuracy.

4. Conclusions

The aero-engine is the heart of one aircraft and its RUL is one of the major issues for condition-based maintenance of the aircraft or the whole fleet. For RUL prediction with multi-sensor signals, one statistical method that appropriately fuses the multi-sensor information into one generic HI and then fits the HI with empirical parametric models is proposed in this study. Case studies are carried out with the simulation data set of one aircraft gas turbine engine. The results demonstrate that the presented method can effectively utilize the selected sensors to model and predict the RUL of the aircraft engine. In this work, the scenario of single operational condition with two failure modes is studied, validation of the presented method for multiple operating conditions under multiple failure modes will be pursued in the future.

Acknowledgments

This work was financially supported by the Science and Technology Research Program of Chongqing Municipal Education Commission (Grant No. KJ1704109) and the Open Found of The State Key Laboratory of Mechanical Transmissions (Grant No. SKLMT-KFKT-201809).

References

[1] H. Hanachi, C. Mechefske, J. Liu, A. Banerjee, Y. Chen. Performance-based gas turbine health monitoring, diagnostics, and prognostics: A survey. IEEE Transactions on Reliability, 2018, 67, 1340-1363.

[2] Z. X. Li, K. Goebel, D. Z. Wu. Degradation modeling and remaining useful life prediction of aircraft engines using ensemble learning. Journal of Engineering for Gas Turbines and Power, 2019, 141, 041008.

[3] S. Al-Dahidi, F. Di Maio, P. Baraldi, E. Zio. Remaining useful life estimation in heterogeneous fleets working under variable operating conditions. Reliability Engineering and System Safety, 2016, 156, 109-124.

[4] J. B. Yu. Aircraft engine health prognostics based on logistic regression with penalization

regularization and state-space-based degradation framework. Aerospace Science and Technology, 2017, 68, 345-361.

[5] C. G. Huang, H. Z. Huang, Y.F. Li. A Bi-Directional LSTM prognostics method under multiple operational conditions. IEEE Transactions on Industrial Electronics, 2019, 66, 8792-8802.

[6] X. S. Si, W. B. Wang, C. H. Hu, D. H. Zhou. Remaining useful life estimation - a review on the statistical data driven approaches. European Journal of Operational Research, 2011, 213, 1-14.

[7] P. F. Wang, B. D. Youn, C. Hu. A generic probabilistic framework for structural health prognostics and uncertainty management. Mechanical Systems and Signal Processing, 2012, 28, 622-637.

[8] K. Le Son, M. Fouladirad, A. Barros. Remaining useful lifetime estimation and noisy gamma deterioration process. Reliability Engineering and System Safety, 2016, 149, 76-87.

[9] S. F. Lasheras, P. J. G. Nieto, F. J. De Cos Juez, R. M. Bayón, V. M. G. Suárez. A hybrid PCA-CART-MARS-based prognostic approach of the remaining useful life for aircraft engines. Sensors, 2015, 15, 7062-7083.

[10] K. B. Liu, S. Huang. Integration of data fusion methodology and degradation modeling process to improve prognostics. IEEE Transactions on Automation Science and Engineering, 2016, 13, 344-354.

[11] B. Zhang, K. Zheng, Q. Q. Huang, S. Feng, S. Q. Zhou, Y. Zhang. Aircraft engine prognostics based on informative sensor selection and adaptive degradation modeling with functional principal component analysis. Sensors, 2020, 20, 920.

[12] A. Saxena, K. Goebel, D. Simon, N. Eklund. Damage propagation for aircraft engine run-to-failure simulation. In Proceedings of the International Conference on Prognostics and Health Management, Denver, CO, USA, 6-9 October 2008.

Research on 3D reconstruction method of wear particle dynamic image based on multi contour space mapping

Han Wang[1], Hongfu Zuo[1,*], Zhenzhen Liu[1], Hang Fei[1], Yan Liu[1], Zhiqiang Jiang[2]

[1] College of civil aviation, Nanjing University of Aeronautics and Astronautics, Nanjing, China

[2] China Aeronautical Radio Electronics Research Institute, Shanghai, China

*Corresponding author: rms@nuaa.edu.cn

Abstract. Aiming at the problems of current image monitoring methods of lubrication oil wear particles, this paper designs and builds a dynamic monitoring system for oil wear particles based on microfluidic microscopic images. A contour-based 3D reconstruction method of debris particle images is proposed. The image sequences of rotating wear particles tracked by a single target are used as data, and the contour of the wear particle is extracted and the data is stored. The minimum area external rectangle method is used to correct the rotation of the particle images for the problem of deflection. And an algorithm based on cylindrical coordinate space conversion is used to convert the discrete contour point data into three-dimensional space. Complete the 3D model reconstruction of microfluidic wear particles. The ability to analyse wear particles in oil online monitoring technology is improved, which also shows new ideas for wear status monitoring and fault diagnosis technology.

Keywords: Lubricating oil monitoring, Image analysis, Wear particle detection, 3-D reconstruction.

1. Introduction

Rotation mechanical equipment wear and failure will produce different wear particles, and the identification and detection of abrasive particles in the lubrication oil is particularly important. Oil wear particles contain a large amount of mechanical wear and failure information. Analysis of their size, quantity, color, morphological characteristics and contour shape, combined with the working status and wear mode of the system, can effectively determine the degree and type of wear, analyze the cause of the failure, and locate the worn area. As an important and efficient wear detection method, oil abrasive particle monitoring can determine the wear status of equipment or key parts after the wear particle information is clear, so as to realize the current performance analysis of the machine and equipment and the prediction of the next failure development trend, and finally make a specific decision.

At present, there are many researches in the direction of wear particle analysis, and there are many studies on the two-dimensional characteristics of wear particles, and they have been used in oil monitoring experiments. However, with the continuous development of image processing technology and high-speed vision technology, the three-dimensional image analysis method of wear particles has gradually become the focus of research in this direction. The three-dimensional reconstruction analysis method of wear particles is still in the preliminary research stage of studying two-dimensional plane parameters. In the aspect of wear particle analysis, systematic research on the methods of 3D reconstruction of wear debris particles has not been carried out, and there is no complete theory to characterize the morphological characteristics of 3D debris particles [1].

In recent years, scholars have begun to study the 3D reconstruction technology of wear particles. In 2005, He [2] studied the method of ordinary optical microscope to reconstruct the three-dimensional surface of abrasive particles, improved the ordinary optical microscope, and conducted preliminary three-dimensional ferrography analysis. However, limited to the speed and effect of the algorithm, the analysis of wear particles has not been carried out, so the follow-up in-depth research is of great significance. In 2016, Wu [3] studied on-line extraction of wear particle features based on video sequences and three-dimensional representation models. A video-based rapid 3D reconstruction system for two-dimensional wear particles is proposed, which completes the dynamic wear particle extraction based on the Gaussian mixture model, uses the acquired feature information to track the wear particles, and realizes the three-dimensional surface reconstruction of the simulated wear particles in multiple views. Because the system is affected by the image quality, image processing methods such as the motion recovery and enhancement of the wear image need to be studied in depth to improve the image quality of the wear image. In 2017, Peng [4] used the advantages of video image processing to propose a system and method for three-dimensional wear particle analysis, and obtained that the wear particles in the oil rotate under the action of torque, and images of different views can be used. Sequence to build the important conclusions of the three-dimensional model. And put forward that the future research direction is to optimize the experimental system, improve the accuracy of wear particle detection, and perform three-dimensional reconstruction in multiple views.

In order to obtain more comprehensive information on wear debris, the technology of on-line visual ferrograph (OLVF) technology is being developed [5]. Li designed a new reflecting light based on LED arrays to solve the problem of ferro-imaging quality that is affected by the lower-brightness and non-uniform illumination of the original reflecting light [6]. In order to automatically acquire three-dimensional information of wear debris for online condition monitoring, Peng developed a microfluidic device consisting of an oil flow channel and a video imaging system. A microchannel is designed to ensure the continuous rotation of particles such that their three-dimensional features can be captured [7]. Liu put the application of magnetic fluid in ferrograph to improve the technique for separating nonferrous wear debris by magnetic means [8].

In image analysis and processing algorithms, Peng applied two-dimensional fast Fourier transform, power spectrum and angular spectrum analyses to describe wear particle surface textures in three dimensions, and developed a software for the image analysis of boundaries and surfaces. [9-11]. Wu proposed the description of wear debris from on line ferrograph images by their statistical color. The method provides a primary exploration on describing the color of wear debris by on-line ferrograph images [12]. Wu also investigated statistical dimension of wear debris in on-line ferrograph images, in order to descript debris images for wear characterization [13]. Wu proposed the method of incorporating gray and boundary based segmentation, and watershed-based morphological separation of wear debris chains for on-line ferrograph analysis [14,15]. Wu present the development of a video-based system to extend the particle information in 3-dimension (3-D). The proposed method contains three main procedures including: particle extraction using a Gaussian mixture model, multiple particle tracking with Kalman filter, and 3-D feature reconstruction by the shape-from-silhouette method [16,17]. Xi had researched the restoration of online video ferrograph images for out-of-focus degradations. The main idea is to extract object edges, magnify with a non-linear gain factor, then combine with the input image to produce an enhanced image to facilitate further analysis [18]. Liu proposed an unsupervised segmentation for wear particle's image using local texture feature [19]. To address the image degradation issue (debris images captured often suffer degradation due to debris motion blur and lubrication contamination), Wu developed a new method of wear debris image restoration to reduce the effect of blur [20]. Peng used a Multi-SVDD method to model dynamic oxides identification [21].

Therefore, the three-dimensional reconstruction analysis of wear particle images has become an important research direction of oil online monitoring. The optimization of the experimental system of oil wear particle online monitoring has become the key to the study, and the processing algorithm of wear particle images also needs further research. In this paper, a set of simple and reliable 3D wear particle reconstruction methods for the whole process is proposed to meet the above needs, which provides a new and important means to obtain abundant debris information.

2. Experimental platform design

In order to collect two-dimensional wear particle image, combined with microfluidic chip technology and high-speed vision technology, an oil dynamic monitoring system based on microfluidic micro image analysis technology is designed. The system structure is shown in the figure 1. The system is mainly composed of dynamic injection system and microscopic imaging subsystem. The main functions of the two subsystems are to realize the oil supply injection of the system and acquisition and analysis of wear particle images in oil. The injection dynamic subsystem mainly includes oil pool, micro pump and related oil pipeline. The microscopic imaging subsystem includes a light source, a high magnification imaging lens, a high-speed visual digital camera, and a computer.

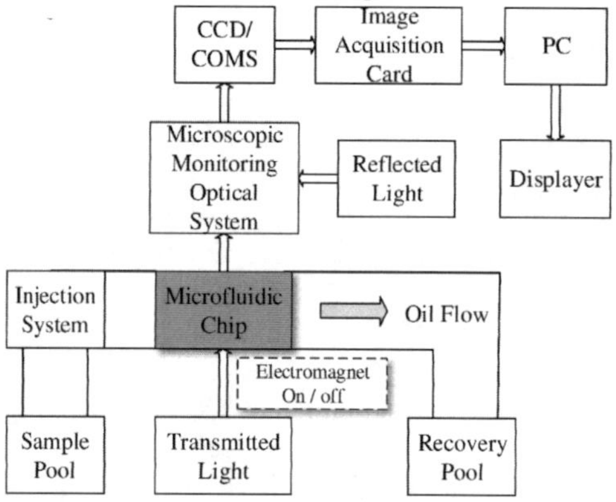

Figure 1. structure of microfluidic microscopic monitoring platform.

When image data is acquired using this system, the uniform linear motion between the sensor and the scene generates the image blurring phenomenon. In this paper, we analyze and study the degradation model of the image, and apply a method based on the degradation function to recover the abrasive grain blurred image; and then use a target detection method based on visual saliency to achieve abrasive grain target detection.

3. Three dimensional reconstruction method of wear particle dynamic image

In this paper, the micro flow oil monitoring system introduced in the previous section is used to obtain the dynamic two-dimensional image of wear particles. A three-dimensional reconstruction method of wear particle dynamic image based on multi contour space mapping is proposed to realize the three-dimensional reconstruction of wear particle. The overall process is shown in the figure.

3.1. Particle profile extraction

Image contour detection is an extremely important step in computer vision. Generally speaking, for the contour detection of irregular object image, firstly, the edge detection operator is used to extract the object edge. Then the redundant pixels are removed according to the contour characteristics of the object. Finally, these edge pixels are connected to determine the contour boundary of the object.

Wear particle contour extraction is also a key process in wear particle 3D reconstruction. Because of the impurities and image background noise in the microchannel, the wear particle contour extraction must be able to effectively suppress the noise and accurately detect the position of the edge. In this paper, the wear particle image edge detection algorithm combined with background difference and Canny operator is used for contour extraction to realize the extraction process of wear particle contour in 3D reconstruction. For the debris particle image in micro flow, firstly, the particle image is segmented to suppress the interference of bubbles, impurity particles in background oil and chip pollutants that may appear in the particle image. And then the canny edge detection operator is used to extract the contour of the abrasive particle image to extract the abrasive particle contour accurately.

3.2. Image deflection correction

Under perfect conditions, the debris particles will rotate stably around a fixed axis. However, in the microfluidic chip channel of the system, the fluid movement is complex, and the abrasive particles will deflect at a certain angle. The deflected wear particle contour image is shown in the figure 2(a).

Before the next step of three-dimensional reconstruction of abrasive particles, the deflected particle image must be corrected first. Before image deflection correction, it is necessary to calculate the angle of abrasive particle deflection. In this paper, a minimum area circumscribed rectangle method is used to calculate the minimum area circumscribed rectangle of the abrasive particle profile, and the abrasive particle deflection angle is calculated through the rectangular axis of symmetry. The specific algorithm is as follows:

Step 1: Store the data, input the abrasive particle contour image, and define the data (x, y) as the coordinates of debris particle boundary points. Define the rectangular coordinates of the boundary as $(rectx, recty)$.

Step 2: Search the boundary rectangle, classify according to the number of coincident points between the boundary rectangle and the debris boundary, round off the boundary rectangle with the number of coincident points of 0, 1 and 2 respectively, keep the boundary rectangle with the number of coincident points of 3 or more, and execute the cycle.

Step 3: Calculate the minimum boundary rectangular area, save $(rectx, recty)$ and obtain the minimum area circumscribed rectangle.

Step 4: Obtain the angle of each boundary of the minimum area circumscribed rectangle, and calculate the deflection angle of the rectangular symmetry axis.

Through the minimum area circumscribed rectangle method, the abrasive particle profile and the minimum circumscribed rectangle are obtained, as shown in the figure 2(b), in which the green rectangle is the minimum area circumscribed rectangle.

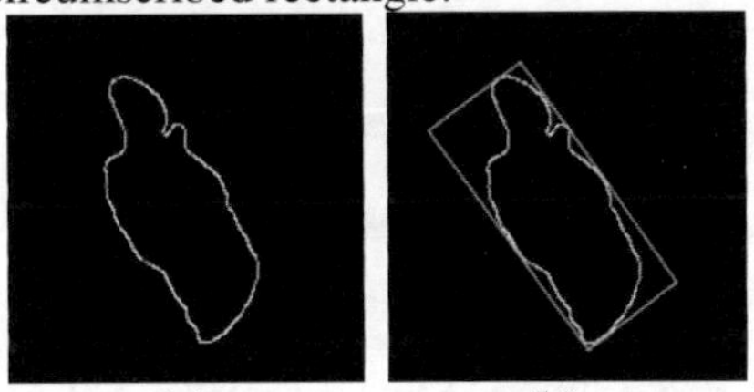

(a) (b)

Figure 2. Particle profile and minimum circumscribed rectangle.

3.3. Particle boundary point extraction

Due to the uncertainty of the flow of wear particles in the oil, in order to accurately store the two-dimensional data $L(x, y)$ of wear particle boundary points in the same coordinate system, it is necessary to establish a unified two-dimensional coordinate system. Firstly, the center of the abrasive image is determined, then the unified two-dimensional coordinate system is determined with the center point of the abrasive image as the reference point, and finally the two-dimensional data $L(x, y)$ of each contour is stored.

In the binary image, the area of wear particles can be determined by the number of pixels in the image particle area. The center of the particle is the area center of the irregular abrasive particle. The calculation formula of the center coordinates $(\overline{x}, \overline{y})$ of the abrasive particle image with different attitudes is shown in the formula.

$$\overline{x} = \frac{\sum \overline{x_i}}{N}, \quad \overline{y} = \frac{\sum \overline{y_i}}{N} \tag{1}$$

Where, N is the number of pixels in the wear particle area in the image. $\sum \overline{x_i}$ is the sum of abscissa of pixels in the particle area, and $\sum \overline{y_i}$ is the sum of ordinate of pixels in the particle area.

3.4. Three-dimensional space mapping

The abscissa and ordinate systems are established by using the contour image stored by the particle points. The two-dimensional debris contour image of the established coordinate system is shown in Figure 3.

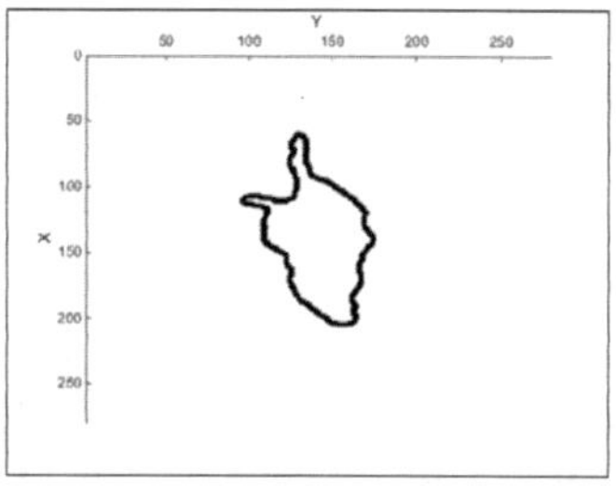

Figure 3. Two dimensional wear particle contour image.

In the established plane rectangular coordinate system XOY, the two-dimensional coordinates $A(x, y)$ of any particle contour point have their corresponding coordinates $A'(x, y, z)$ in three-dimensional space. Because the particle rotates relative to the rotating axis, the abrasive particle contour is different in three-dimensional space. Therefore, this paper adopts an algorithm based on cylindrical coordinate transformation to realize the mapping from two-dimensional contour to three-dimensional space.

Firstly, a plane rectangular coordinate system is established on the contour image stored by the particle boundary points, and the coordinate origin is located in the upper left corner of the image. The coordinate data of particle contour points are determined by the plane rectangular coordinate system, and the two-dimensional coordinates of all particle contour points are stored.

Then convert particle $A'(x, y, z)$ coordinates to three-dimensional polar coordinate system, and store particle polar radius r, polar angle θ and height value z. The cylindrical coordinates (r, θ, z) are used to determine the position of each contour point of particles in three-dimensional space.

The cylindrical coordinate system is transformed into Cartesian coordinate system, and the mapping from polar coordinates and cylindrical coordinates to Cartesian coordinates is shown in the figure 4. The stored polar radius and polar angle of particle contour points are mapped to Cartesian coordinates, and the height value remains unchanged to form three-dimensional abrasive particle contour point coordinates (x, y, z).

IOP Conf. Series: Materials Science and Engineering 1207 (2021) 012017 doi:10.1088/1757-899X/1207/1/012017

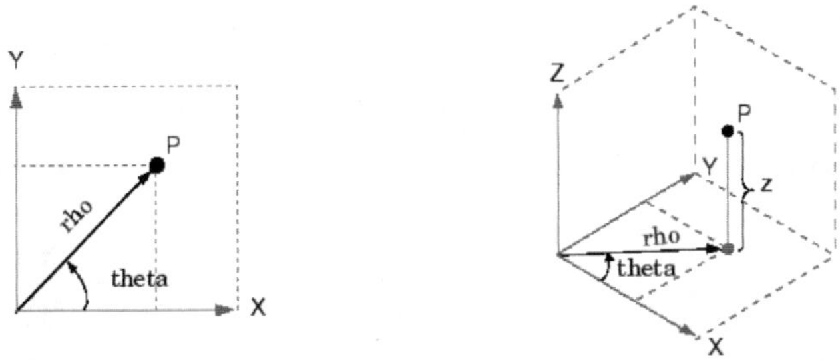

Figure 4. Mapping polar and cylindrical coordinates to Cartesian coordinates.

In this paper, the above algorithm based on cylindrical coordinate transformation can quickly and accurately map the wear particle from two-dimensional contour to three-dimensional space. The two-dimensional abrasive particle contour image in Figure 4 is mapped to the particle contour image in three-dimensional space, as shown in Figure 5. In the figure, the edge pixels of abrasive particle contour in three-dimensional space are represented by blue dots.

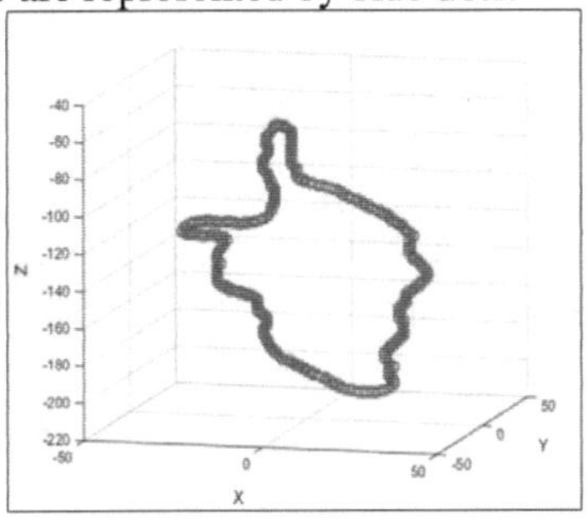

Figure 5. Wear particle contour image in three-dimensional space.

4. Experimental test

The collected lubricating oil samples were monitored on the experimental platform described in Section II, and nineteen original wear particle images were obtained as shown in the figure 6.

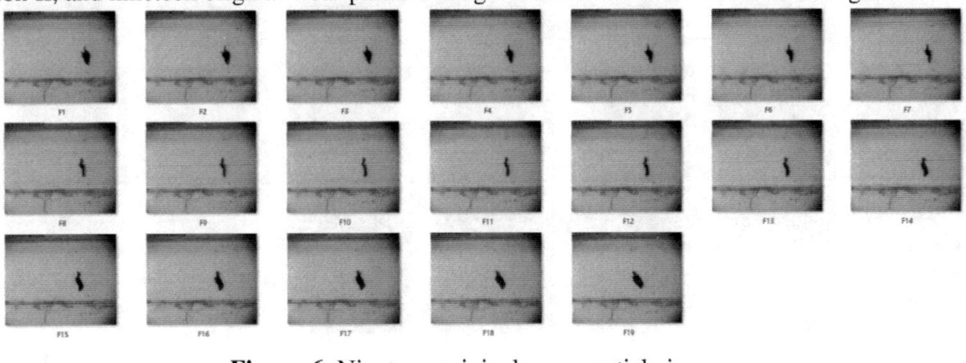

Figure 6. Nineteen original wear particle images.

Through the wear particle segmentation method based on oil background difference, 19 images of the same wear particle are segmented. Canny edge detection algorithm is used to detect the edge of 19 segmented images of the same wear particle, and finally realize the contour extraction of particle image. After edge detection, 19 contour image sequences of the same wear particle are shown in the figure 7.

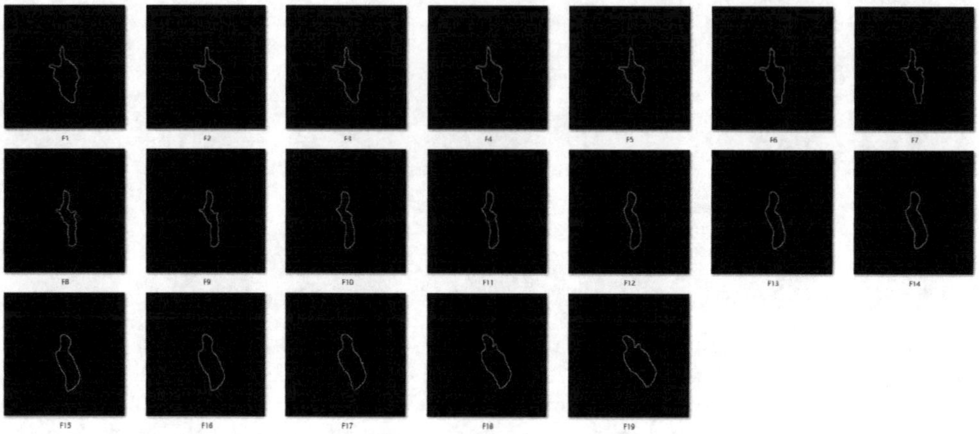

Figure 7. Nineteen contour image sequences of debris particles

It can be seen from the figure that after the wear particle image edge detection algorithm based on oil background difference and Canny operator is used for contour extraction, the contour line of particles is very clear and fine, the boundary is completely closed. However, due to the deflection of wear particle contour image, further particle image processing needs to be completed.

After the minimum area circumscribed rectangle is obtained, the wear particle deflection angle can be obtained by using the rectangular symmetry axis, and then the wear particle contour image can be corrected according to the deflection angle. Then, the wear particle contour image is corrected according to the deflection angle to obtain the wear particle contour image after deflection correction.

After determining the wear particle image center, taking the particle image center as the reference point, a unified particle boundary point stored contour image is established for nineteen corrected wear particle contour images. The corrected contour images used for the storage of debris boundary points are shown in the figure 8, and these contours are stored in two-dimensional data.

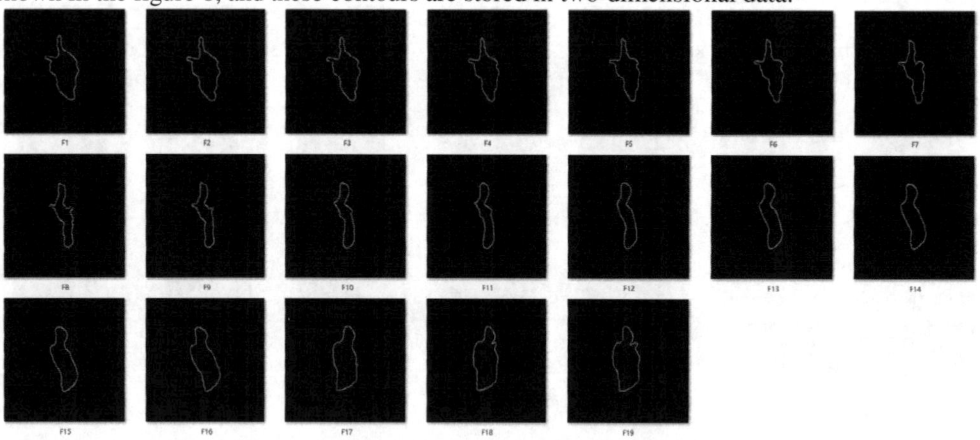

Figure 8. Contour image stored by particle boundary points.

The two-dimensional contour to three-dimensional space mapping method is used to convert the wear particle image in Figure 8. After conversion, nineteen contour images of the same wear particle in the three-dimensional space are shown in Figure 12.

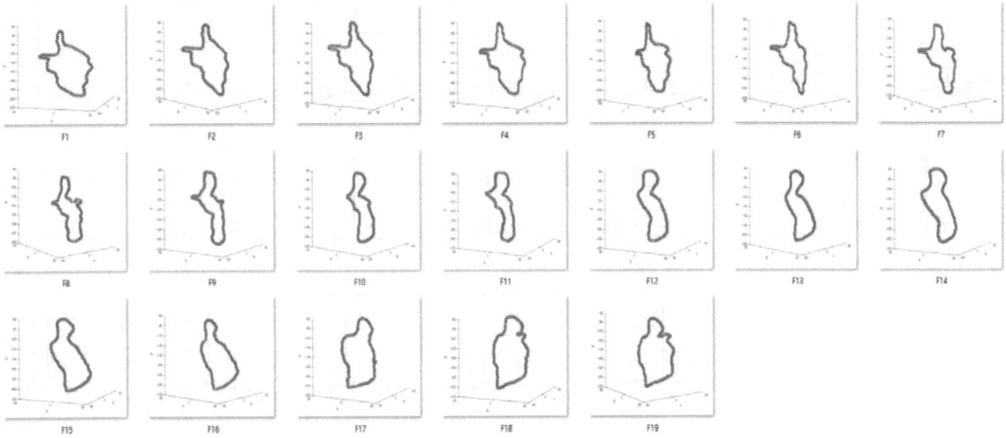

Figure 9. Nineteen contour images of the particle in three-dimensional space

The nineteen contour images are combined into a three-dimensional scatter diagram. The front view, oblique view and top view of the three-dimensional scatter diagram of wear particles are shown in Figure 10.

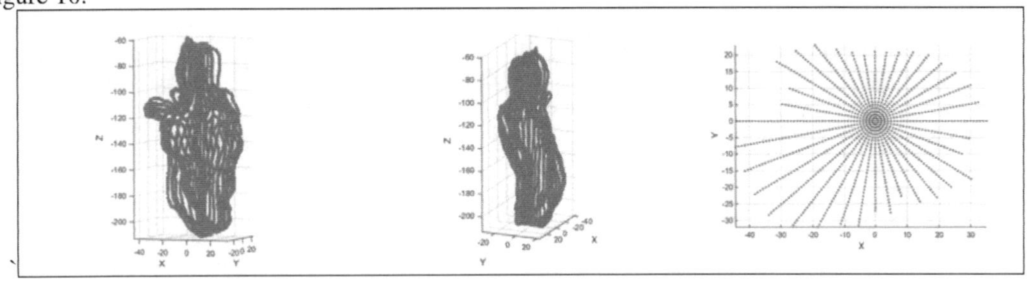

Figure 10. Front view, oblique view and top view of three-dimensional composite diagram of particles.

Finally, use Geomagic Design X software for rendering to form the three-dimensional model surface of wear particles. The front view, side view, top view and oblique view of the three-dimensional model of debris particles are shown in Figure 11 (a), (b), (c) and (d).

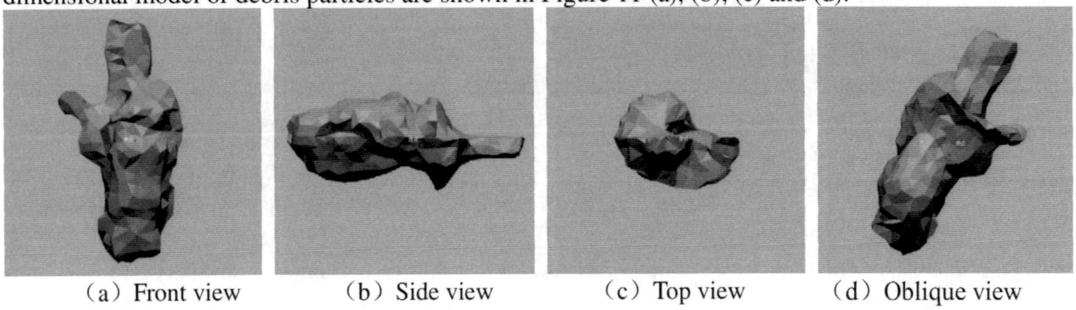

(a) Front view (b) Side view (c) Top view (d) Oblique view

Figure 11. Views of 3D model of wear particles.

From the reconstruction results of the three-dimensional model of wear particles, it can be concluded that most of the morphological information of particles can be reflected, including the thickness, width, length and aspect ratio of abrasive particles. Under a certain measurement error, deeper parameters such as the volume of abrasive particles can be estimated.

5. Conclusions

In this paper, a three-dimensional reconstruction method of wear particle image based on contour is proposed. For the same wear particle tracked in microchannel, the three-dimensional model reconstruction based on its contour image sequence is realized.

Firstly, the image sequence of rotating wear particles tracked by single target is extracted as data, and the relationship between the frame sequence of the particle image and the corresponding particle rotation step is analyzed in detail. Then, the particle image edge detection method combined with oil background difference and Canny operator is used to segment and detect the edge of 19 images of the same wear particle, and the particle contour is extracted. Aiming at the problem of particle deflection, a minimum area circumscribed rectangle method is used to calculate the minimum area circumscribed rectangle of abrasive particle contour. The deflection angle of abrasive particle image is calculated through the rectangular symmetry axis, and the deflection correction of abrasive particle image is completed. After determining the abrasive particle image center in the binary image, taking the particle image center as the reference point, a uniform size contour image for boundary point storage is established, and the two-dimensional data of particle boundary points are stored. An algorithm based on cylindrical coordinate transformation is used to convert the image data to three-dimensional space, realize the mapping from two-dimensional contour to three-dimensional space, and establish the three-dimensional scatter diagram of wear particles. Finally, the three-dimensional model reconstruction of micro flow abrasive particles is completed, which provides more abundant information for wear particle analysis technology.

In future work, the authors will investigate how to acquire clearer and more complete images of wear particles and construct more sophisticated 3-D models of the particles. At the same time, the quantitative evaluation of reconstruction results should be completed. Finally, continue to study the real impact of three-dimensional reconstruction of wear particles in lubricating oil on condition monitoring during mechanical rotation, such as condition prediction and wear analysis of mechanical structure.

Acknowledgments

This work was financially supported by the China National Natural Science Foundation (Grant No. U1933202).

References

[1] Shaochen Li. Research on oil wear particle monitoring technology based on electrostatic induction and micro image [D]. Nanjing University of Aeronautics and Astronautics, 2009.

[2] Xiaoyun He. Analysis and 3D reconstruction of wear particle surface morphology [D]. Wuhan University of Technology, 2005.

[3] Hongkun Wu,Ngai Ming Kwok,Sanchi Liu,Tonghai Wu,Zhongxiao Peng. A prototype of on-line extraction and three-dimensional characterisation of wear particle features from video sequence[J]. Wear,2016,368-369.

[4] Yeping Peng, Tonghai Wu, Shuo Wang, Ying Du, Ngaiming Kwok, Zhongxiao Peng. A microfluidic device for three-dimensional wear debris imaging in online condition monitoring[J]. Proceedings of the Institution of Mechanical Engineers, 2017, 231(8).

[5] Wu T H, Mao J H, Wang J T, et al. A new on-line visual ferrograph[J]. Tribology Transactions, 2009, 52(5): 623-631.

[6] Li B, Feng S, Mao J. Design and implementation of online image visible ferrographic LED array reflector [J].Journal of Xi'an Jiaotong University, 2014, 48 (10): 29-34.

[7] Peng Y, Wu T, Wang S, et al. A microfluidic device for three-dimensional wear debris imaging in online condition monitoring[J]. Proceedings of the Institution of Mechanical Engineers, Part J: Journal of Engineering Tribology, 2017, 231(8): 965-974.

[8] Liu T, Zhao J, Liu S, et al. Magnetization mechanism of nonferrous wear debris by magnetic fluid in ferrography[J]. Proceedings of the Institution of Mechanical Engineers, Part J: Journal of Engineering Tribology, 2016, 230(7): 827-835.

[9] Peng Z, Kirk T B, Xu Z L. The development of three-dimensional imaging techniques of wear particle analysis[J]. Wear, 1997, 203: 418-424.
[10] Peng Z, Kirk T B. Two-dimensional fast Fourier transform and power spectrum for wear particle analysis[J]. Tribology International, 1997, 30(8): 583-590.
[11] Peng Z, Kirk T B. Automatic wear - particle classification using neural networks[J]. Tribology letters, 1998, 5(4): 249-257.
[12] Wu T, Wang J, Peng Y, et al. Description of wear debris from on-line ferrograph images by their statistical color[J]. Tribology Transactions, 2012, 55(5): 606-614.
[13] Wu T, Peng Y, Du Y, et al. Dimensional description of on-line wear debris images for wear characterization[J]. Chinese Journal of Mechanical Engineering, 2014, 27(6): 1280-1286.
[14] Wu T, Wu H, Du Y, et al. Imaged wear debris separation for on-line monitoring using gray level and integrated morphological features[J]. Wear, 2014, 316(1-2): 19-29.
[15] Wu H, Wu T, Peng Y, et al. Watershed-based morphological separation of wear debris chains for on-line ferrograph analysis[J]. Tribology Letters, 2014, 53(2): 411-420.
[16] Wu H, Kwok N M, Liu S, et al. A prototype of on-line extraction and three-dimensional characterisation of wear particle features from video sequence[J]. Wear, 2016, 368: 314-325.
[17] Wu T, Peng Y, Wang S, et al. Morphological feature extraction based on multiview images for wear debris analysis in on-line fluid monitoring[J]. Tribology Transactions, 2017, 60(3): 408-418.
[18] Xi W, Wu T, Yan K, et al. Restoration of online video ferrography images for out-of-focus degradations[J]. EURASIP Journal on Image and Video Processing, 2018, 2018(1): 31.
[19] Liu H, Wei H, Xie H, et al. Unsupervised segmentation of wear particle's image using local texture feature[J]. Industrial Lubrication and Tribology, 2018, 70(9): 1601-1607.
[20] Wu H, Li R, Kwok N M, et al. Restoration of low-informative image for robust debris shape measurement in on-line wear debris monitoring[J]. Mechanical Systems and Signal Processing, 2019, 114: 539-555.
[21] Peng Y, Wu T, Wang S, et al. Oxidation wear monitoring based on the color extraction of on-line wear debris[J]. Wear, 2015, 332: 1151-1157.

An EMI radiation grouping approach based on spectral cluster in near filed scanning for assessment

Quan Huang[1,*], Yuxin Wu[1], Yan Gao[2,*], Wenxiao Fang[1,*], Zhiqiang Yi[1], Yun Huang[1], Guoguang Lu[1], Kaiwei Wang[1]

[1]China Electronic Product Reliability and Environmental Testing Research Institute, Guangzhou, China
[2]Institute of systems engineering, AMS, Beijing, China

*Corresponding author: fangwx@ceprei.com, gaomajiayi@163.com

Abstract. With the ever-increasing operating frequency in integrated circuit, it is very essentialto assess the radiation used to help the IC designer. Based on the similarity of electromagnetic patterns obtained from the radiation of ICs and their nonlinear edge, we develop a post-processing technique to group the electromagnetic patterns. A near field scanning is performed to obtain and extract the electromagnetic pattern that is used to validate the technique. Experiment results show that it can accurately group the electromagnetic patterns by which radiation assessment can be performed.

Keywords: EMI; grouping; radiation; near field scanning

1. Introduction

Near field scanning is an established technique that is usually used to detect the electromagnetic (EM) field distribution on a plane above the device under test (DUT), such as an integrated circuit [IC] or printed circuit board (PCB). The electromagnetic field distribution on the plane can be visualized, which is called electromagnetic pattern. The pattern's mechanism is currently needed to study further based on the strength of the emitted frequency, the form of the package, etc., it has been evidenced that electromagnetic patterns are constructive for assessing radiation quantity, localizing emission source[1] and designing for electromagnetic compatibility (EMC) of sub-system, module or sub-module [2], such as PCBs [3] and ICs [4]. In recent years, near field scanning was widely employed to measure the characterization of IC electromagnetic radiation [5], evaluate patch antenna radiation pattern [6] and observe magnetic domain in magnetic material with a high-resolution probe [7]. Near field scanning plays an important role in electromagnetic interference diagnosis. In [8], a near-field-to-near-field transformation method for the EMI radiation is proposed to obtain magnetic field on multi-planes by the expansion of the probe output based on the theory of plane wave spectrum.

Nonintrusive electromagnetic interference (EMI) detection is a powerful method for EMI diagnosis in high-speed links and multiple signal PCB designs. A noncontact time-domain voltage measurement method is used to detect the EMI signals of two adjacent on a PCB by a near field electric probe, and a reconstruction algorithm is developed to recover the voltage without contact [9]. From statistical view, a high order statistics (HOS) method is used to detect the number of electromagnetic interference sources [10], which is performed in frequency domain considering the characteristics of EMC problem. The EMIs , which are detected from a normal and a faulty circuits, are compared to recognize the switch fault in DC-DC converters [11]

The electromagnetic pattern is extracted after near field scanning and the EMI source is located at the position with higher radiation strength, which is applied to a chip [12], and the flash memory is found to be the major EM emission of the chip. The probe's position is correctly localized by a 2D optical tracking system in near field measurement [13], and an emission source microscopy (ESM) algorithm inpired from synthetic aperture radar (SAR) technique is applied to localize radiating sources.

Machine learning has been applied to many fields such as face recognition, data mining, object detection and recommendation systems. With the ever-increasing operating frequency of IC, electromagnetic interference is becoming more complex, which is not coped with by conventional analysis. In [14], the EMI radiation is predicted by combining the deep neural network (DNN) and Bayesian optimization algorithm (BOA). Besides, using the power supply and temperature variation, the RF integrated interference limitation is also estimated by elaborately designing a DNN, which is helpful to improve the working-effectiveness between the IC designer and the EMC designer [15] and promote the exchange of the designers' issues and solutions. To our understanding, the shape of electromagnetic pattern can reveal much information about the emission source and is usually irregular due to the structure of the device under test (DUT) and the scanning distance between probe and its surface [16]. The scanning step has an important impact on the gradient of the shape, which plays an essential role in localization accuracy.

In this paper, using the electromagnetic patterns, an EMI source diagnosis algorithm based on spectral cluster is investigated, which is used to group electromagnetic patterns based on similarity and diagnose their corresponding radiation. This paper is organized as follows. The EMI source diagnosis and grouping algorithm are developed in section II. Section III introduces the near field scanning experiment. Experiment results and discussion are arranged in section IV. And Section V draws the conclusion.

2. EMI Source Grouping and Assessment Technique on Electromagnetic Pattern

Near field scanning is performed by a mover which is controlled by a computer. An electromagnetic probe that is connected to a spectrum analyzer is controlled to sense the electromagnetic field emitted by the DUT along its surface in a specific scanning area. A spectrum matrix with same size as the scanning area would be obtained for a single frequency after scanning. As shown in Figure 1, for an $m \times n$ scanning area with $m * n$ sense points, if the number of scanning frequency points of the spectrum analyzer is N , which is denoted as $f_1, f_2, \ldots, f_N$, the spectrum matrix for each $f_i, i = 1, 2, \ldots, N$ is expressed as,

$$S = \begin{bmatrix} s_{11} & s_{12} & \cdots & s_{1n} \\ s_{21} & s_{22} & \cdots & s_{2n} \\ \vdots & \vdots & \vdots & \vdots \\ s_{m1} & s_{m2} & \cdots & s_{mn} \end{bmatrix} \quad (1)$$

Where each s_{ij} is sensed by the probe at each scanning point in the scanning area.

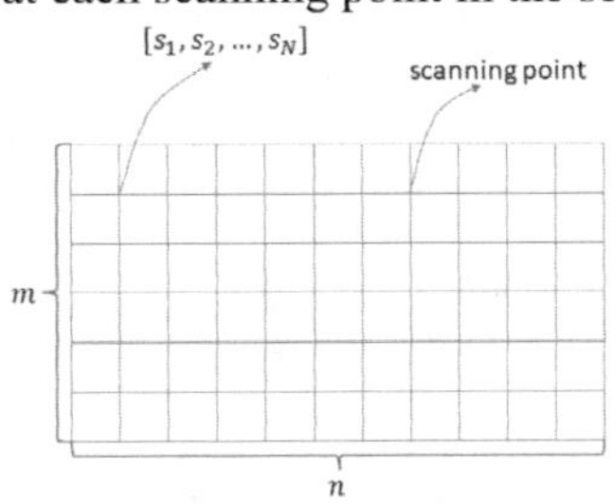

Figure 1. A scanning area with size of $m \times n$ in near field scanning and s_i denotes the spectrum at frequency of f_i .

Then, electromagnetic pattern is obtained by visualizing the spectrum matrix as a grey image or an RGB image.

2.1. Frequency Extraction Algorithm

In near field scanning, frequencies that have obvious electromagnetic pattern are seldom. In order to select these frequencies, an extraction algorithm is developed based on statistical variance or statistical standard deviation (std), which is computed as follows,

$$var(f_i) = \sum_{i=1}^{m}\sum_{j=1}^{n}(s_{ij} - \mu)^2 \tag{2}$$

Or,

$$std(f_i) = \sqrt{var(f_i)} \tag{3}$$

Where μ is the mean of $\boldsymbol{S}$, computed as,

$$\mu = \frac{1}{m * n}\sum_{i=1}^{m}\sum_{j=1}^{n} s_{ij} \tag{4}$$

It is known that signal variance is greater than that of noise. By setting a σ, those frequencies that have electromagnetic pattern can be extracted by comparing their variances with σ.

2.2. Grouping and Assessment Algorithm

For an integrated circuit, its clock usually generates many harmonics due to the nonlinear property of components such as diode, transistor and field effect tube (FET). Radiation at high frequency will be easy to be coupled by another circuit and has a significant impact on its operation stability. For a frequency and its harmonics, their electromagnetic patterns are similar in shape, which indicates they may be distributed in the same area of the IC. It is convenient to assess radiation and analyze those frequencies that have the same electromagnetic patterns after grouping. Grouping is performed based on the similarity of electromagnetic pattern. It is very natural that we can think of all spectrum matrices as a full connection network due to the harmonic relationship of the majority of frequencies. The adjacent matrix is denoted as,

$$\boldsymbol{W} = \begin{bmatrix} w_{11} & w_{12} & \cdots & w_{1n} \\ w_{21} & w_{22} & \cdots & w_{2n} \\ \vdots & \vdots & \cdots & \vdots \\ w_{n1} & w_{n2} & \cdots & w_{nn} \end{bmatrix} \tag{5}$$

Its element w_{ij} is computed as,

$$w_{ij} = w_{ji} = e^{-\frac{1}{2}(\boldsymbol{v}_i^T \Sigma^{-1} \boldsymbol{v}_j)} \tag{6}$$

Where $\boldsymbol{v}_i$ and $\boldsymbol{v}_j$ are column vector which are obtained by flattening the spectrum matrix of f_i and f_j, respectively. Σ is an invertible covariance matrix of the matrix formed by placing all flattened spectrum matrices as its columns.

Then the degree matrix can be obtained from adjacent matrix,

$$\boldsymbol{D} = \begin{bmatrix} d_1 & 0 & 0 & 0 & \cdots & 0 \\ 0 & d_2 & 0 & 0 & \cdots & 0 \\ 0 & 0 & d_3 & 0 & \cdots & 0 \\ 0 & 0 & 0 & d_4 & \cdots & 0 \\ \vdots & \vdots & \vdots & \vdots & \cdots & \vdots \\ 0 & 0 & 0 & 0 & \cdots & d_n \end{bmatrix} \tag{7}$$

Where d_i is computed as follows,

$$d_i = \sum_{j=1}^{n} w_{ij} \tag{8}$$

The Laplacian matrix is computed by subtracting $\boldsymbol{W}$ from $\boldsymbol{D}$,

$$\boldsymbol{L} = \boldsymbol{D} - \boldsymbol{W} \tag{9}$$

Where $\boldsymbol{L}$ is a symmetric matrix due to the full connection of all spectrum matrices, which is grouped by spectral cluster.

To our understanding, the grouped electromagnetic patterns have the same properties. First, its shape is similar in each group, which means the EMI sources formed by these frequencies distributed in the same physical structure and area. Second, the strength of each electromagnetic pattern in a group is usually different, and the gradient between them can be used to assess the rate of attenuation of harmonic wave.

To sum up, the grouping and assessment algorithm followed after the frequency algorithm, it can be summarized as following steps:

1) Compute the variance or standard variance of each spectrum matrix for each frequency by Eq. (2).
2) Flatten each spectrum matrix to column vector.
3) Construct the adjacent matrix by Eq. (5).
4) Construct the degree matrix by Eq. (7).
5) Compute the Laplacian matrix by Eq. (8).
6) Make eigen-decomposition of Laplacian matrix, and get the eigen-vectors corresponding to the first k smallest eigen-values to compose a matrix $\boldsymbol{Q} = [q_1, q_2, q_3, \dots, q_k]$.
7) Group each rows of $\boldsymbol{Q}$ by k-means or other cluster algorithm.

It can be seen that there is a dimension-reduced operation on the symmetric Laplacian matrix. We can treat $\boldsymbol{L}$ as a new sample matrix but it only has half information due to its symmetry. Assuming that its rows represent number of samples and columns represent features, then the dimension-reduced operation can be explained to reduce dependency or noise of the original data that is polluted when detecting and transmitting at wire.

3. Near Field Scanning on An FPGA

As described before, near field scanning experiment consists of the DUT, probe (electric or magnetic), mover, spectrum analyzer and computer, as shown in Figure 2.

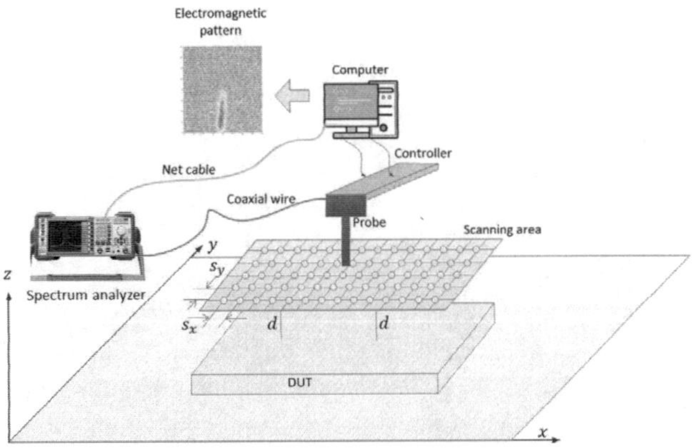

Figure 2. The near field scanning setup.

It is known that the smaller the distance between the surface of the DUT and the scanning plane, the more accurate the scanning data (magnetic or electric field data). The scanning steps denoted as s_x and s_y shown in Figure 2 decide scanning density that is a key factor to describe details of the electromagnetic pattern. However, high scanning density is not necessary for scanning area that its electromagnetic field stays constant or not changes dramatically. Scanning with high density and wide frequency is time-consumed.

Figure 3 shows the practical experiment setup on an FPGA shielded by a circle metal clamp. The DUT is placed above a ground plane designed according to IEC61967-3 [17], and the RF magnetic field is probed in direction parallel to the scanning surface.

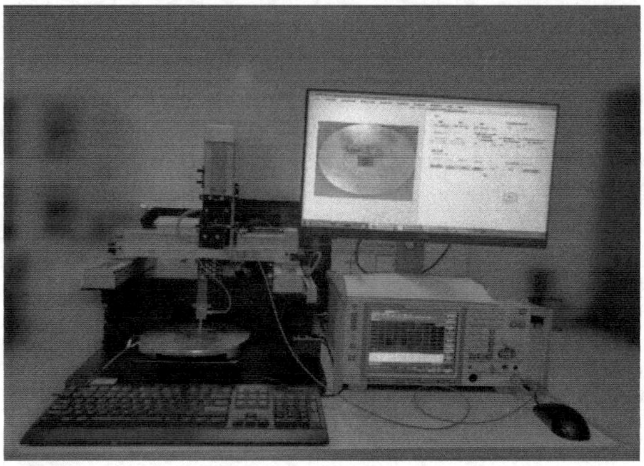

Figure 3. The near field scanning system with a 3-axis mover.

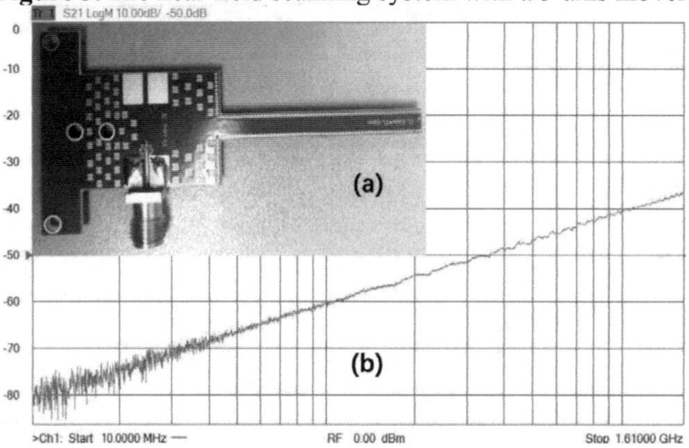

Figure 4. (a) Magnetic probe. (b) Its S_{21} from 10MHz to 2GHz.

Figure 5. The top view (left) and bottom view (right) of the FPGA test board. The width and height of the FPGA chip are both 17 mm.

The input clock frequency of the board is configured as 50MHz injected via pin E16 and its GPIO is running at 2.5V LVTTL standard.

4. Results and Discussion

The near field scanning is running in the same step of 106 μm in x-axis and y-axis directions forming a scanning area of 160×160 points, that is $m = 160$ and $n = 160$ in Figure 1. The spectrum analyzer is configured at frequency range of 10 MHZ to 500 MHz with 1001 sweep points. The probe is placed with a distance of 1 mm above the surface of the FPGA chip, i.e. $d = 1$ mm in Figure 2.

4.1. Frequency Extraction

Figure 6 shows the variances of each frequency. It can be seen that the variances of only 15 frequencies are greater than the threshold ($\sigma = 0.15$). They are the frequencies of 50MHz, 100MHz, 150MHz, 190MHz, 230MHz, 250MHz, 300MHz, 350MHz, 377MHz, 400MHz, 417MHz, 438MHz, 450MHz, 480MHz and 500MHz, which includes the harmonic waves of clock frequency (50 MHz).

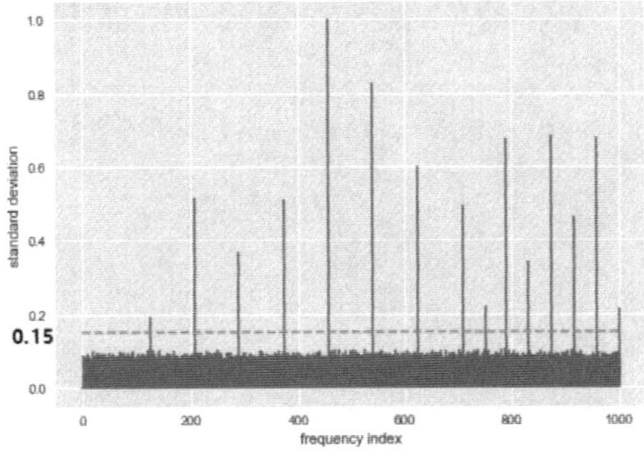

Figure 6. Variances of each frequency.

Frequencies that are not harmonic waves of clock frequency may be generated by nonlinear properties of components such as diode and transistor. It is also caused by the third-order cross modulation.

4.2. Group the Electromagnetic Patterns of An FPGA

We apply the grouping algorithm to the spectrum matrix of the 15 frequencies extracted by their variance with $k = 4$ groups as shown in Figure 7 to Figure 10. It can be seen that electromagnetic patterns in each group are similar in shape, while dissimilar in different groups such as Group-0 and Group-2, Group-1 and Group-2. But the grouping is not perfect. It is obvious that Group-0 and Group-1 are similar, to a certain extent. The strength of EMI radiation in each group for different frequency is intuitive by observing the electromagnetic patterns.

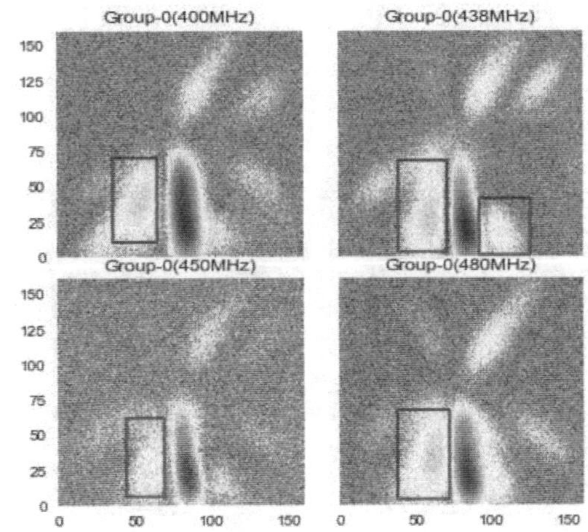

Figure 7. Results of Group-0

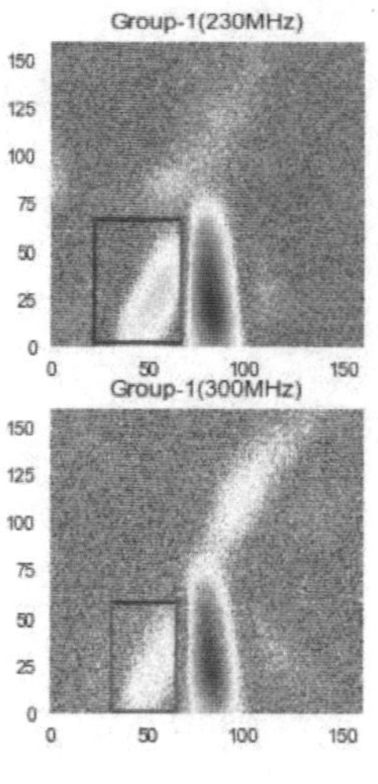

Figure 8. Results of Group-1

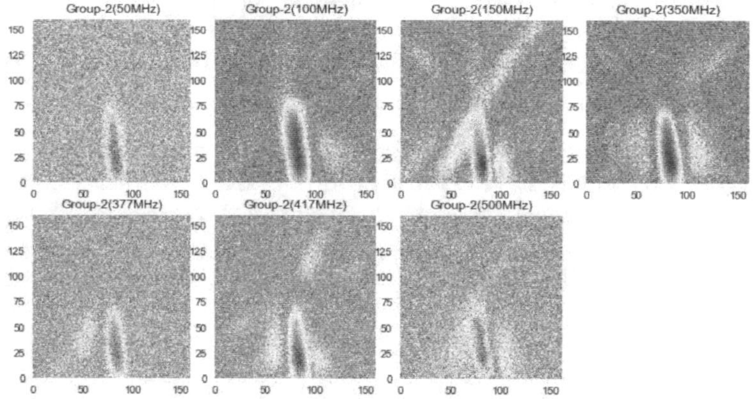

Figure 9. Results of Group-2

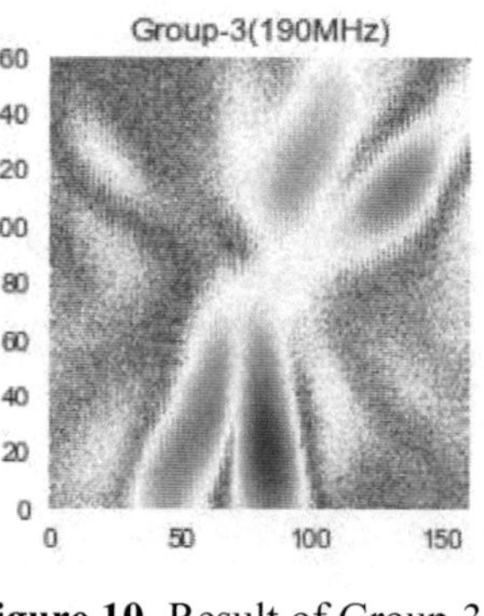

Figure 10. Result of Group-3

As seen from Figure 7 to Figure 10, the redder the electromagnetic pattern is, the higher the radiation is. The electromagnetic patterns of harmonic wave of clock (50 MHz) are grouped in the same one as shown in Figure 9, for instance, the electromagnetic pattern of 100 MHz, 150 MHz, 350 MHz and 500 MHz. But there is exception about that, the 400 MHz and 250 harmonic waves are not grouped in the same one, so do the 450 MHz and 300 MHz. It may be explained that the signal of a trace in the FPGA running at high frequency is coupled by its adjacent trace with low coupling efficiency, which leads to electromagnetic pattern with lower strength, such as Figure 7 and Figure 8 marked in red rectangle.

5. Conclusions

In this work, we developed a post-processing method to electromagnetic patterns of integrated circuits in near field scanning. Based on the fact that variance of signal is always greater than that of noise, frequencies that have electromagnetic patterns are extracted to be used to group. Spectral cluster is introduced to group the patterns due to their similarity in shape and nonlinear edges. The grouping technique is applied to an FPGA and has a good result. However, the number of groups has to be set artificially, which is a drawback for many cluster algorithms. Besides, interference always exists due to nonlinear properties of components such as diodes and transistors, which results in many frequencies that are not harmonics of clock frequency. These frequencies can be assessed by their electromagnetic patterns.

Acknowledgments

This work was supported in part by the Natural Science Foundation of Guangdong Province (2018A030310360), the Key Area R&D Program of Guangdong Province (2019B010141002), the National Natural Science Foundation of China (No.61801124) and the Science and Technology Program of Guangzhou, China (No. 201904010210). The authors would like to thank the editors and reviewers for valuable suggestions and comments for improving the original manuscript.

References

[1] T. -H. Song, X. -C. Wei, Z. -Y. Tang and R. X. -K. Gao, "Broadband Radiation Source Reconstruction Based on Phaseless Magnetic Near-Field Scanning," in IEEE Antennas and Wireless Propagation Letters, vol. 20, no. 1, pp. 113-117, Jan. 2021.

[2] S. M. M. Mirtalaei, S. H. H. Sadeghi, and R. Moini, “A combined method-of-moments and near-field measurements for EMI evaluation of switched-mode power supplies”, IEEE Trans. Ind. Electron., vol. 61, no. 4, pp. 1811-1818, Apr. 2014.

[3] L. Beghou, B. Liu, L. Pichon, and F. Costa, “Synthesis of equivalent 3-D models from near field measurements–Application to the EMC of power printed circuit boards,” IEEE Trans. Magn., vol. 45, no. 3, pp. 1650–1653, Mar. 2009.

[4] Z. Yu, J. A. Mix, S. Sajuyigbe, K. P. Slattery and J. Fan, "An Improved Dipole-Moment Model Based on Near-Field Scanning for Characterizing Near-Field Coupling and Far-Field Radiation From an IC," in IEEE Transactions on Electromagnetic Compatibility, vol. 55, no. 1, pp. 97-108, Feb. 2013.

[5] M. A. chahine, R. Perdriau, M. Ramdani and A. Saleh, "A calibrated near field scanning system applied to the characterization of IC electromagnetic radiation," 2012 2nd International Conference on Advances in Computational Tools for Engineering Applications (ACTEA), 2012, pp. 124-127.

[6] Y. Shu, X. Wei and Y. Yang, "Patch Antenna Radiation Pattern Evaluation Based on Phaseless and Single-Plane Near-Field Scanning," 2019 IEEE International Conference on Computational Electromagnetics (ICCEM), 2019, pp. 1-3.

[7] M. Aoyagi, S. Niratisairak, T. Sioda and T. Ishibashi, "Extinction Ratio in Apertureless Reflection-Mode Magneto-Optical Scanning Near-Field Optical Microscopy," in IEEE Transactions on Magnetics, vol. 48, no. 11, pp. 3670-3672, Nov. 2012.

[8] J. Zhang, X. Wei, R. Yang, R. X. Gao and Y. Yang, "An Efficient Probe Calibration Based Near-Field-to-Near-Field Transformation for EMI Diagnosis," in IEEE Transactions on Antennas and Propagation, vol. 67, no. 6, pp. 4141-4147, June 2019, doi: 10.1109/TAP.2019.2902741.

[9] H. Qiu, C. Luo, W. Fang and L. Jiang, "Noncontact Time-domain EMI Measurement of Two Adjacent Traces on a PCB," 2018 IEEE Symposium on Electromagnetic Compatibility,

Signal Integrity and Power Integrity (EMC, SI & PI), 2018, pp. 634-639, doi: 10.1109/EMCSI.2018.8495320.

[10] S. Zhenfei, S. Donglin and X. Shuguo, "Detecting the Number of EMI Sources Based on Higher Order Statistics," 2009 20th International Zurich Symposium on Electromagnetic Compatibility, 2009, pp. 465-468, doi: 10.1109/EMCZUR.2009.4783491.

[11] M. A. Rahman, E. Pazouki, Y. Sozer and J. A. De Abreu-Garcia, "Fault Detection of Switch Mode Power Converters Based on Radiated EMI Analysis," 2019 IEEE Energy Conversion Congress and Exposition (ECCE), 2019, pp. 2968-2972, doi: 10.1109/ECCE.2019.8913272.

[12] T. Ordas, M. Lisart, E. Sicard, P. Maurine, and L. Torres, "Near-field mapping system to scan in time domain the magnetic emissions of integrated circuits," Proc. Int. Workshop Power Timing Modeling, Optim. Simul., Berlin, Germany, 2009, pp. 229–236, vol. 5349.

[13] H. He, V. Khilkevich and D. Pommerenke, "2D imaging system with optical tracking for EMI source localization," 2015 IEEE Symposium on Electromagnetic Compatibility and Signal Integrity, 2015, pp. 107-110, doi: 10.1109/EMCSI.2015.7107668.

[14] H. Jin, L. Zhang, H. Ma, S. Yang, X. Yang and E. Li, "Machine learning for complex EMI prediction, optimization and localization," 2017 IEEE Electrical Design of Advanced Packaging and Systems Symposium (EDAPS), 2017, pp. 1-3, doi: 10.1109/EDAPS.2017.8276967.

[15] M. Chang, "A Machine Learning Approach to Predict Supply and Temperature Variation Aware RF Integrated Circuit Interference Limitation," 2020 International Workshop on Electromagnetics: Applications and Student Innovation Competition (iWEM), 2020, pp. 1-2, doi: 10.1109/iWEM49354.2020.9237389.

[16] Z. Hong et al., "Electromagnetic Pattern Extraction and Grouping for Near-Field Scanning of Integrated Circuits by PCA and K-Means Approaches," in IEEE Transactions on Electromagnetic Compatibility, vol. 61, no. 6, pp. 1811-1822, Dec. 2019, doi: 10.1109/TEMC.2018.2890026.

[17] Integrated Circuits Measurement of Electromagnetic Emissions, 150 kHz to 1 GHz, Part 3: Measurement of Radiated Emissions—Surface Scan Method, IEC TS 61967, 2005.

Adversarial flow-based model for unsupervised fault diagnosis of rolling element bearings

Jun Dai[1], Jun Wang[1,*], Linquan Yao[1], Juanjuan Shi[1], Lei Wang[2] and Guolei Wang[2]

[1] School of Rail Transportation, Soochow University, Suzhou, Jiangsu, China
[2] Xiangyang BOYA Precision Industrial Equipments Co., Ltd., Xiangyang, Hubei, China

*Corresponding author: junking@mail.ustc.edu.cn

Abstract. Nowadays, numerous supervised deep learning models have been applied to bearing fault diagnosis. However, labelling the health states of the bearing vibration data is a time-consuming work and dependent on expert experience. In order to tackle this problem, a novel unsupervised bearing fault diagnosis method named adversarial flow-based model is explored in this paper. Flow-based model is a type of generative models that is proved to be better than other types in many aspects. This paper introduces the flow-based model into the field of machinery fault diagnosis, and designs an appropriate model architecture so as to train the model in unsupervised and adversarial ways. The proposed model contains an autoencoder (AE), a flow-based model, and a classifier. Firstly, the AE maps the vibration data from signal space to latent vector space. Then, the flow-based model aligns the distributions of the latent vectors of different bearing states with specific prior distributions. Finally, the classifier tries to discriminate the aligned latent vectors from the vectors sampled from the prior distributions. With the help of distinguishable prior distributions and the adversarial training mechanism between the classifier and the flow-based model together with the AE, the bearing data with the same health states are clustered into the same areas. The good clustering performance of the adversarial flow-based model is verified by a dataset with 10 health states from a bearing test rig.

Keywords: Fault diagnosis, flow-based model, adversarial training, unsupervised learning, rolling element bearings.

1. Introduction

As a critical rotating part of mechanical systems, rolling element bearings are widely applied in transportation systems, industrial systems, and electrical systems. Owing to the harsh operating environments, such as high speed and heavy dynamic load, bearings are prone to failure that are difficult to be recognized [1]. Bearing fault may affect the normal operation of mechanical systems and lead to serious accidents. To ensure the healthy operation of rolling bearings, massive data are collected, which promoted the evolution of data-driven based fault diagnosis techniques [2].

Nowadays, many supervised learning models, such as deep neural networks (DNN) [3], convolutional neural networks (CNN) [4], deep belief networks (DBN) [5], have been used to bearing fault diagnosis. Latent features in the bearing vibration signals can be extracted adaptively by these models, and used for recognition of the bearing health states. However, these models are all supervised model, a large amount of historical data and corresponding labels are required for model training. In practical diagnosis task, labelling massive bearing vibration signals is time consuming and requires expert experience. Hence, unsupervised learning models, such as autoencoder (AE) [6], K-nearest neighbor (KNN) [7], and generative model [8], have attracted much attention to tackle the absence of labels in bearing fault diagnosis.

Generative adversarial network (GAN) and variational autoencoder (VAE) are typical generative models which are designed to learn the data distributions in an unsupervised way, and have been gradually employed in fault diagnosis. Chen et al. [9] constructed a sliding-window convolutional variational autoencoder (SWCVAE) to realize real-time anomaly detection. However, the AE is only an approximate method for estimation of data distributions. Dai et al. [10] proposed a novel GAN to capture the distribution of normal samples, which adopts adversarial training mechanism between a generator constructed via the AE and a discriminator for identification of data authenticity. A health indicator was established for monitoring the real-time health states of a machine. The results have demonstrated impressive performance on unsupervised fault diagnosis. However, training the GAN model is a challenging task owing to the training instability, such as mode collapse, gradient vanishing, and gradient explosion.

Flow-based model is a new type of generative models, which is popularized by Dinh et al. [11] for accurately learn of data distributions through a sequence of invertible transformations. The flow-based model has been utilized in many fields, including image generation [12], voice conversion [13], and speech synthesis [14]. Yang et al. [15] introduced the flow-based model to the AE to construct a sequence-to-sequence learning model to realize remaining useful life prediction of a mechanical system. State-of-the-art results have been achieved by the flow-based models in these literatures. However, the flow-based models have not been used to recognize the health states of mechanical systems until now.

In this paper, an adversarial flow-based model is constructed for unsupervised fault diagnosis, by embedding the flow-based model into the AE and classification networks. The proposed model aligns the latent vectors obtained by the AE from the bearing vibration signals of each health states with a specific prior distribution through the flow-based model and a classifier, which is trained in unsupervised and adversarial ways. The AE and flow-based model try to trick the classifier into obtaining a certain category assignment for the aligned latent vectors, while the classifier tries to discriminate the aligned latent vectors from the prior distributions and gives them vague category judgments. After fully trained, the proposed model will reach a balance, which means the distribution characteristics of the bearing vibration signals have been fully captured by the AE and flow-based model, and the classifier cannot discriminate the aligned latent vectors from the prior distributions. As a result, the aligned latent vectors of each health state strictly follow one specific prior distribution, which exhibits great clustering and unsupervised classification performance. A bearing vibration dataset with 10 health states is analyzed in this paper. The experiment verified the superiority of the proposed adversarial flow-based model over traditional AE model.

2. Flow-based model

Flow-based model is a kind of generative networks, which aims to accurately capture the probability distribution of input samples. Given an observed data $x \in X$ that follow the distribution p_X, and a prior probability distribution p_Z on a variable $z \in Z$, the probability density follows the formula below:

$$\int p_X(x)dx = \int p_Z(z)dz = 1 \tag{1}$$

The flow-based model designs a bijection function $f: X \rightarrow Z$ with $g = f^{-1}$. The function $f(\cdot)$ maps the x from original latent distribution p_X to the prior distribution p_Z. The change of variable formula defines a model distribution on X by:

$$p_X(x)dx = p_Z(f(x))df(x)$$
$$\Rightarrow p_X(x) = p_Z(f(x))\frac{df(x)}{dx}. \tag{2}$$

Generalization to higher dimensions, the probability density can be expressed as:

$$p_X(x) = p_Z\left(f(x)\right)\left|\det\left(\frac{\partial f(x)}{\partial x^T}\right)\right| \tag{3}$$

$$\log\left(p_X(x)\right) = \log\left(p_Z\left(f(x)\right)\right) + \log\left(\left|\det\left(\frac{\partial f(x)}{\partial x^T}\right)\right|\right) \tag{4}$$

where $\frac{\partial f(x)}{\partial x^T}$ is the Jacobian matrix of $f(\cdot)$ at x. To simplify the calculation of the Jacobi matrix, a flow network named real-valued non-volume preserving (Real NVP) [16] is applied, whose architecture is shown as figure 1.

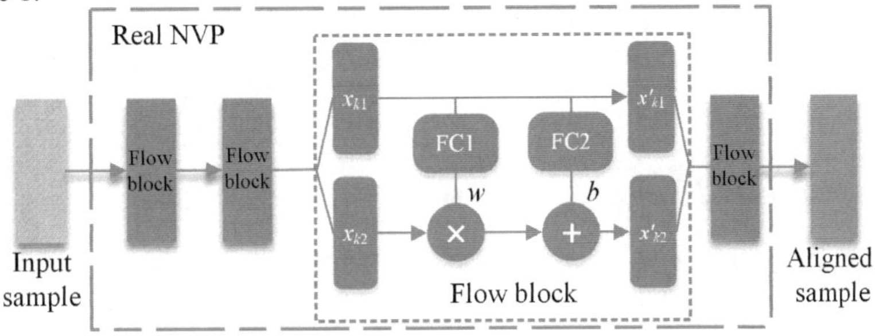

Figure 1. Architecture of the Real NVP.

The architecture of the Real NVP is stacked by several flow blocks. In each block, the input data are divided into two parts, x_{k1} and x_{k2}. The first part x_{k1} is transmitted directly to the output, i.e., $x'_{k1} = x_{k1}$. For the second part, x'_{k2} is calculated according to the following formula:

$$x'_{k2} = w \cdot x_{k2} + b \tag{5}$$

where w, b represent the weight and bias, respectively, which are output by the fully connected networks. The Jacobi matrix of this structure is a diagonal matrix and can be formulated as the following equation:

$$Jacobi\ matrix = \begin{bmatrix} E & O \\ M & W \end{bmatrix} \tag{6}$$

where E, O, W represent the identity matrix, null matrix, and weight matrix, respectively. M is a matrix which will not influence the determinant. The determinant of the Jacobi matrix is calculated by successive multiplication of the elements at the main diagonal.

3. Adversarial flow-based model

The flow-based model described above is trained through an unsupervised approach. That is to say, the model is trained without using label information. In order to enhance the clustering performance in unsupervised rolling element bearing fault diagnosis, an adversarial flow-model is proposed, whose architecture is illustrated in figure 2. The proposed model contains an AE, a flow-based model, and a classifier. In the forward propagation process, the signal $\boldsymbol{x}$ after simple preprocessing using Fourier transform is first input to the AE, by which the latent feature z of the signal in latent space are learned by reconstructing the signal $\boldsymbol{x}$ as $\boldsymbol{x}'$ in the frequency domain. Then the latent vector z is input into a flow-based model, by which the vector will be mapped to another space after aligning to a specific prior distribution. Thereafter, the vector $\boldsymbol{k}$ sampled from one of the K prior distributions and the aligned latent vector z' are input into a classifier, which is trained in an adversarial way and predicts the labels for both of the vectors.

During the training process, the vectors sampled from the prior distributions are treated as real samples. The classifier will give these vectors certain classification assignments. On the contrary, the aligned latent vectors are considered as fake samples and vague judgments will be given by the classifier. However, the AE and flow-based model are trained to align the latent vector to follow the prior distributions as much as possible, so that the classifier could not tell the real and fake samples apart. This is an adversarial training process. To further enhance the adversarial performance in classification, both the encoder and the classifier are required to assign classification equally to the samples (assume the amount of samples per class is the same for all categories). After fully trained, the original signal with the same healthy state will be clustered by the proposed model in both the latent space and the space of the designed prior distributions.

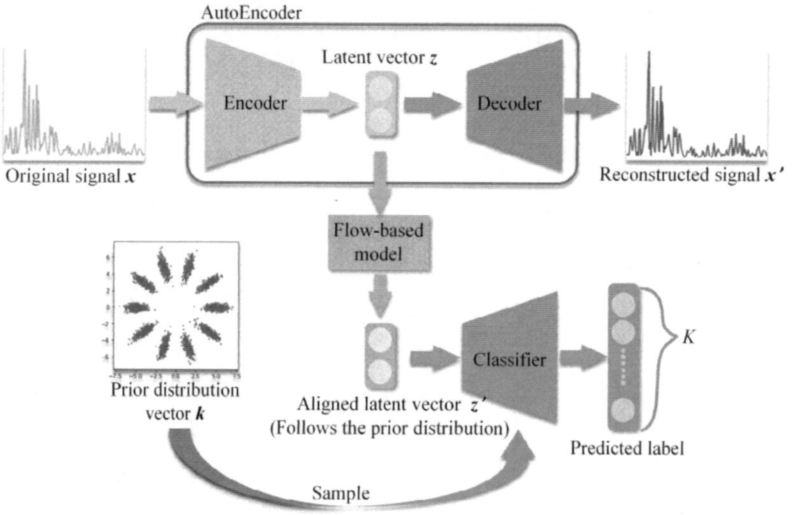

Figure 2. Architecture of the proposed adversarial flow-based model.

In this paper, three parts of functions, including reconstruction function, flow-based model function and adversarial function are applied to train the proposed model. The reconstruction function measures the dissimilarity between the original signal and the reconstructed signal, as formulated by:

$$F_{rec} = \min_{En,De} \left\| x - x' \right\|_2 \tag{7}$$

The flow-based model function measures the probability density of the latent vectors, as defined by:

$$F_{flow} = \max_{flow} \left[\log p_{latent}(z) \right] = \max_{flow} \left[\log p_{prior}\left(f(z)\right) + \log \left| \det \left(\partial f(z) \big/ \partial z^T \right) \right| \right] \tag{8}$$

The adversarial function is used to optimize the encoder and classifier respectively. For the classifier, the function aims to discriminate the source of input vector and assign classification equally. The objective function for the classifier is shown as:

$$F_{adversarial-Classifier} = \max_{C} \left\{ H_{prior}\left[p(y|C) \right] - E_{\boldsymbol{k}\sim prior}\left[H\left[p(y|\boldsymbol{k}, C) \right] \right] + E_{\boldsymbol{x}\sim data}\left[H\left[p(y|f(En(\boldsymbol{x})), C) \right] \right] \right\} \tag{9}$$

where $H[\cdot]$represents the Shannon entropy. C represents the classifier. En represents the encoder

For the encoder, the adversarial function is optimized to obtain certain and equal category assignment from the classifier. The objective function is shown as:

$$F_{adversarial-Encoder} = \min_{En} \left\{ -H_f\left[p(y|C) \right] + E_{\boldsymbol{x}\sim data}\left[H\left[p(y|f(En(\boldsymbol{x})), C) \right] \right] \right\} \tag{10}$$

4. Experiment validation

In this section, a bearing dataset containing 10 types of health states is analyzed by proposed method. The experimental platform is shown in figure 3. The test bearing (Type: NSK 6205) was installed at the right end of a shaft driven by an AC motor. The rotating speed was kept at 900 RPM and the radial load was kept at 1 kN. The sampling frequency was 10 kHz. The complete dataset contains a normal state (N) and three fault states, including inner race fault (IR), outer race fault (OR), and ball element fault (B). According to the width of the fault slit, each fault state is divided into three sub-states, corresponding to the width of 0.2 mm, 0.3 mm, and 0.4 mm, respectively. For convenience, one fault state is denoted by the fault location combining the slit width. For example, IR02 represents the fault state with a 0.2 mm width of slit located in the bearing inner-race way. Each fault state contains 500 samples and each sample has 2048 points in the time domain.

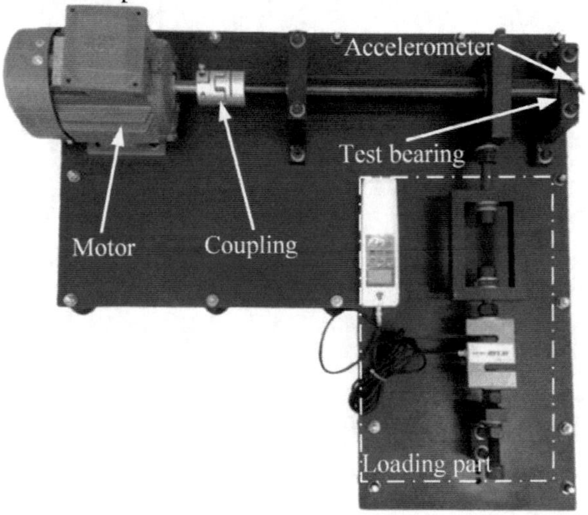

Figure 3. Experimental platform for data acquisition of rolling element bearings with 10 types of health states.

In the experiment, the designed prior distributions are K (= 10) the two dimensional Gaussian distributions, which corresponds to 10 types of the bearing health categories. Each sub-distribution is generated from an initial Gaussian distribution $X \sim N\left(\begin{bmatrix} a \\ 0 \end{bmatrix}, \begin{bmatrix} b_1 & 0 \\ 0 & b_2 \end{bmatrix}\right)$ by rotating an angle θ around the origin and can be formulated as:

$$X \sim N\left(\begin{bmatrix} a*\cos\theta \\ a*\sin\theta \end{bmatrix}, \mathbf{U}\begin{bmatrix} b_1 & 0 \\ 0 & b_2 \end{bmatrix}\mathbf{U}^T\right) \quad \theta = \frac{2\pi}{K}\cdot i, \ \{i \mid 0 \le i \le K-1, \ i \in N\} \tag{11}$$

where $\begin{bmatrix} a*\cos\theta \\ a*\sin\theta \end{bmatrix}$ is the mean value matrix of each sub-Gaussian distribution. $\boldsymbol{U}\begin{bmatrix} b_1 & 0 \\ 0 & b_2 \end{bmatrix}\boldsymbol{U}^T$ is the covariance matrix, where $\begin{bmatrix} b_1 & 0 \\ 0 & b_2 \end{bmatrix}$ is the covariance matrix of initial Gaussian distribution and $U = \begin{bmatrix} \cos\theta & -\sin\theta \\ \sin\theta & \cos\theta \end{bmatrix}$ is the rotation matrix.

In this paper, a=5, b_1=0.5, and b_2=0.02. The 10 prior distributions are displayed in the lower left quarter of figure 2. The time domain samples are transformed into frequency domain by fast Fourier transform (FFT) at first. Then, the min–max scaling is performed on the frequency domain samples. The samples are all scaled within the range of [0,1]. The model is optimized by Adam optimization algorithm (Learning rate = 0.0005). The structure details of the proposed adversarial flow-based model are summarized in table 1.

After fully trained by the bearing dataset, the aligned latent vectors by the flow-based model in the proposed method are showed in figure 4(a). In the figure, the samples belonging to the same bearing health state are assembled at the same area in the two dimensional space, which strictly follows a specific prior distribution shown in the lower left quarter of figure 2. Therefore, the 10 types of bearing health states are accurately classified to different categories in an unsupervised way.

Table 1. The structure details of the proposed adversarial flow-based model

Modules	Parameter name	Parameters	Activation function
	Encoder-layer-1	1024 to 512	Tanh
Encoder	Encoder-layer-2	512 to 128	Tanh
	Encoder-layer-3	128 to 2	/
	Decoder-layer-1	2 to 128	Tanh
Decoder	Decoder-layer-2	128 to 512	Tanh
	Decoder-layer-3	512 to 1024	Tanh
Flow-based	Number of flow layer	4	/
model	Dimension of hidden layer	8	/
	Classifier-layer-1	2 to 100	Tanh
Classifier	Classifier-layer-2	100 to 100	Tanh
	Classifier-layer-3	100 to K	Softmax

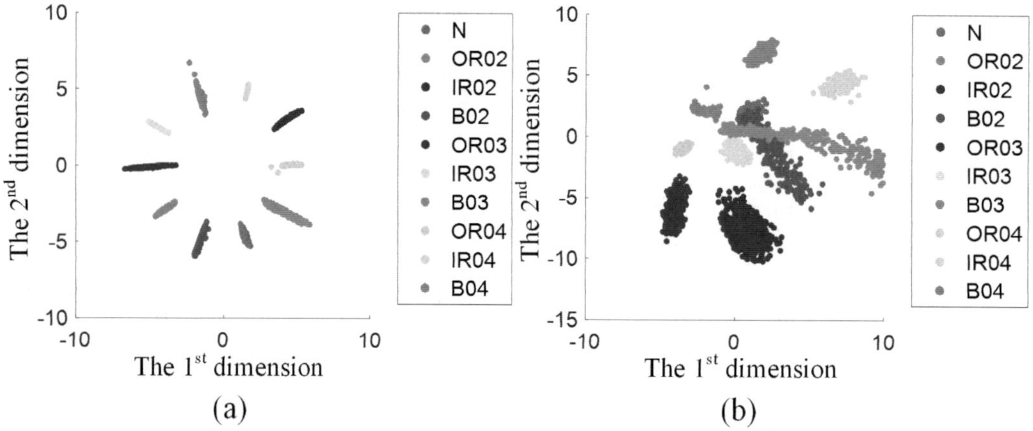

Figure 4. Visualization of (a) the aligned latent vectors in the proposed method and (b) the latent vectors of the AE.

As a comparison, an AE model with the same structure as that of the proposed model is constructed to analyse the same dataset. The latent vectors obtained by simple AE model are displayed in figure 4(b). Some bearing health states cannot be discriminated from another by the latent vectors of the AE, especially the states of N, B02, and B03. And the clustering performance of every health state is not satisfactory, because the assembled area has much larger covariance value than proposed method. The comparison results prove the outstanding clustering ability of the adversarial flow-based model in unsupervised rolling element bearings fault diagnosis.

5. Conclusion

In this paper, a novel unsupervised bearing fault diagnosis model called adversarial flow-based model is presented. The contribution is that the flow-based model is embedded in the AE and classification networks for mapping of the latent vectors on prior distributions, which is implemented in unsupervised and adversarial ways. Two dimensional Gaussian distributions with different mean and covariance values were designed as the prior distributions, which benefits the clustering performance of the proposed model. The signals with the same health states can be clustered in the same areas, whereas those with different health states can be separated distinctly. A bearing vibration dataset containing 10 types of health states was used to validate the clustering effectiveness of the proposed adversarial flow-based model. The results proved the superiority of the proposed model in clustering and unsupervised classification of bearing health states. More dataset will be applied to validate the proposed model in the future work.

Acknowledgments

This work was financially supported in part by the National Natural Science Foundation of China under Grant 51805342, in part by the Natural Science Foundation of Jiangsu Province under Grant BK20180842, in part by the China Postdoctoral Science Foundation under Grant 2021M692354, and in part by the Postgraduate Research & Practice Innovation Program of Jiangsu Province under Grant KYCX20_2661.

References

[1] Liu R, Yang B, Zio E and Chen X 2018 Artificial intelligence for fault diagnosis of rotating machinery: A review *Mech. Syst. Signal Process.* **108** 33–47
[2] Lei Y, Yang B, Jiang X, Jia F, Li N and Nandi A K 2020 Applications of machine learning to machine fault diagnosis: A review and roadmap *Mech. Syst. Signal Process.* **138** 106587
[3] Teng W, Cheng H, Ding X, Liu Y, Ma Z and Mu H 2018 DNN-based approach for fault detection in a direct drive wind turbine *IET Renew. Power Gener.* **12** 1164–71
[4] Wen L, Li X, Gao L and Zhang Y 2018 A new convolutional neural network-based data-driven fault diagnosis method *IEEE Trans. Ind. Electron.* **65** 5990–8
[5] Yu J and Liu G 2020 Knowledge extraction and insertion to deep belief network for gearbox fault diagnosis *Knowledge-Based Syst.* **197** 105883
[6] Shao H, Jiang H, Zhao H and Wang F 2017 A novel deep autoencoder feature learning method for rotating machinery fault diagnosis *Mech. Syst. Signal Process.* **95** 187–204
[7] Lu J, Qian W, Li S and Cui R 2021 Enhanced K-nearest neighbor for intelligent fault diagnosis of rotating machinery *Appl. Sci.* **11** 919
[8] Liu H, Zhou J, Xu Y, Zheng Y, Peng X and Jiang W 2018 Unsupervised fault diagnosis of rolling bearings using a deep neural network based on generative adversarial networks *Neurocomputing* **315** 412–24
[9] Chen T, Liu X, Xia B, Wang W and Lai Y 2020 Unsupervised anomaly detection of industrial robots using sliding-window convolutional variational autoencoder *IEEE Access* **8** 47072–81
[10] Dai J, Wang J, Huang W, Shi J and Zhu Z 2020 Machinery health monitoring based on unsupervised feature learning via generative adversarial networks *IEEE/ASME Trans. Mechatronics* **25** 2252–63
[11] Dinh L, Krueger D and Bengio Y 2014 NICE: Non-linear independent components estimation *arXiv Prepr. arXiv1410.8516*
[12] Lugmayr A, Danelljan M, Van Gool L and Timofte R 2020 SRFlow: Learning the super-resolution space with normalizing flow *European Conference on Computer Vision* pp 715–32
[13] Serrà J, Pascual S and Segura C 2019 Blow: A single-scale hyperconditioned flow for non-parallel raw-audio voice conversion *arXiv Prepr. arXiv1906.00794*
[14] Kim H, Lee H, Kang W H, Cheon S J, Choi B J and Kim N S 2020 WaveNODE: A continuous normalizing flow for speech synthesis *arXiv Prepr. arXiv2006.04598*
[15] Yang H, Ding K, Qiu R C and Mi T 2020 Remaining useful life prediction based on normalizing flow embedded sequence-to-sequence learning *IEEE Trans. Reliab. Preprint* 10.1109/TR.2020.3010970
[16] Dinh L, Sohl-Dickstein J and Bengio S 2016 Density estimation using real NVP *arXiv Prepr. arXiv1605.08803*

IOP Conf. Series: Materials Science and Engineering 1207 (2021) 012020 doi:10.1088/1757-899X/1207/1/012020

Joint feature enhancement mapping and reservoir computing for improving fault diagnosis performance

L J Kong[1], Y W Huang[2], Q B Yu[1], J Y Long[2] and S Yang[1*]

[1]School of Management Science and Engineering, Chongqing Technology and Business University, Chongqing 400067, China
[2]School of Mechanical Engineering, Dongguan University of Technology, Dongguan 523808, China

E-mail: yangsmart@hotmail.com

Abstract. Complicated industrial robot structure and harsh working conditions may cause signal features collected in the condition monitoring process to be seriously disturbed. In this paper, a joint feature enhancement mapping and reservoir computing (FEM-RC) method is presented to handle the industrial robot fault diagnosis problem. Firstly, a feature enhancement mapping (FEM) method is proposed to achieve intraclass distance minimization and interclass distance equalization to obtain an enhanced feature matrix. Then, the first reservoir computing (RC) network is adopted to map the original feature matrix to the feature enhancement matrix, and the second RC network is for fault type classification. The results of the experiment carried out on a six-axial industrial robot demonstrate that compared with other peer models, the present FEM-RC has better fault diagnosis performance and robustness.

1. Introduction

Industrial robots often have complex structures and extreme working environments, which are prone to various forms of failure and corresponding costs [1, 2]. Therefore, it is of long-term significance to carry out condition monitoring and fault diagnosis on them. In recent years, the fault diagnosis of engineering system is mainly based on three ideas [3]: simulating the vibration response of the gearbox, fault simulation modelling under different loads; time-domain methods, frequency-domain methods, time-frequency domain methods, and other signal processing methods; data-driven intelligent diagnosis method. Specific signal processing analysis methods include short-time Fourier transform [4], wavelet analysis [5], wavelet packet transform [6], empirical mode decomposition [7], Winger-Ville distribution [8], matching pursuit method [9] and so on. However, for large-scale industrial rotating machinery systems (including industrial robots) with variable operating conditions, various fault types and unclear fault mechanisms, it is often difficult to ensure that the built diagnosis model can accurately extract features and mine rich potential information in big data. Thus, deep learning has become a research hotspot in the field of fault diagnosis. In actual work, feature selection and feature extraction [10, 11] are both methods to find the most effective features from the original data. Guo et al. [12] proposed a novel hierarchical learning rate adaptive deep convolution neural network based on the improved algorithm to achieve bearing fault pattern recognition and fault degree evaluation. Tang et al. [13] used a new fault diagnosis method named Fisher discriminative sparse representation based on deep belief network to obtain superior performance for feature extraction and classification in the field of complex system fault diagnosis. Shao et al. [14] proposed to construct a new deep auto encoder with noise reduction auto encoder and compressed auto encoder to enhance feature learning ability. He et al. [15] presented a new method that combines the convolutional neural network and long short-term memory network to perform a gradual changing fault classification.

Although these feature processing methods are powerful in some cases, there are still challenges in dealing with mechanical fault diagnosis. Especially deep learning models have the limitation of high computational complexity.

In this study, we proposed a feature enhancement mapping method to improve the performance of fault diagnosis. Firstly, a set of activation function mapping method is used between the original data and the class center to reduce the intraclass decision distance and form a new feature matrix. Then, each type of fault is distributed equally in the feature space to increase the decision distance between different categories to form a feature-enhanced matrix. Thirdly, reservoir computing (RC) is adopted to map the original feature matrix to the feature-enhanced matrix. Finally, put the feature-enhanced matrix into the other RC network for classification, which can achieve high class separability with a small computational complexity. In order to verify the performance of FEM-RC, it was applied to a six-axial industrial robot gearbox fault diagnosis experiment, and compared with the peer fault diagnosis models based on RC, stack autoencoder (SAE), and deep Boltzmann machine (DBM), respectively.

2. Model definition

This section describes the proposed FEM-RC model in detail. The detailed methods of the intraclass distance minimization and the interclass distance equalization are given in subsection 2.1. Mapping via RC network is realized in subsection 2.2. Application to intelligent fault diagnosis is implemented in subsection 2.3.

2.1. Intraclass distance minimization and interclass distance equalization

A feature enhancement mapping (FEM) approach is proposed for strengthening original characteristics expression. Establish feature enhancement mapping to reduce the intraclass distances of the same class in each dimension of the feature matrix to form a new class center matrix, and then adopt a balanced distribution algorithm to increase the distance between different classes in each dimension of the class center matrix. The transformation of training samples is shown in equation (1).

$$O = w_1 G \tag{1}$$

$$G = a(x) = \frac{1}{1+e^{-x}} \tag{2}$$

where $w1 \in R^{d \times d}$ represents the weight between raw sample data and the feature matrix reconstructed by FEM. G is the sigmoid function in activation functions shown in equation (2), $O \in R^{d \times 1}$ is the output layer and $x \in R^{d \times 1}$ is the input layer consisting of original data.

To achieve feature enhancement mapping, the process of intraclass distance minimization and interclass distance equalization in each dimension is shown in equation (3) and (4) separately.

$$q_i = \frac{1}{l}\sum_{j=1}^{l} x(j) \tag{3}$$

$$\bar{x}_i^{desired} = min\, q_i + (min\, q_i - max\, q_i) \times \frac{i-1}{c-1} \tag{4}$$

where $q_i$is the center of the i-th category, l is the sample of the i-th category and suppose that the input layer contains c categories (i= 1, 2, …, c). $\bar{x}_i^{desired} \in R^{dx1}$represents the sample of reconstructed feature matrix with even distribution in vector space. To acquire the w_1, the raw sample data and the feature matrix reconstructed by FEM can be represents by equations (5) and (6).

$$\bar{x}_i^{desired} = w_1 G \tag{5}$$

$$w_1 = (G^T G)^{-1} G^T\, \bar{x}_i^{desired} \tag{6}$$

To avoid under fitting and over fitting, the regularization factor is added in equation (7). The regression model with different λ value is trained to control the fitting trend of the model.

$$w_1 = (G^T G + \lambda e)^{-1} G^T \bar{x}_i^{desired} \quad (7)$$

where λ is the regularization factor and e is the unit diagonal matrix. Then, the feature matrix reconstructed by FEM can be calculated by equation (1).

2.2. *Mapping via reservoir computing*

The RC network is composed of input layer, reservoir layer and output layer. Figure 1 shows the network structure of the classic RC network.

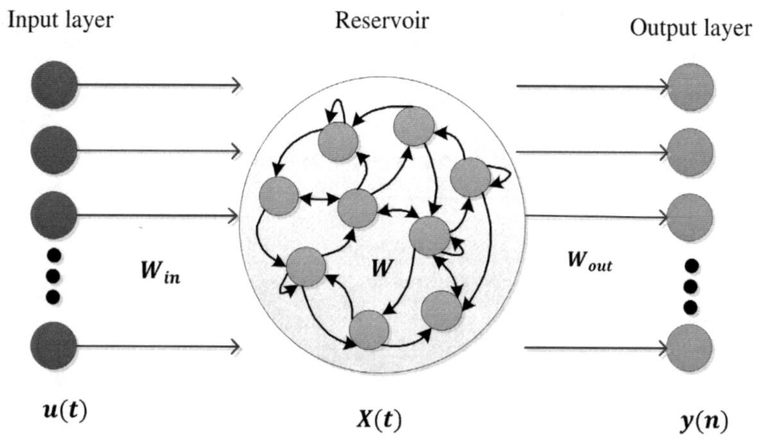

Figure 1. Structure of reservoir computing network.

At time t, the input vector, the reservoir state, and the output of the network are represented by $u(t) \in R^K$, $x(t) \in R^N$ and $y(t) \in R^L$ respectively. Correspondingly, the connection weights between the input layer and the reserve pool, the recursive connection weights inside the reservoir, and the connection weights between the reservoir and the output layer are represented by $W_{in} \in R^{N\times K}$, $W_r \in R^{N\times N}$ and $W_{out} \in R^{L*(N+K)}$ respectively. The state updating equation of the reservoir is shown in (8), and the output state updating equation of RC is shown in equation (9).

$$x(t) = f(W_r \times x(t-1) + W_{in} \times u(t)) \quad (8)$$

$$y(t) = F\left(W_{out} * \left(u(t); x(t)\right)\right) \quad (9)$$

where the activation functions $f(\cdot)$ and $F(\cdot)$ are tanh and identity respectively.

As shown in the optimization objective function of equation (10), the network output weight is obtained by minimizing the mean square error of the output of the RC network and the expected output. Suppose X denotes the network state matrix and Y denotes the desired output sequence matrix.

$$min\|W_{out}X - Y\|_2^2 + \gamma\|W_{out}\|_2^2 \quad (10)$$

where γ is regularization coefficient, ridge regression is used to solve for W_{out}:

$$W_{out} = YX^T(XX^T + \gamma I)^{-1} \quad (11)$$

2.3. *Application to intelligent fault diagnosis*

The application procedure of the FEM-RC model for industrial fault diagnosis is shown in figure 2, and the detailed steps are as follows:

Step1: Different data are collected according to the input variables of different fault categories to construct an input feature matrix.

Step2: Divide data into a training set and a testing set.

Step3: Adopt FEM to enhance the features of the training set and test set.

Step4: Perform the first RC network to map the training set to a uniformly distributed feature matrix.

Step5: Perform the second RC network to map the uniformly distributed feature matrix to the corresponding label and evaluate the model's classification performance.

Step6: Fed testing data into the trained FEM-RC model to classify fault state mode.

Step7: Output the classification result. End.

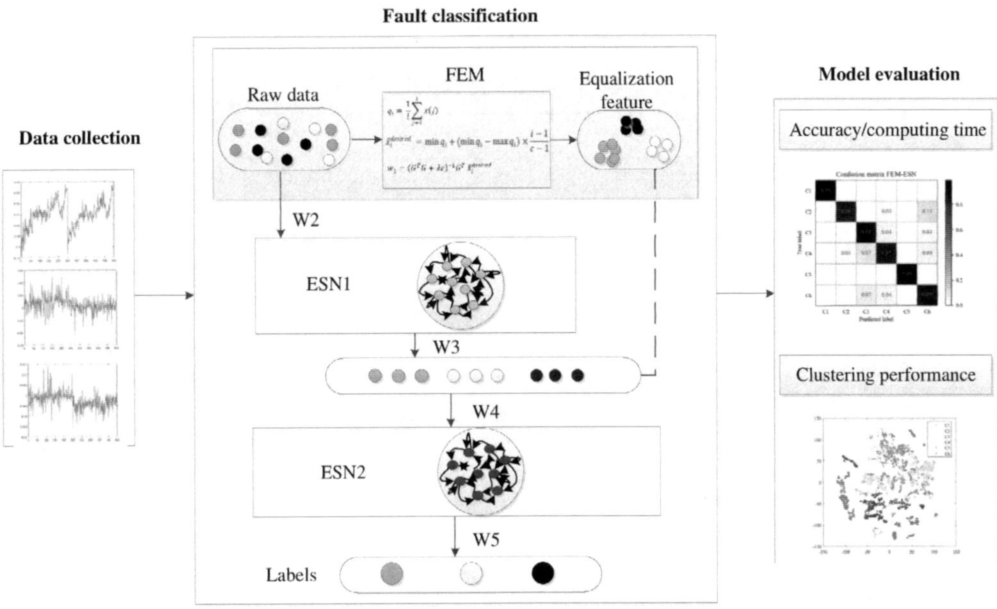

Figure 2. Application flow chart using FEM-RC model for intelligent fault diagnosis.

3. Experiments

To verify the performance of the proposed FEM-RC in fault diagnosis, it was applied to the health state diagnosis of a six-axial industrial robot and compared those with the other three models. Firstly, the confusion matrix is adopted to express the detailed classification accuracy of each model. Secondly, t-SNE (t-Distributed Stochastic Neighbor Embedding) is adopted to visualize the data distribution before and after FEM processing.

3.1. Dataset

The attitude signal of the industrial robot was collected through the WT901C-485 attitude sensor. The sampling frequency of the attitude sensor was 100 Hz, and the sampling time was 20s. Six failure state modes were defined in table 1. The data source information of every failure was detailed in table 2.

Finally, 13500k data points were collected in total, and each sample length was 900 points. In this way, 15000 samples can be obtained and every 2500 samples correspond to one kind of fault. Among them, half of the acquired samples were randomly selected as the training samples and the rest were used as the test samples.

Table 1. Different fault pattern settings of industrial robots.

Label	Fault type	Fault location	Fault degree

C1	Healthy	/	/
C2	Tooth breakage	The second axis RV40E-121 sun gear	Full broken tooth
C3	Tooth breakage	The second axis RV40E-121 planetary gear	Full broken tooth
C4	Tooth breakage	The third axis RV40E-121 sun gear	Full broken tooth
C5	Tooth breakage	The third axis RV40E-121 planetary gear	Full broken tooth
C6	crackle	The second axis RV40E-121 planetary gear	Width 0.5mm, depth 0.5mm

Table 2. Sample information.

Fault type	Number	Label
C1	2500	100000
C2	2500	010000
C3	2500	001000
C4	2500	000100
C5	2500	000010
C6	2500	000001

3.2. Model parameter settings

All the training and testing processes of the above models were programmed with matlab2018a and executed on a PC with Intel® Core (TM) i5-4590 CPU @ 3.3GHz processors and 16 GB RAM.

The main parameter settings of different models are in detailed in table 3 below. Especially, *I(.), H(.)* and *O(.)* represent the neurons in input layer, hidden layer and output layer.

Table 3. Different fault diagnosis models.

Model	Parameter setting	Number of neurons in each layer
RC	Spectral radius=0.1; activation='tanh'; sparse connectivity=0.01	I(900)-H(900)-O(6)
FEM-RC	Activation='sigmoid'; penalty coefficient=0.1	I(900)-H(900)-H(100)-H(900)-H(900)-O(6)
SAE	Learning rate=0.01; momentum=0.1; activation='sigmoid'; batch size=50	I(900)-H(200)-H(100)-H(50)-O(6)
DBM	Learning rate=0.001; iteration=100; activation='sigmoid'; batch size=50	I(900)-H(500)-H(200)-H(50)-O(6)

3.3. Performance evaluation

This section describes the performance of fault diagnosis using the proposed model from two aspects. The comparison of classification accuracy and confusion matrix of different fault diagnosis models are shown in table 4 and figure 3 respectively. Figure 4 introduces the evaluation and visualization of the clustering effect.

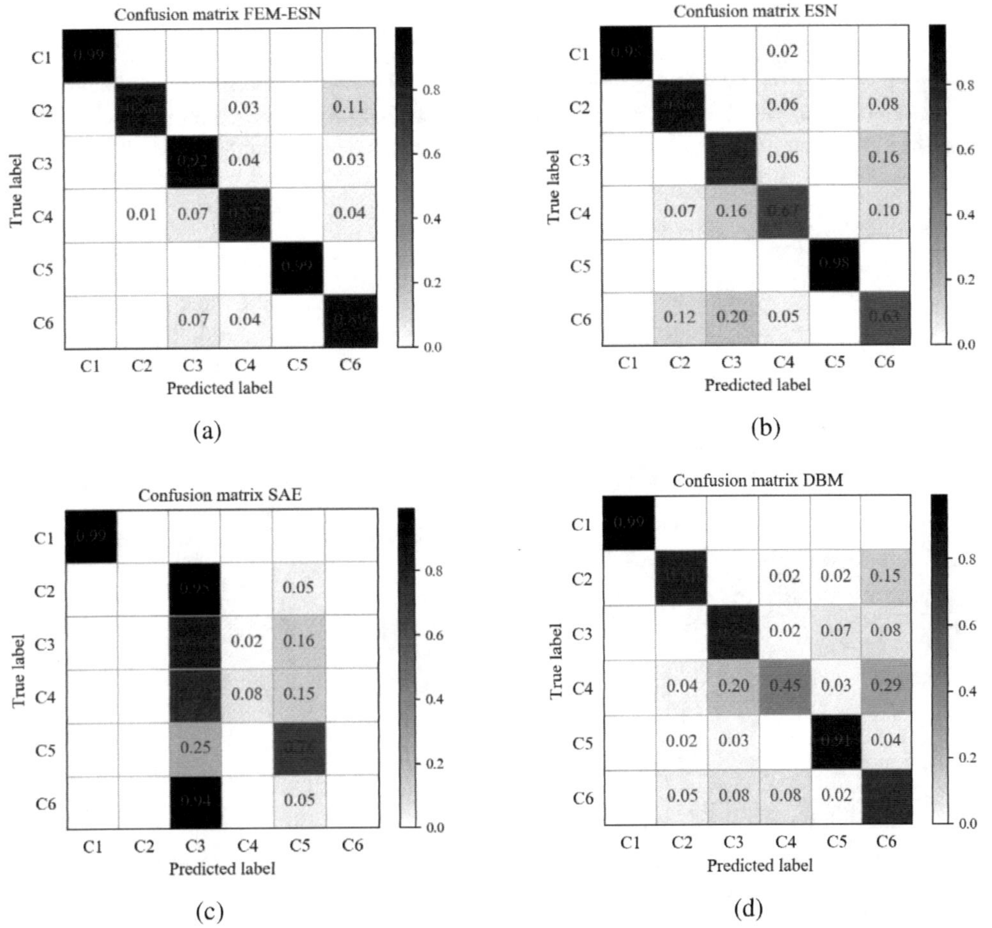

Figure 3. The confusion matrix for: (a)FEM-RC; (b)RC; (c)SAE; (d)DBM

Table 4. Different fault diagnosis models.

Model	Average accuracy (%)	Standard deviation	Computing time(s)
RC	81.70	0.53	196
FEM-RC	94.09	0.50	245
SAE	49.60	6.59	321
DBM	78.95	3.17	749

For each model, twenty trials are carried out under the same parameters, and the classification results are detailed as follows.

Table 4 shows the mean and standard deviation of classification accuracy and computing time. The fault diagnosis results given by the proposed FEM-RC model achieved the highest average accuracy with 94.09%. The average fault diagnosis accuracy is 81.7% by RC, 49.6% by SAE and 78.95% by DBM. Meanwhile, FEM-RC acquires the best robust with the lowest standard deviation among the five approaches. Table 4 also lists the time for FEM-RC model and other comparative models to perform a single fault diagnosis experiment. The results show that the average operation time of the FEM-RC model is 245s, and the average operation time of the SAE and DBM is 321s and 749s, respectively. It shows that the computational efficiency of FEM-RC reduces the complexity of fault diagnosis model.

Figure 3 shows the confusion matrix of each model at the highest accuracy in 20 experiments. Each row of the confusion matrix corresponds to the real value of the label, and each column corresponds to the predicted value of the model. When the ratio on the diagonal of the confusion matrix is higher, it means that the classification of each type of data is better. It can be seen from figure 3 that FEM-RC has a relatively high classification accuracy among five diagnosis models.

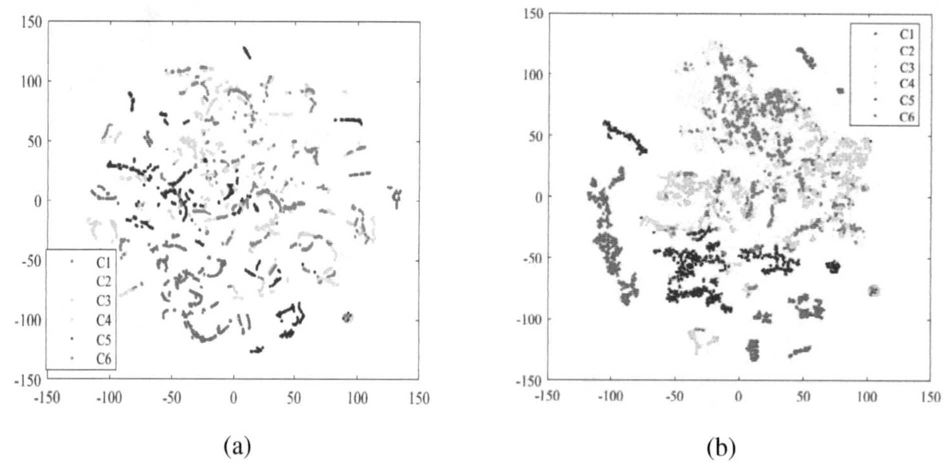

Figure 4. Visualization of sample feature data distribution: (a) raw data; (b) feature data after FEM processing

As shown in figure 4, t-SNE is used to analyze the feature enhancement effect of FEM on the original data more intuitively. Each color pattern in the picture represents a fault state mode (there are six different faults in this experiment). It can be seen that the original samples corresponding to each type of fault in figure 4(a) are irregularly mixed together. However, in figure 4(b), the data of the same data type begin to gather. Among them, samples of C0, C1, C4, C5 are obviously separated, and there is a clear boundary between each category. In conclusion, FEM can optimize the feature data and enhance the feature representation of the essential information of the data.

4. Conclusions

In this paper, a joint feature enhancement and reservoir computing (FEM-RC) approach has been proposed for industrial robot diagnosis. FEM was proposed to enhance the class features of the original data by minimizing the intraclass distance and equalizing the interclass distance, and then used the reserve pool to calculate the mapping relationship between the original data and the enhanced matrix. Finally, RC was used for simple and efficient classification to solve the problems of data redundancy and slow running speed caused by high-dimensional data.

Experiments have been conducted from two perspectives to evaluate the performance of the proposed FEM-RC. Firstly, three peer machine learning approaches were selected to compare and verify the fault diagnosis performance of FEM-RC. Secondly, t-SNE was adopted to evaluate the clustering effect that FEM could enhance the feature representation. Thirdly, confusion matrix detailed diagnostic accuracy to further validate the robustness of the proposed method. In these comparison experiments, the proposed FEM-RC approach achieved better performance.

Acknowledgments

This work was supported in part by the National Natural Science Foundation of China (51905058, 52175080), the Chongqing Natural Science Foundation (cstc2019jcyj-zdxmx0013, cstc2020jcyj-msxmX0182), the Guangdong Basic and Applied Basic Research Foundation (2019B1515120095), the Graduate Innovation Project of Chongqing Technology and Business University (yjscxx2021-112-27), and the Intelligent Manufacturing PHM Innovation Team Program (2018KCXTD029, TDYB2019010).

References

[1] Day C P 2018 Robotics in industry-their role in intelligent manufacturing. *Engineering* **4** 440-445

[2] Miao Y H, Zhang B Y, Yi Y G and Lin J 2021 Application of improved reweighted singular value decomposition for gearbox fault diagnosis based on built-in encoder information. *Measurement* **168** 108295

[3] Lei Y, Lin J, Zuo M J and He Z J 2014 Condition monitoring and fault diagnosis of planetary gearboxes: A review. *Measurement* **48** 292-305

[4] Ivan G, Hrvoje P and Damir N 2021 Fault detection in dc microgrids using short-time fourier transform. *Energies* **14** 277

[5] Zhu Y, Li G P, Wang R, Tang S N, Su H and Cao K 2021 Intelligent fault diagnosis of hydraulic piston pump based on wavelet analysis and improved alexnet. *Sensors* **21** 549

[6] Heidari M, Homaei H, Golestanian H and Heidari A 2016 Fault diagnosis of gearboxes using wavelet support vector machine, least square support vector machine and wavelet packet transform. *J. Vibroeng.* **18** 860-875

[7] Malik H and Sharma R 2017 EMD and ANN based intelligent fault diagnosis model for transmission line. *J. Intell. Fuzzy. Syst.* **32** 3043-50

[8] Xie P, Yang Y X, Jiang G Q, Du Y H and Li X L 2012 A new fault detection and diagnosis method based on Wigner-Ville Spectrum Entropy for the rolling bearing. *Appl. Mech. Mater.* **197** 346-350

[9] Tropp J A and Gilbert A C 2017 Signal recovery from random measurements via Orthogonal Matching Pursuit. *IEEE Trans. Inf. Theory* **53** 4655-466

[10] Nargesian F, Samulowitz H, Khurana U, Khalil E B and Turaga D 2017 Learning feature engineering for classification. *Int. Conf. on Ijcai* **352** 2529-35

[11] Hira Z M, Gillies D F 2015 A review of feature selection and feature extraction methods applied on microarray data. *Adv. Bioinform.* **2015** 1-3

[12] Guo X J, Chen L and Shen C Q 2016 Hierarchical adaptive deep convolution neural network and its application to bearing fault diagnosis. *Measurement* **93** 490-502

[13] Tang Q, Chai Y, Qu J F and Ren H 2018 Fisher discriminative sparse representation based on DBN for fault diagnosis of complex system. *Appl. Sci.* **8** 795

[14] Shao H D, Jiang H K and Wang F 2017 An enhancement deep feature fusion method for rotating machinery fault diagnosis. *Knowl. Based Syst.* **119** 200-220

[15] He Y F, Liu Y L, Shao S, Zhao X H, Liu G J, Kong X J, Liu L and Stoican F 2019 Application of CNN-LSTM in gradual changing fault diagnosis of rod pumping system. *Math. Probl. Eng.* **2019** 1-9

Simulation and experimental investigation of contact spot temperature for electrical contact components

Chao Zhang, Zien Zhao, Wanbin Ren*

School of Electrical Engineering and Automation, Harbin Institute of Technology, Harbin, China

*Corresponding author: renwanbin@hit.edu.cn

Abstract. The contact spot temperature of electrical contact components substantially affects the reliability and electrical life of any electrical connections within the electrical engineering. In this paper, finite element model of typical spring structure components is built by using COMSOL Multiphysics software. Furthermore, the transient process of contact temperature is simulated by taking account of film resistance on the contact surface. Moreover, a test rig is introduced that makes it possible to measure the electrical contact resistance and temperature within the electrical contact components simultaneously. Finally, correlation between contact resistance and contact spot temperature with different contact force and current levels are investigated explicitly.

Keywords: electrical contact; contact temperature; contact resistance; finite element simulation

1. Introduction

Current flowing through two separable electrical components is a typical electrical connection method within electrical apparatus. The features of contacts surface, rough and unclean, form the constriction resistance and film resistance of mated components. Contact resistance and contact temperature rise coexist for a long time and are influenced each other. Abnormal increase of contact resistance and contact temperature can induce electrical contact welding, short circuit, fire casualty and so on. The correlative factors include contact force, current, material physical properties and interface characteristics. Demand on low and stable contact temperature rise and contact resistance is increasing with the trend of system being adaptive in more complicated and extreme environments.

Relationship between temperature and contact voltage drop was firstly described by φ–θ relationship in Holm's classical theory in condition of symmetric contact, and Timsit proved the validity of the theory in last century [1,2]. In recent years, analysis about contact resistance gradually enriched the electrical contact theory, which has considered factors including constriction radius [3], *a*-spot thickness [4] and heterogeneous materials [5] etc. The essential character that conductive spots are hidden in contacts outside surfaces poses challenges to identify the conductive surface and to measure the contact temperature. Swingler [6,7] obtained pictures of conductive area of bolted contacts using CT scan technology. Sawada [8] researched the current density distribution in contact surface by introducing semiconductor wafer experiment. Although there are already reports about measuring instruments that can measure temperature and contact resistance, factors which correlative with temperature rise and contact resistance in electrical contact components are still insufficient.

Firstly, a new designed test rig for measuring the temperature rise and loop resistance of typical contact components simultaneously is introduced. The transient process of temperature rise and loop resistance is recorded with different contact forces and currents. In addition, the contact spot temperature rise is determined by comparing the finite element simulation results and the experimental results. The research method in this paper provide an alternative for evaluate the reliability of electrical contact components.

2. Numerical simulation of multi-physics field coupling

In order to accurately estimate temperature changes within the whole electrical contact components and contact spot temperature rise, this paper applies COMSOL Multiphysics 5.3a finite element software to establish a physical simulation model. 3D symmetric geometry and meshes are shown in Fig. 1.

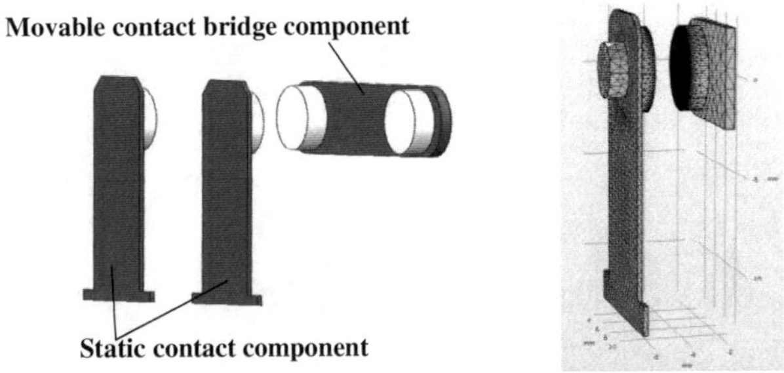

(a) Illustration of electrical contact components (b) Simulation model

Figure 1. 3D symmetric geometry and mesh

The material properties used in the simulation are shown in Table 1. Firstly, structure mechanics module is called for determining the real contact radius. On this basis, an electric-thermal coupling analysis is performed sequentially. The initial ambient temperature is 28℃, which is room temperature of real experimental conditions. Set heat transfer coefficient of the interface of model. Thereby electric potential and the contact spot temperature T_m within the contact region will be obtained, as well as the entire potential distribution and the temperature distribution within the model. To analyze the effect of film resistance on the temperature during the increasing process, the interface element of contact resistance between the contact surfaces is added. The film resistance can be simulated by setting the surface resistivity. The complete multi-physics coupling simulation process is shown in Fig.2.

Table 1. Material properties used in simulation calculation

Material	Beryllium bronze	$AgSnO_2$
Conductivity(S/m)	4.02×10^{-8}	2.4×10^{-8}
Density(kg/m^3)	8960	9800
Elastic modulus (GPa)	26.8	79
Poisson's ratio	0.35	0.37
Heat capacity (J/(kg·K))	385	272
Thermal conductivity (W/(m·K))	195	325
Melting Point(K)	1356	1233

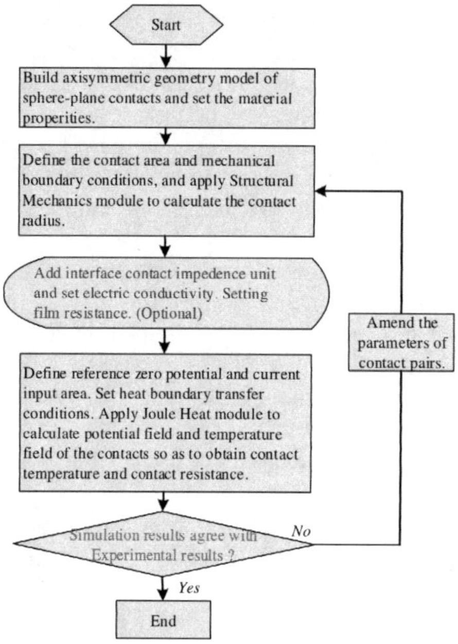

Figure 2. Simulation flow chart

The steady mechanical stress, electrical potential and temperature distribution of electrical contact components with carrying current of 20A and movable contact bridge over travel of 1.8mm is plotted in Fig.3. As shown, the maximum von mises stress reaches 53.34MPa, which is below the allowable stress of Beryllium bronze material. The maximum stress of contact region is only 3.06MPa, and the mechanical contact area of 0.27mm^2 could be exported from the results. Also, the distribution of electrical potential and current density flow route of cross section in Plane XY is illustrated in Fig.3(c). As expected, current density lines are obviously constricted into the mechanical contact zone, which results into the substantial potential drop. So, the maximum temperature of 317.69K also appears the center of mechanical contact zone. It should be noted that the contact spot temperature relates closely to the electrical contact situation. Once the temperature approaching the melting point of electrode material, the welding failure occurs certainly.

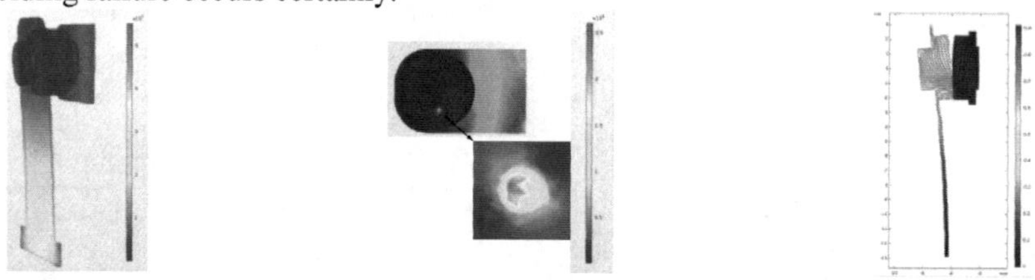

(a) Von Mises Stress (b) Zoom-in contact surface (c) Electrical current and potential

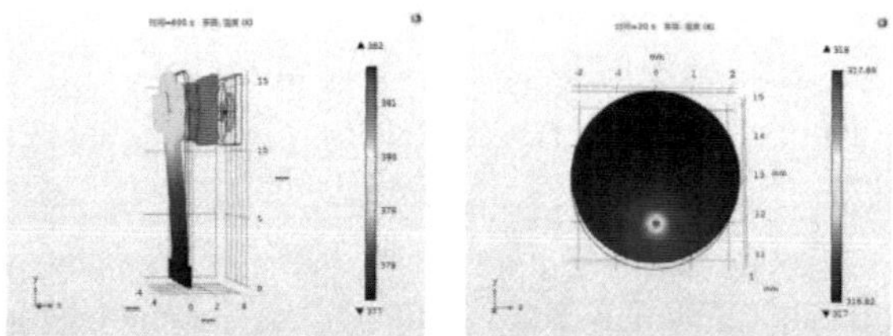

(d) Temperature distribution contour (e) Temperature situation of contact surface

Figure 3. Simulation results (convection coefficient is 14W/(K·m), and film resistivity is 0.1Ω·cm^2)

Fig.4 shows the variation in contact spot temperature rise as a function of time at 20A current loads. The steady temperature rise of contact spot is 61.75K. The changing rule of transient temperature rise follows negative exponent function, and the transient temperature rise of heating element can generally be written as

$$\tau = \tau_w - (\tau_w - \tau_0)exp(-t / T) \quad (1)$$

where τ_w is stable temperature rise, τ_0 is initial temperature rise, T is thermal time constant.

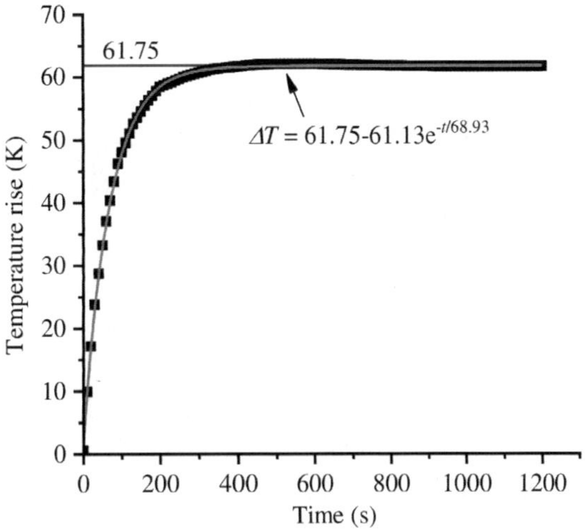

Figure 4. Variation in contact spot temperature rise as a function of time at 20A current loads

3. Experimental method

The new test rig is mainly composed of mechanical structure and measurement unit. The mechanical structure is mainly used for clamping real size electrical contact components and flexible mechanical loading in one dimensional direction. The maximum of loading cell is 5000cN and the sensitivity is 1.5 mV/V. Four-wire method is used to measure contact resistance with external controllable constant current source, and the current accuracy is 1%. Contact voltage and current are recorded by our designed signal amplification circuit. Temperature is measured by the miniature thermocouple electrical contacts. T end thermocouple is chosen, of which temperature measuring range is from -200℃ to 200℃ and the uncertainty is 0.2℃.

As shown in Fig.1(a), the typical electrical contact component includes two static contact components and one movable contact bridge component. Rivets are silver tin oxide material, and the leaf springs are Beryllium bronze material. Specimens were degreased using acetone, methyl alcohol and distilled water in an ultrasonic cleaner in order, dried and carefully mounted in the test rig. The thermocouple is installed at the bottom of leaf spring. In order to improve the stability of the measurement, and reduce disturbance by environment factors like airflow, and temperature change and micro vibration, the mechanical structure part adopts is under protection of vibration isolation platform and transparent organic glass housing. Schematic diagram of test rig is shown in Fig.5. Experimental condition is shown in Table 2.

Table 2. Experimental condition

Contact force	0.01~1N
Contact current	5~50A
Temperature	25±1℃
Humidity	61±2%RH

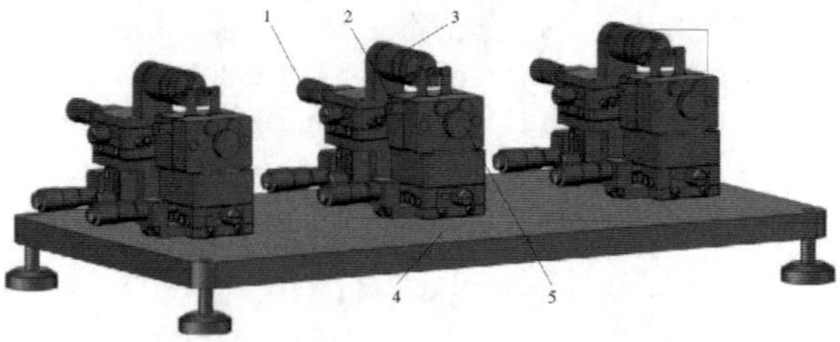

(a) Three work positions

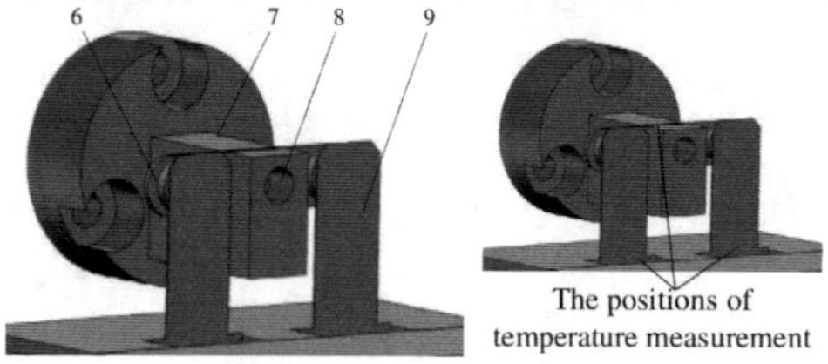

(b) Zoomed in view

Figure 5. Mechanical structure of test rig. 1. Linear translation stage for moveable bridge component 2. Copper terminal 3. Force transducer 4. Base plate 5.Clamping structure for static contact component 6. Moveable bridge component 7. Support 8. Jackscrew 9. Static contact component.

The mechanical load force is flexible adjusted by using the displacement control of movable bridge component during experiment condition setting process. The displacement reference zero point is defined as the contact force is 0.01N, meanwhile the loop resistance is 319mΩ.

4. Experimental results and discussion

Fig.6 illustrates the relationships between contact force, loop resistance and movable contact bridge displacement. As shown in Fig.6(a), the contact load force increases almost linearly with the displacement. However, the loop resistance drops significantly when the displacement increases from 0 to 0.2mm. After that threshold value, the decent rate slows down. The simulation results agree well with the experimental results.

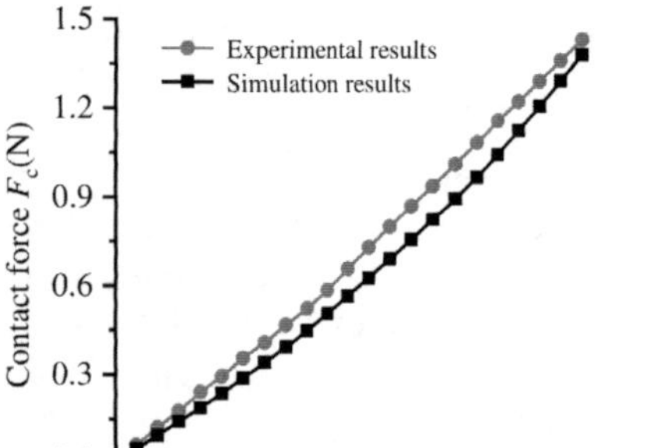

(a) Contact force

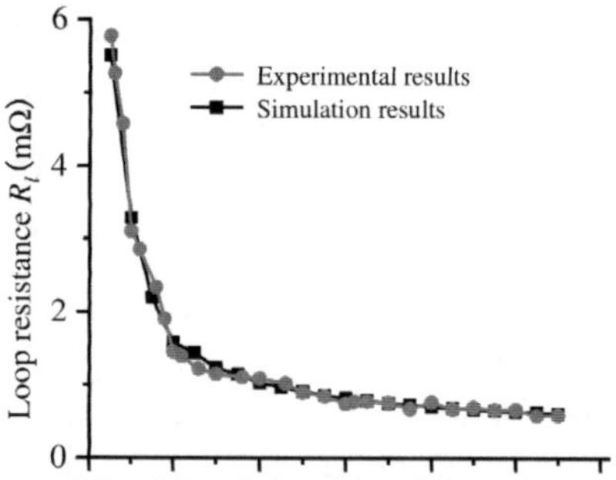

(b) Loop resistance

Figure 6. Variations in contact force and loop resistance as a function of movable contact bridge displacement

Fig.7(a) shows the measured transient temperature rise of static spring and movable spring. The steady temperature rise of movable spring end and static spring end are 19.9K and 14.8K individually. The corresponding thermal time constant is 265.8s and 469.2s respectively according to the exponential function fitting results. It indicates that the heat dissipation of movable spring component becomes easier.

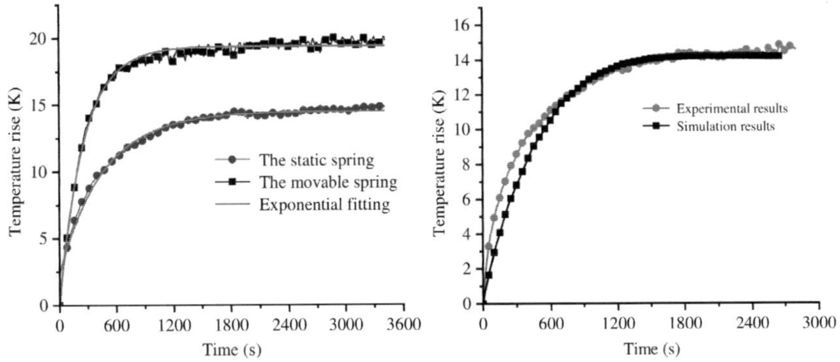

(a) Comparison of two measured positions (b) Comparison of experiment and simulation (static spring)

Figure 7. Variations in temperature rise of electrical contact components as a function of time

In order to reoccur the experimental results in simulation model, and further investigate the influencing factors with the use of FEM model, we build the same geometry model and use the associated material prosperities. However, the suitable convection coefficient is needed to improve the simulation accuracy. After several simulation attempts, the convection coefficient of 8.38W/(m^2·K) is determined as the optimal value. The simulation results of temperature rise with the use of such value is compared with the experimental results. And the fitting function of temperature rise with the use of experimental results is expressed as

$$\tau = 14.39 - 13.12e^{-\frac{t}{451.1}} \tag{2}$$

The fitting function of temperature rise with the use of simulation results is expressed as

$$\tau = 14.41 - 14.34e^{-\frac{t}{443.9}} \tag{3}$$

So, the simulation model and associated method is accurate enough to further analyze the influence relationship.

Variations in temperature rise as a function of time for current load of 20A and contact load force of 0.6N, 0.8N and 1N are plotted in Fig.8. As shown, the steady temperature rise is 17.75K, 16.08K and 14.46K individually when the contact force is 0.6N, 0.8N and 1N. And the thermal time constant remains about 450s. Therefore, the contact load force has negligible influence in the heat transfer process. Variations in temperature rise as a function of time for contact load force of 0.8N and current load of 10A, 20A and 30A are plotted in Fig.9. As shown, the steady temperature rise is 7.04K, 20.72K and 49.92K individually when the current load is 10A, 20A and 30A.

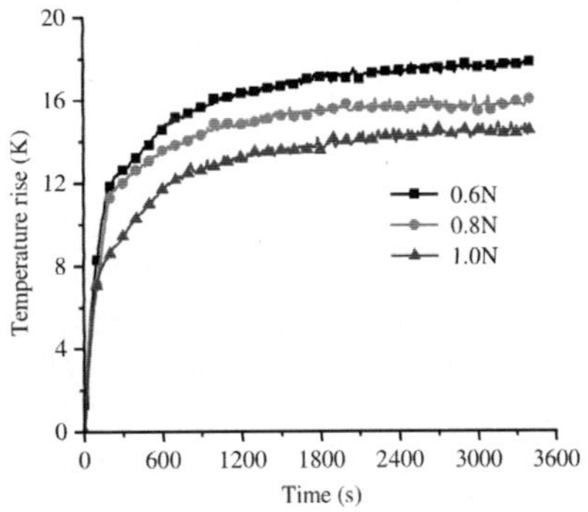

Figure 8. Effect of contact load force on temperature rise of the static spring end (current load is 20A)

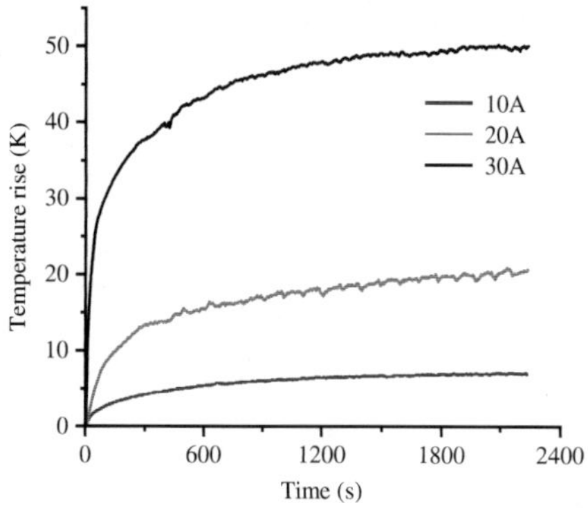

Figure 9. Effect of current load on temperature rise (contact load force is 0.8N)

5. Conclusions

This investigation mainly aims to determine the effect of contact force and current on contact spot temperature and contact resistance of spring contact component accurately. A new test rig, which could measure contact resistance and contact temperature simultaneously between contact pairs with changeable static contact force is presented. The direct measuring temperature method, that is implanting the thermocouple within the real size rivet contacts, is realized successful. A multi-physics field coupling simulation method is proposed based on the finite element software COMSOL Multiphysics 5.3a. Considered both experimental results and simulation results, the highest temperature that is the contact spot temperature is determined accurately. The heat dissipation of movable spring component is easier than the static one. The steady temperature and the thermal time constant of the static contact pair is negatively correlated with the contact force. Meanwhile, the stable temperature and thermal time constant is positively correlated with the contact current. Increase of contact force and current will result in decrement of the contact resistance.

Acknowledgments

The authors express their gratitude for the kind support provided by The National Natural Science Foundation of China (Contract Number 51777039 and 51377029).

References

[1] R.S. Timsit. On the evaluation of contact temperature from potential drop measurements. Proceedings of Holm conference on electrical contacts. 1982: 147-154.

[2] R.S. Timsit. The 'melting' voltage in electrical contacts. IEEE Transactions Comp., Hybrids, Manufact, Technol., 1991, 14: 285-292.

[3] P. Zhang, Y.Y. Lau, R.S. Timsit. On the spreading resistance of thin-film contacts. IEEE Transactions on electron devices, 2012, 59(7): 1936-1940.

[4] R.S. Timsit. Constriction resistance of thin-film contacts. IEEE Transactions Compon. Packag. Technol, 2010, 33(3): 636-642.

[5] Y. Sano, T. Kaneko. Applicability and correction of temperature-voltage relation in the case of point contact. IEEE Transactions on components, packaging, and manufacturing technology-Part A, 1998, 21(2): 345-351.

[6] A.V. Lalechos, J. Swingler, J. Crane. Visualization of the contact area using the x-ray computer tomography technique. Proceedings of IEEE Holm conference on electrical contacts. 2008: 263-269.

[7] J. Swingler, A Lalechos. Visualization and size distribution of contact spots at a real undismantled electrical contact interface. Journal of physics D: Applied physics, 2009, (42), 085304.

[8] S. Sawada, S. Tsukiji, T. Tamai, Y. Hattori. Constriction current behavior of oxide film effect observed by using LED wafer. Proceeings of ICEC-ICREPEC, 2012-13. 4

Exploiting graph neural network with one-shot learning for fault diagnosis of mechanical equipment

Shuai Yang [1,*], Xu Chen [1], Yun Bai [2]

[1]School of Chongqing Technology and Business University, Chongqing, China

*Corresponding author: jerryyang@ctbu.edu.cn

Abstract. For the classification of mechanical fault diagnosis, a graph neural network (GNN) method with one-shot learning is proposed. Convolutional Neural Network (CNN) is used to extract the feature vectors and One-Hot coding from images of Fault diagnosis of mechanical equipment. Inputting feature vectors and One-Hot coding into GNN, according to the Adjacency Matrix between vertices in the Graph, and is used for classification and inference. The method with one-shot learning is used for fault diagnosis classification. Through the fault classification for the industrial robot RV reducer and public data set CWRU pictures, the effectiveness of the method is verified. Five categories are used for fault diagnosis and classification in RV Reducer of the industrial robots. 80 categories are used in the public data set CWRU, and 55 categories are used as the training set. GNN is employed to spread the label information from the supervised sample of the unlabeled query data. The large-scale dataset can then be used as baseline classes to learn transferable knowledge for classifying novelties with one-shot samples. The one-shot learning with graph neural network GNN significantly improves the classification accuracy. The results show that the proposed method is superior to other similar methods and has a substantial potential for improvement in Fault diagnosis of mechanical equipment.

Keywords: Fault diagnosis; Graph neural network; One-shot learning; bearing; Industrial robot

1. Introduction

The early detection and diagnosis of mechanical equipment failure are essential to avoid catastrophic accidents[1]. For the classification of fault diagnosis of mechanical equipment, a large number of labeled image data is needed in the past, but sometimes the field is challenging, and experts with relevant experience are needed to label the data. This process is time-consuming and laborious. In this way, only a few pictures of each class need to be labeled by using one-shot Learning. Marking a few images in each category can automatically mark the remaining large numbers of images with high accuracy [2].

Because of the advantages of GNN in representing the sample relationship within and between classes, it has been widely concerned by researchers in recent years. The general graph structure cannot be directly applied to CNN, and a special graph convolution operation is required to get the hidden graph structure behind it. GNN is able to shift distance measures by the Euclidean Space and Non-Euclidean Space. Its structure is composed of nodes and edges, where each node has said a sample image, produced by the node adjacency matrix is expressed as the side, the relationship between the sample and by iteratively the execution node characteristics and the relationship between node updates, from the label information on the supervised sample will be spread to the unlabeled query data, In this way, the sample image without the label can be predicted to achieve image classification task[3, 4].

This paper is based on the deep feature learning of CNN and GNN to obtain more representation attribute features. The GNN is used to understand the node features and the adjacency matrix between nodes (the measure of similarity) and other relevant information. Based on the deep learning of multi-source high-level representation features, the fault causes of mechanical equipment can be accurately located [5]. Experimental results show that our method is more competitive in fault diagnosis by comparing with the other one-shot learning methods. In this paper, the GNN of the one-shot learning method is proposed to carry out mechanical fault diagnosis and classification [6].

There are two main contributions to the present work. On the first hand, one-shot learning is applied in mechanical fault classification and diagnosis. In One-Shot Learning, there is less training sample data for each category. If a multi-classification model is trained directly, it will be unable to train adequately due to the small number of samples of each category of mechanical equipment. One of the advantages of GNN is that it can do information diffusion through the connection between nodes. If each sample is regarded as a node in the graph, and the edge between nodes is a measure of their distance, then the label information of the existing label samples can be selectively spread to the most similar samples that need to be predicted according to the similarity between nodes. On the other hand, this paper is based on the one-shot learning method of meta-learning and then uses GNN to spread the label information from the supervised sample to the unlabeled query data. The large-scale dataset can then be used as base classes to learn transferable knowledge for classifying novelties with one-shot samples.

2. Methodology

In the first section, the one-dimensional signals of mechanical equipment are transformed into two-dimensional pictures by short-time Fourier transform (STFT). In the second section, feature vectors are extracted by CNN. The third section is to input the feature vector into the GNN and to apply the GNN to the learning with the one-shot sample. Finally, the prediction results are obtained for the classification of Fault diagnosis of mechanical equipment.

2.1. Transforming one-dimensional signal into two-dimensional image by STFT

Because one-dimensional data lacks frequency-domain signal, it is necessary to use short-time Fourier transform to transform one-dimensional data into two-dimensional data [7, 8].

The steps of STFT for the original signal x(t) are as follows:

First move the window to the starting point of the signal. At this time, the center position of the window function is at $t=\beth\_0$, and the signal is windowed:

$$y(t) = x(t) * w(t - \beth_0) \tag{1}$$

Then perform the Fourier transform:

$$X(\omega) = \mathcal{F}\big(x(t)\big) = \int_{\infty}^{+\infty} x(t) * w(t - \beth_0)e^{-jwt}dt \tag{2}$$

Thus, the spectrum distribution X(ω) of the first segmented sequence is obtained. In real applications, since the signal is a discrete point sequence, what we get is a spectrum sequence X[N].

For the convenience of expression, we define the function here, which represents the spectrum result after transformation of the original function at the center of the window function, namely:

$$S(\omega, \beth) = \mathcal{F}\big(x(t) * w(t - \beth)\big) = \int_{\infty}^{+\infty} x(t) * w(t - \beth_0)e^{-jwt}dt \tag{3}$$

As shown in Fig. 1, Repeat the above operation by constantly sliding the window and FFT, and finally get the spectrum results of all the segments from $\beth_0 \sim \beth_N$, and finally we get S, which is the result of STFT transformation. Thus, the one-dimensional data is converted into the picture form of the two-dimensional data.

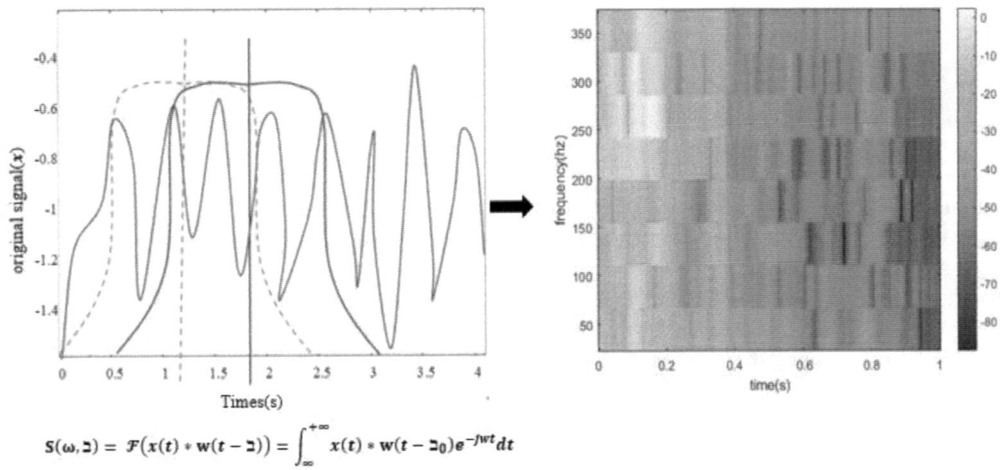

Fig 1. Principle of short-time Fourier transform

2.2. Using CNN to extract feature vectors

The actual process of feature extraction (training) and matching feature (recognition) by CNN is the dot product of filters with different weight parameters and the input region block. The convolutional layer and the pooling layer can extract image features, and finally determine the parameters of the convolutional kernel through backpropagation to obtain the final features, which is a rough process of CNN feature extraction [9], The size of one batch is [image-channel, batch-size, image-h], each sample is an image, including n-way * n-shot supported samples and a query sample.

2.3. Fault classification using graph neural network learned with one-shot learning

GNN is a graph model formed by many nodes and edges. It maps every image in the training data to a vertex on the graph. Through training, it obtains the adjacency matrix between the vertices in the graph and uses it to classify and infer. In this paper, each node represents an image, and the weight of each edge represents the relationship between the two images (distance or similarity) [10]. Specific weight calculation process:

$$\tilde{A}_{i,j}^{(k)} = \varphi_{\tilde{\theta}}(x_i^{(k)}, x_j^{(k)}) = \mathrm{MLP}_{\tilde{\theta}}(abs(x_i^{(k)} - x_j^{(k)})) \quad (4)$$

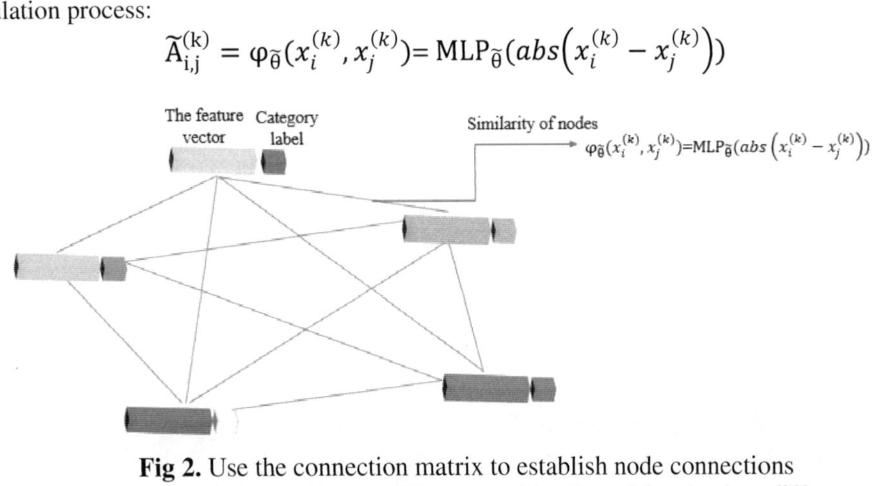

Fig 2. Use the connection matrix to establish node connections

As shown in Fig. 2, It is learned by a nonlinear combination of the absolute differences between two node features. By this architecture, the symmetric distance characteristic is satisfied by construction $\varphi_{\tilde{\theta}}(a, b) = \varphi_{\tilde{\theta}}(b, a)$, and it is easy to learn its own distance characteristic $\varphi_{\tilde{\theta}}(a, a) = 0$, and output the corresponding weight value.

Softmax each row of the computed pass the data and transformation to Graph Conv. Gc apply these transformations to embedding and then go through a fully connected layer and LeakyRelu Generate a new embedding to ensure that the sum of the weights between each node and all other nodes is 1, and then after the adjacency matrix is obtained, GNN can be used to calculate the next layer of the network to complete the GNN transfer, The calculation process is as follows.

$$x_l^{(k+1)} = Gc\left(x^{(k)}\right) = \rho\left(\sum_{B \epsilon A} Bx^{(k)}\, \theta_{B,l}^{(k)}\right), l = d_1 \dots d_{k+1} \quad (5)$$

The accumulation symbol indicates that the adjacency matrix B can adopt a variety of calculation methods, and add them together. According to this formula, the update rule of the node feature can be obtained.

As shown in Fig. 3, due to the denseness of the edges in the graph, the depth is simply interpreted as giving the model more expressive power. However, there are only two cases in this paper: the weight of the connection between the node and itself is 1, and the weight of the connection with other nodes is $\widetilde{A}_{i,j}^{(k)}$. In the training process, it is also necessary to change the weight of each layer of the network. The input $V^{(k)}$ and the output of the Gc block are cascaded to form the lower-level network input $V^{(k+1)}$, (this step is not required for testing).

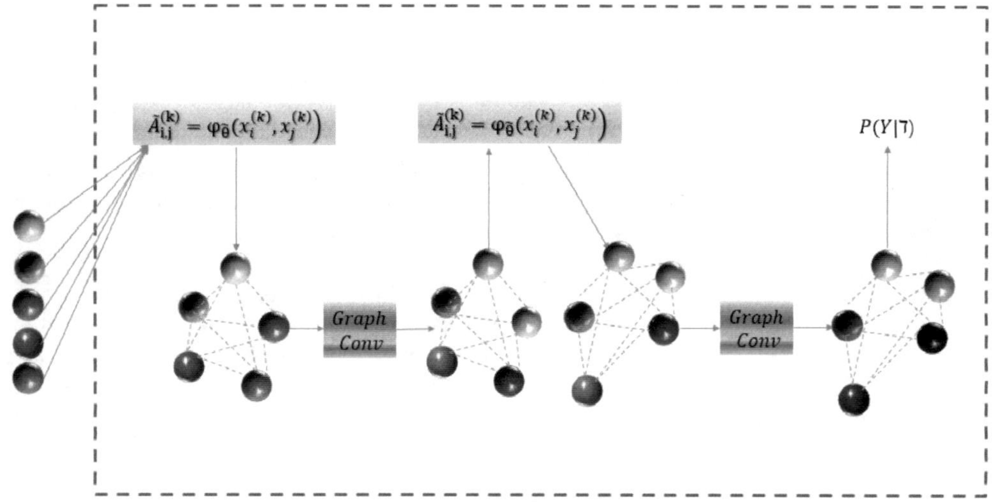

Fig 3. Graph Neural Network illustration

The initial point feature is defined as

$$x_i^{(0)} = (\phi(x_i), h(l_i)) \quad (6)$$

Where ϕ(·) (a CNN) is a way to extract semantic features from the image, and h(·) represents the translation of the tag into a one-hot vector.

The final loss function

$$\min \frac{1}{L} \sum_{i \leq L} \ell(\Phi(ך; \theta), Y_l) + \mathcal{R}(\theta) \quad (7)$$

$$\ell(\Phi(ך; \theta), Y) = -\sum y_k log P(Y_* = y_k | ך) \quad (8)$$

Where $\Phi(ך; \theta) = P(Y|ך)$, the predicted label is obtained through maximum likelihood estimation.

By using GNN and small one-shot learning in meta-learning, it can be applied to the fault diagnosis of mechanical equipment. The details are illustrated in Fig. 4 and summarized below.

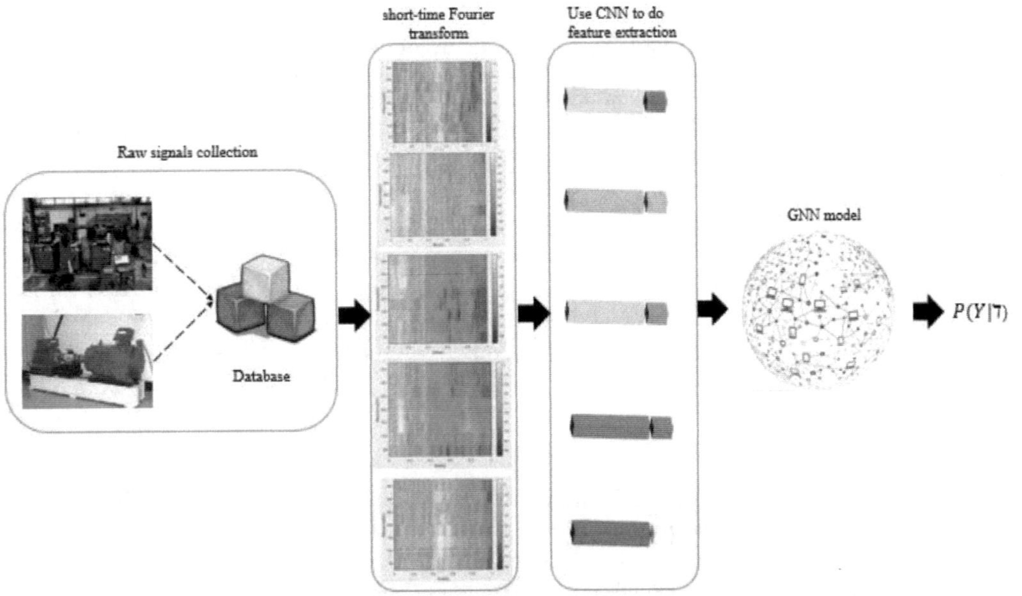

Fig 4. An overview of the proposed approach for classification of Fault diagnosis of mechanical equipment

3. Results and discussion

Siamese Net, SAE+RF and GNN were used to classify the industrial robot data. Tab. 1 shows the comparison of the different models. Obviously, GNN had the highest classification rate each time (96.43% on average). The accuracy of Siamese Net was 80.86%, and the performance of SAE+RF was the worst, only 31.17% on average. It was proved that GNN is the best classification method for industrial robot fault diagnosis. Conversely, SAE+RF was worst at classifying one-shot samples of data for fault diagnosis.

Table 1. Fault diagnosis accuracy of the comparisons on the industrial robot dataset

Model	Accuracy of each test (%)					Mean (%)
	1	2	3	4	5	
Siamese Net	79.40	83.25	79.30	82.05	80.30	80.86
SAE+RF	29.87	29.63	31.43	32.00	32.93	31.17
GNN	**96.56**	**96.63**	**96.26**	**95.91**	**96.81**	**96.43**

As shown in Tab.2, Similarly, Siamese Net, SAE+RF and GNN were also classified for CWRU dataset. The absolute advantage of GNN can also be found.

Table 2. Fault diagnosis accuracy of the comparisons on the CWRU dataset

Model	Accuracy of each test (%)					Mean (%)
	1	2	3	4	5	
Siamese Net	81.20	82.10	81.60	81.05	81.90	81.57
SAE+RF	24.67	24.89	25.56	22.22	WRU	24.62
GNN	**95.42**	**95.39**	**96.29**	**96.27**	**96.54**	**95.98**

4. Conclusions

A classification method for fault diagnosis of mechanical equipment is proposed in this paper. This method can improve the classification accuracy of fault diagnosis. The main contributions of this paper can be summarized as follows: First, the application of small sample research and learning methods in mechanical fault classification and diagnosis. In one-shot Learning, the training sample data of each category is less. One of the advantages of GNN is that information can be diffused through the connections between nodes. If each sample is regarded as a node in the graph, and the edge between the nodes is a measure of their distance, then the existing label can be used. According to the similarity between the nodes, the label information of the samples is selectively diffused to the most similar samples that need to be predicted. In the second part, a novel fault recognition model is constructed. Two different networks (CNN, GNN) carry out joint training examples. One is the general image convolutional neural network to extract the image feature expression; The other is the convolutional graph neural network, which completes the graph learning by constructing Adjacency Matrix on the characteristic difference of nodes.

Acknowledgments

This work is supported by the National Natural Science Foundation of China (51905058), Natural Science Foundation of Chongqing(cstc2020jcyj-msxmX0182,cstc2019jcyj-zdxmX0013), Key Project of Technology Innovation and Application of Chongqing(cstc2019jscx-fxydX0077), Research Start-Up Funds of Chongqing Technology and Business University (1856018, 1752042), Key Research Platform Project of Chongqing Technology and Business University (ZDPTTD201918). The authors also gratefully acknowledge the helpful comments and suggestions of the reviewers, which have improved the presentation.

References

[1] ZHANG X T，QIANG Y T，SUNG F，et al. RelationNet2：deep comparison columns for few-shot learning, 2019-11-10.
[2] Aesup Kim, Jaesik Yoon, Ousmane Dia, Sungwoong Kim,Yoshua Bengio, and Sungjin Ahn. Bayesian model-agnostic meta-learning. In NIPS, 2019.
[3] Koch G, Zemel R, Salakhutdinov R. Siamese neural networks for one-shot image recognition[OL]. 2020-08-31.
[4] Rao K R, Kim D N, Hwang J J. Fast fourier transform: algorithm and application [M]. China Machine Press, 2013:41-45.
[5] V. Garcia and J. Bruna. Few-shot learning with graph neural networks. arXiv preprint arXiv:1711.04043, 2017
[6] Y. LeCun, Y. Bengio, and G. Hinton. Deep learning. nature,521(7553):436, 2015.
[7] Gao, Tianyu, et al. "Hybrid Attention-Based Prototypical Networks for Noisy Few- Shot Relation Classification." (2019)
[8] MUNKHDALAI T, HONG Yu. Meta networks[C]//2017 International Conference on Machine Learning. Sydney, Australia.PMLR, 2017: 2554-2563.
[9] Finn, Chelsea, Pieter Abbeel, and Sergey Levine. "Model-agnostic meta-learning for fast adaptation of deep networks." Proceedings of the 34th International Conference on Machine Learning-Volume 70. JMLR. org, 2017.
[10] Santoro A, Bartunov S, Botvinick M, et al. One-shot learning with memory-augmented neural networks[J]. arXiv preprint arXiv:1605.06065, 2016.

A hybrid approach for modelling multidimensional degradation processes

Renyan Jiang and Binbin Xiong

School of Science and Technology, Changsha University, Changsha, China

Email: jiang@csust.edu.cn

Abstract. Degradation processes are often multidimensional. Modeling such degradation processes needs to address two key issues: indicator fusion and degradation model selection; and they have been separately addressed in the literature. This paper proposes a hybrid approach to jointly address these two issues. The proposed approach first fuses multiple degradation indicators into a composite degradation indicator. This fusion step involves data normalization, aggregation model selection and determination of indicator weights. After the fusion step, the problem becomes one-dimensional, and the existing method to select the degradation model for a one-dimensional degradation process can be applied. The resulting model obtained from the proposed approach can be a two-phase model; and the model for the second phase has a closed-form expression. This considerably facilitates residual life prediction. A real-world example is included to illustrate the proposed approach and its appropriateness.

1. Introduction

Degradation processes are often multidimensional [1-3], and a fusion technique is usually applied to combine all the degradation indicators into a composite degradation indicator [4]. The indicator fusion needs to address several issues such as data normalization, aggregation model selection and determination of indicator weights. To reduce the subjectivity, the weights are usually derived using a mathematical method [5]; and this may require estimating the time to the failure threshold of a degradation path using interpolation or extrapolation.

The stochastic process models for modeling one-dimensional degradation processes can be age-, state- and age-state-dependent. The traditional methods to build such models are usually to directly fit a set of degradation data to an assumed model. Since the assumption on the degradation model can be unrealistic, the resulting model can be unreliable. To improve, Jiang presents an innovative approach to aid model selection and the basic idea is as follows [6]. A one-dimensional degradation process can be graphically displayed in a two-dimensional plane, where the horizontal axis (or simply h-axis) can be age or state variable and the vertical axis (or simply v-axis) can be degradation rate or its reciprocal. Thus, there are four coordinate systems for selection. For each coordinate system, the sample correlation coefficient (CC) and the average of five-point moving coefficients of variation (CV) of the ordered data can be calculated. The best coordinate system has the largest |CC| (i.e., good linearity) and the smallest CV (i.e., small variability). In such a coordinate system, the degradation rate or its reciprocal can be approximately represented by a simple function of age or state variable so that the mean degradation function has a closed-form expression. This will considerably facilitate residual life prediction and ensure good prediction accuracy.

According the above discussion, the indicator fusion and coordinate system selection have been separately addressed in [5-6]. The purpose of this paper is to propose a hybrid approach to jointly address these two problems. The proposed approach consists of three main steps as shown in figure 1.

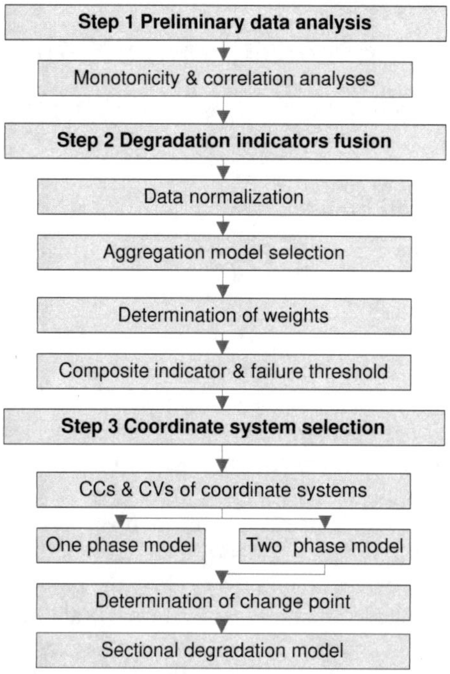

Figure 1. Proposed three-step modeling framework.

In the first step, a preliminary data analysis will be carried out to examine the monotonicity of degradation indicators and their dependence. The second step combines all the degradation indicators into a composite degradation indicator. This will involve several issues such as data normalization, aggregation model selection, determination of indicator weights and failure threshold modeling. The weights are mathematically derived and the dependence is handled according to a competing risk approach. Finally, the coordinate system for displaying the degradation process of composite indicator is selected based on the value of |CC/CV| in the third step. The resulting model obtained from the proposed approach can be a two-phase model, which is typically a mixture of the delay time model and stochastic model [7], and analytically tractable. A real-world example is included to illustrate the proposed approach and its appropriateness. Thus, the main contribution of the paper is the three-step modelling framework shown in figure 1, which is generally applicable to multidimensional degradation processes. In addition, a weighted linear method will be developed to estimate pseudo-failure time, which is applicable for the situation where the extrapolation is needed; and a method to estimate the phase boundary will be developed, which is applicable for the situations where the degradation rate curve in the best coordinate system is S-shaped or inverse S-shaped.

The paper is organized as follows. The proposed approach is presented in Section 2 and illustrated in Section 3. The paper is concluded in Section 4.

2. PROPOSED APPROACH

2.1. Preliminary data analysis

Let

$$y_{ij}^{(k)};\ i = 1, 2, \ldots, m, j = 1, 2, \ldots, n, k = 1, 2, \ldots, K \tag{1}$$

denote the observed value of degradation indicator Y_i of the kth unit at time point t_j. Without loss of generality, assume that $Y_i(0) = 0$ and $Y_i(t)$ has an increasing trend though it may locally decrease. Let L_i denote the failure threshold of Y_i.

A preliminary data analysis can be carried out based on the information in equation (1). For any unit and t_j, if $y_{ij}^{(k)} - y_{i,j-1}^{(k)} \geq 0$, Y_i can be considered to be monotonic.

Let r_{il} denote the CC between Y_i and Y_l. Its critical value is given by [8]

$$r_c = t_c / \sqrt{1+t_c^2} \tag{2}$$

where t_c is the Student's t-value associated with the 95% level, one tail, and the degrees of freedom nK - 1. If $r_{il} > r_c$, Y_i can be thought to be linearly correlated with Y_l; otherwise, they can be thought to be mutually independent.

2.2. *Indicator fusion*

According to [4-5], the indicator fusion deals with specifying the two relations in equation (3):

$$z = \varphi(y_i^*; w_i),\ L(z) = \phi(z; \mu_z, \sigma_z) \tag{3}$$

where z is the composite indicator, $\varphi(.)$ is an aggregation model to be specified, y_i^* is the normalized value of y_i, and w_i is the weight of Y_i, which meets

$$w_i \in (0, 1),\ \sum_{i=1}^{m} w_i = 1 \tag{4}$$

$L(.)$ is the normal random failure threshold of composite indicator Z, $\phi(.)$ is the normal density function with parameters μ_z and σ_z. The indicator fusion involves the following four issues

- Normalization of the data;
- Determination of indicators' weights;
- Selection of the aggregation model;
- Determination of parameters of the random failure threshold model.

Since the method to normalize the data may change the weights in an implicit way [9-10], it is necessary to ensure their compatibility.

2.2.1. Equal-failure-threshold transformation. For the above setting, the data is normalized by:

$$y_i^* = y_i / L_i \tag{5}$$

This transformation has the following features:

- y_i^* is dimensionless;
- all the normalized indicators have a common failure threshold 1;
- the time for Y_i to reach L_i (denoted as T_i) is equal to the time for Y_i^* to reach 1, implying that the transformation does not impact the distribution of T_i, and hence does not impact the weights derived from the probability method of [9].

2.2.2. Determination of indicators' weights. Let $F_i(t)$ and $f_i(t)$ denote the cumulative distribution function and density function of T_i, respectively. The probability of event ($T_i < T_l$) is given by

$$p_{il} = \int_0^{\infty} [1 - F_l(t)] f_i(t) dt \tag{6}$$

Specially, when $i = l$, $p_{il} = 0.5$. When T_i is small, Y_i is an important indicator. Therefore, a small T_i corresponds to a large p_{il} and a large w_i. For a fixed l value ($l = 1, 2, \ldots,$ or m), w_i is calculated by

$$w_i = p_{il} / \sum_{l=1}^{m} p_{ll} \tag{7}$$

The problem is now to determine $F_i(t)$. This needs to estimate the pseudo-failure time of each degradation path [11-12]. Reference [12] uses a weighted least squares method (LSM) to fit a degradation path to a power-law model and calls the model as the local mean degradation model. The weight function is given by:

$$w(t) = \exp[-(t-\mu)^2 / (2\sigma^2)] \tag{8}$$

where $\mu = t_n$ and $\sigma = \mu/3$. The pseudo-failure time is the intersection between the fitted power-law model and the degradation limit. Since the estimation for the pseudo-failure time is generally based on extrapolation, a large weight should be assigned to the recent observations. Equation (8) can meet this requirement well. Jiang and Huang use a cross-validation approach to optimize the value of σ [13].

To make the estimated pseudo-failure time robust, the local mean degradation model is taken as a linear model, given by

$$y_p(t) = a + bt \tag{9}$$

To make the estimated pseudo-failure time accurate, the fitted local mean degradation model should be based on the last three observations. Thus, the value of σ is taken as

$$\sigma = (t_n - t_{n-3}) / 3 \tag{10}$$

The values of a and b are determined through minimizing the following

$$SSE = \sum_{j=1}^{n} w(t_j)[y_j - y_p(t_j)]^2 \tag{11}$$

and the pseudo-failure time is estimated as

$$t_p = (L-a)/b \tag{12}$$

Thus, $F_i(t)$ can be obtained through fitting m pseudo-failure times to a proper distribution; and the weight of indicator can be obtained from equation (6) and equation (7).

2.2.3. Aggregation model. Reference [4] presents four categories of aggregation models; and the most widely used model is the generalized weighted mean model, given by [14]

$$z_p = (\sum_{i=1}^{m} w_i y_i^{*p})^{1/p},\ p \in (-\infty, \infty) \tag{13}$$

When $p = 0$, 1 and 2, the model reduces into the weighted geometric mean, weighted arithmetic mean and weighted quadratic mean, respectively.

Aggregation model selection needs to consider the compensation among the indicators [4]. If the indicators are highly correlated, the weighted arithmetic mean model with a full compensability is proper; otherwise, an aggregation model with a relatively weak compensability is desired, and the aggregated value should be closer to the value of the poorer indicator. For example, for a non-decreasing degradation path, the aggregated value should be larger than the weighted arithmetic mean value. This requires $p > 1$.

2.2.4. Determination of random failure threshold. The time to failure for a unit can be defined as t_f, which meets

$$t_f = \min_i(t_i^*) \tag{14}$$

where t_i^* corresponds to $y_i^* = 1$. Equation (14) is actually a competing risk model and applicable for both independent and dependent cases. The value of Z at $t = t_f$ corresponds to the failure threshold of the composite indicatior, which is generally not a constant. This implies that the failure threshold of the composite indicatior is a random variable. Without loss of generality, it can be assumed to follow a normal distribution $\Phi(z; \mu_z, \sigma_z)$, where μ_z and σ_z are distribution parameters.

To estimate μ_z and σ_z, the data of $Z(t_f)$, (z, s), are extracted from equation (1). If $\max_i(y_i^*) > 1$ [$\max_i(y_i^*) < 1$], then $s = 1$ [$s = 0$] and the corresponding z is a left [right] censored observation of $Z(t_f)$. The log-likelihood function is given by

$$\ln(L) = \sum_{s=1} \ln[\Phi(z)] + \sum_{s=0} \ln[1-\Phi(z)] \tag{15}$$

The parameters are determined through maximizing ln(L).

2.3. Coordinate system selection for composite indicator

The data of composite indicator can be written as

$$z_j^{(k)}; j = 1, 2, \ldots, n, k = 1, 2, \ldots, K \tag{16}$$

The data are transformed into degradation rate data given by [6]

$$r_j^{(k)} = [z_j^{(k)} - z_{j-1}^{(k)}]/(t_j^{(k)} - t_{j-1}^{(k)}),\ z_j = [z_j^{(k)} + z_{j-1}^{(k)}]/2,\ t_j = (t_{j-1}^{(k)} + t_j^{(k)})/2 \tag{17}$$

Here, $r_j^{(k)}$ is the degradation rate at age t_j and state z_j, and the sample size of the dataset given by equation (17) is nK. The data can be displayed in a two-dimensional plane with the h-axis being t_j or z_j and the v-axis being r_j or $1/r_j$. For each coordinate system, calculate the values of CC and CV. The best coordinate system corresponds to the largest value of |CC/CV|. The degradation model is built based on the information of the data plot in the best coordinate system. This is illustrated in the next section.

3. Illustration

3.1. a Data and earier model

Xu et al. deal with the reliability evaluation of a certain product. The degradation indicators are capacitance (Y_1) and wear amount (Y_2), increase with time and have zero initial values [15]. The failure thresholds are $L_1 = 14$ and $L_2 = 100$, respectively. Table 1 shows the observed degradation paths.

TABLE 1. Degradation paths of Y_1 and Y_2.

Time	Unit	1	2	3	4	5	6	7	8	9	10
1	Y_1	0.45	0.76	0.64	0.40	0.68	0.83	0.34	0.27	0.68	0.40
	Y_2	0.55	0.80	0.46	0.20	0.10	2.96	0.42	0.74	1.53	3.21
2	Y_1	2.19	0.79	0.56	0.56	1.49	1.53	0.95	1.51	1.33	1.41
	Y_2	1.60	3.42	7.92	16.85	11.88	8.93	8.91	10.97	12.21	5.29
3	Y_1	4.14	2.45	1.21	2.87	3.03	2.62	1.24	3.50	2.21	2.28
	Y_2	7.11	9.43	24.63	18.3	14.94	12.35	14.92	11.97	18.65	21.98
4	Y_1	5.28	3.72	3.81	4.90	4.10	4.81	2.12	3.91	3.03	3.00
	Y_2	21.70	19.09	38.08	20.81	19.85	18.44	36.10	16.40	20.12	26.29
5	Y_1	6.46	4.95	5.68	5.67	6.09	5.86	3.76	6.49	4.40	4.63
	Y_2	29.66	22.53	44.60	28.56	52.27	26.41	43.97	19.63	34.71	33.80
6	Y_1	7.35	6.53	6.49	6.95	6.96	7.34	4.51	8.54	6.00	5.07
	Y_2	38.71	32.09	59.35	34.24	64.06	30.18	55.57	24.26	41.65	41.57
7	Y_1	7.35	6.88	7.29	8.47	8.79	7.92	4.63	9.68	6.19	6.42
	Y_2	49.04	38.16	65.21	45.86	76.69	45.30	61.80	26.53	45.37	48.30
8	Y_1	8.05	8.88	8.41	10.16	9.65	10.97	6.35	11.10	6.76	8.33
	Y_2	54.07	47.54	67.70	51.80	87.62	52.07	80.72	36.77	51.89	52.53
9	Y_1	8.88	9.65	9.57	11.36	10.88	14.11	8.64	13.98	6.27	9.85
	Y_2	59.83	61.57	91.90	66.10	100.51	62.16	84.55	42.86	82.09	60.53
10	Y_1	10.60	11.23	12.84	12.76	12.03	16.23	9.45	17.23	8.99	13.71
	Y_2	65.14	73.18	114.72	79.39	112.76	93.54	96.23	48.47	92.74	68.34

Under the assumption that Y_1 and Y_2 are dependent, Xu et al. use the t-Copula function to model their dependence [15]. The estimated reliability functions under both the independent and dependent assumptions are displayed in figure 2. As seen, the different assumptions result in different estimates.

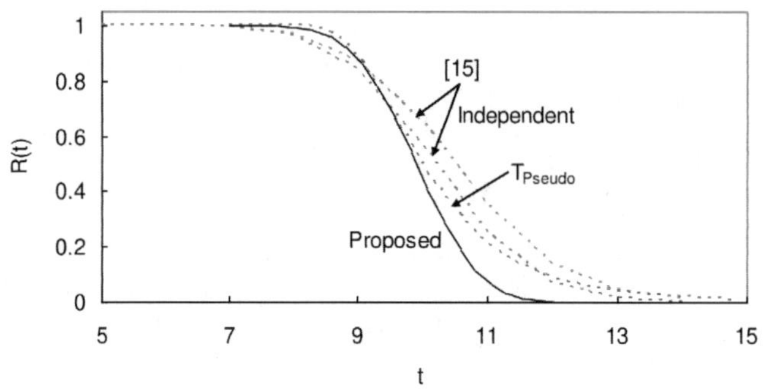

Figure 2. Reliability functions obtained from different approaches.

3.2. Preliminary data analysis

A preliminary data analysis shows:

- Y_1 is non-monotonic and Y_2 is monotonic.
- The CC between Y_1 and Y_2 is 0.7922 and the critical value of CC associated with the 95% level, one tail, and the degrees of freedom $nK - 1 = 99$ is 0.8566, implying that Y_1 and Y_2 can be thought to be mutually independent.

3.3. Aggregation of indicators

The data are normalized by equation (5), and the weighted LSM is used to find the pseudo-failure time of each degradation path. The parameters of the weight function are $\mu = 10$ and $\sigma = 1$ time unit for the current example. The results are shown in the second and third columns of table 2, where T_1 and T_2 are times for Y_1 and Y_2 to reach their failure thresholds, respectively. The 4th column of table 2 shows the values of $T = \min(T_1, T_2)$, which is the pseudo-failure time of a unit.

The pseudo-failure time data are fitted to the Weibull, inverse Weibull, gamma, normal and lognormal distributions, respectively, and the results from the best model are shown in table 3. It is noted that the inverse Weibull distribution (IWD) is the best model for T_2 and T.

Table 2. Pseudo-failure times, X_1 and X_2.

Unit	T_1	T_2	T	X_1	X_2
1	12.45	16.41	12.45	5.07	7.38
2	12.09	12.18	12.09	5.87	6.22
3	10.52	9.35	9.35	4.75	4.60
4	10.93	11.55	10.93	5.61	5.32
5	11.69	8.97	8.97	4.42	4.55
6	9.06	10.34	9.06	5.54	3.51
7	13.46	10.44	10.44	5.28	5.16
8	8.96	18.70	8.96	5.19	3.77
9	13.11	10.37	10.37	5.33	5.03
10	10.16	14.10	10.16	6.16	4.00

Table 3. Distributions of T_1, T_2, T and X_1.

	T_1	T_2	T	X_1
	Weibull	*IWD*	*IWD*	*Normal*
β or μ	8.720	5.769	10.96	5.322
η or σ	11.90	10.71	9.673	0.4854

From equation (6) yields $\Pr(T_1<T_2) = 0.5426$. This yields $(w_1, w_2) = (0.5426, 0.4574)$. Since Y_1 and Y_2 are independent, the generalized weighted mean with $p = 2$ is selected as the aggregation model. figure 3 shows the degradation paths of composite indicator in the coordinate system of $z(t)$ vs. t. The mean degradation function can be approximated by the power-law model with parameters $(\beta, \eta) = (1.353, 10.95)$. Since $\beta > 1$, $z(t)$ vs. t is non-linear.

Using the approach outlined in Section II.B.4 yields the normal random failure threshold with parameters $\mu_z = 0.8627$ and $\sigma_z = 0.01851$. It is noted that $\mu_z + 3\sigma_z = 0.9182 < 1$, implying that the failure limit of the composite indicator is smaller than the failure limit of individual indicators.

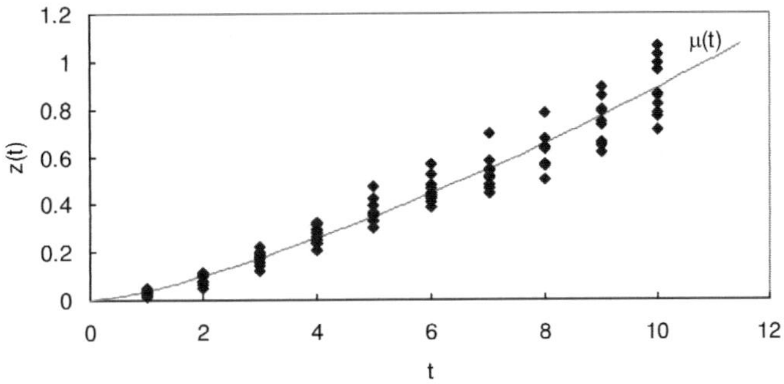

Figure 3. Degradation paths of composite indicator.

3.4. *Degradation model of composite degradation indicator*

3.4.1. Coordinate system selection. Table 4 shows the values of CC, CV and |CC/CV| under 4 coordinate systems. Clearly, the best coordinate system is the one that the *h*-axis is *z* and the *v*-axis is the degradation rate *r*. Figure 4 shows the plot of data under the best coordinate system. As seen, the plot has an inverse S-shaped trend, which implies that the plot is concave [convex] when *z* is small [large]. In other words, the process consists of two phases and their boundary is called the change point (denoted as z_0).

Table 4. Values of CC, CV and |CC/CV| under different coordinate systems.

v-axis	*r*			1/*r*		
h-axis	|*CC*|	*CV*	|*CC/CV*|	|*CC*|	*CV*	|*CC/CV*|
t	0.5415	0.3626	1.493	0.4227	0.3959	1.068
z	0.5829	0.3600	1.619	0.4215	0.3986	1.057

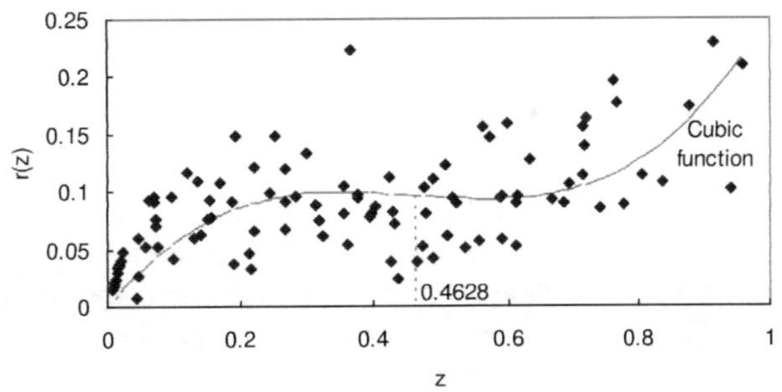

Figure 4 Plot of r vs. z.

3.4.2. Determination of change point. The left tail of the plot can be approximated by a power-law model given by

$$R_1(z) = az^b \tag{18}$$

Fitting the data in (0, $z_{max}/3$) to equation (18) yields $(a, b) = (0.1684, 0.4249)$.

The right tail of the plot can be approximated by

$$R_2(z) = a + bz \tag{19}$$

Fitting the data in ($2z_{max}/3$, z_{max}) to equation (19) yields $(a, b) = (-0.0273, 0.2133)$. The asymptotical relations are displayed in figure 5, and look reasonable.

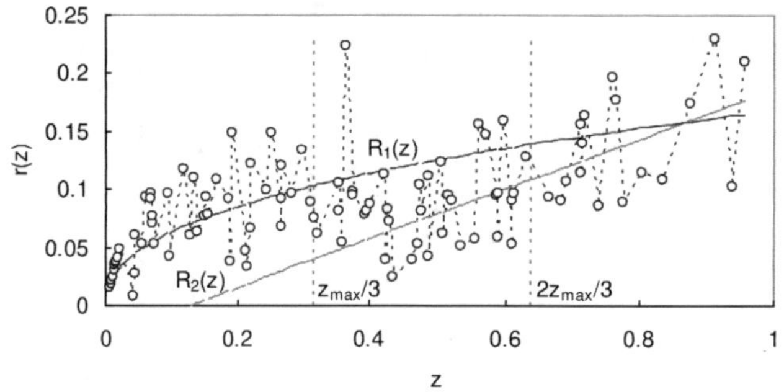

Figure 5. Plots of asymptotical relations.

Absolute errors between the data and asymptotical relations are given by

$$\varepsilon_1(z) = |R_1(z) - r(z)|, \varepsilon_2(z) = |R_2(z) - r(z)| \tag{20}$$

Figure 6 shows the error curves. As seen, $\varepsilon_1(z)$ [$\varepsilon_2(z)$] has an increasing [decreasing] trend so that there exists a point where $\varepsilon_1(z) \approx \varepsilon_2(z)$. This point is defined as the change point of $r(z)$. For the current example, $z_0 = 0.4026 \approx 0.40$, which is close to the inflection point of the cubic function shown in figure 4.

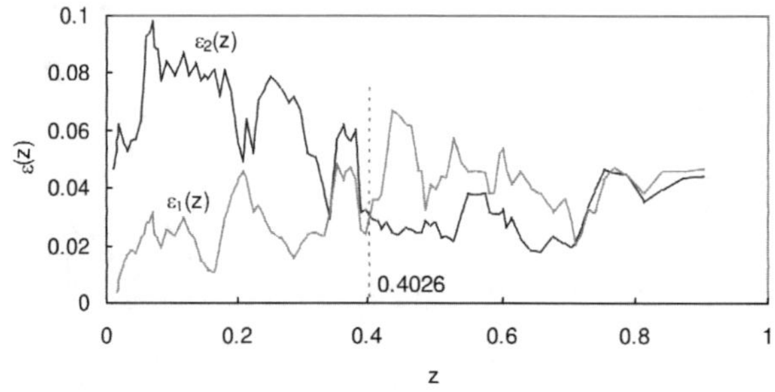

Figure 6. Plots of $\varepsilon_1(z)$ and $\varepsilon_2(z)$.

3.4.3. Degradation model in the first phase. The first phase is the interval of $z < z_0$. Generally, the unit in this phase is safe and there is not a need to make a failure prediction. Therefore, we model the degradation in this phase using a distribution of time to change point (denoted as X_1) where $z(t) = z_0$. For each unit, the value of X_1 can be obtained using interpolation and the results are shown in the 5th column of table 2. For the data of X_1, the normal distribution provides the best fitting and the parameters are shown in the last column of table 3.

The second phase is the interval of $z > z_0$. Let X_2 denote the residual life at X_1, given by

$$X_2 = T - X_1 \tag{21}$$

The values of X_2 are shown in the last column in table 2.

3.4.4. Degradation model in the second phase. Since *z(t)* is monotonic, we model the degradation in the second phase using a gamma process model with shape parameter *u(z)* and scale parameter *v*. The mean function is given by equation (19) with $R_2(z)$ being replaced by *dz*/*dt*. For a given initial condition (t_0, z_0), solving equation (19) yields

$$z(t) = [(a + bz_0)e^{b(t-t_0)} - a]/b \tag{22}$$

Equation (22) is actually the mean degradation function. Therefore, we replace "$z(t)$" with $\mu(t)$. Thus, the shape parameter at t is given by

$$u(t) = [\mu(t) - \mu(t_0)]/v \tag{23}$$

Model parameter set, (a, b, v), can be obtained using the maximum likelihood method. For the current example, $(a, b, v) = (0.0000, 0.1688, 0.01478)$. Since $a \approx 0$, we have

$$\mu(t) = z_0 e^{b(t-t_0)}, \ \mu(t) - \mu(t_0) = z_0[e^{b(t-t_0)} - 1] \tag{24}$$

This implies that the mean increment of Z is proportional to z_0 and exponentially increases with time.

3.4.5. Distribution of residual lifetime. Suppose that the current observation of a path is (t^*, z^*) and $z^* > z_0$. The distribution of residual life, $X = T - t^*$, is given by

$$F(x) = 1 - G[1 - z^*; z^*(e^{b(t-t^*)} - 1)/v, v] \tag{25}$$

where $G(.)$ is the gamma distribution function. Clearly, the residual life depends on both age and state.

To illustrate, consider the last observation point of degradation path of Unit 1 as the current observation, which is $(t^*, z^*) = (10, 0.7109)$. Figure 7 shows the degradation path, failure threshold and residual life distribution. The residual life distribution can be approximated by a normal distribution with parameters $(\mu, \sigma) = (1.181, 0.3441)$. That is, the mean residual life at $t = 10$ is 1.181 or the mean life of Unit 1 is 11.18. The pseudo-failure time obtained from the weighted LSM is 12.45. The relative error between the two estimates is 11.4%, implying that the two estimates are fairly close to each other.

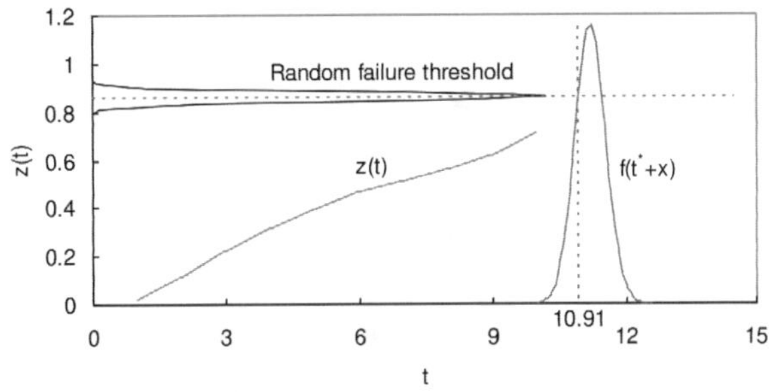

Figure 7. Degradation path and residual life distribution of Unit 1.

The tradeoff Bx life of X is 0.9135 [16], implying that Unit 1 can be preventively replaced at t = 10.91. This illustrates the condition-based maintenance application of the resulting model.

3.4.6. Distribution of time to failure for unit population. From equation (21), the life of unit population is given by

$$T = X_1 + X_2 \tag{26}$$

Here, X_2 is the residual life at z_0 = 0.40. The distribution of X_2 can be approximated by the normal distribution with parameters $(\mu, \sigma) = (4.581, 0.5778)$.

From table 2 yields CC(X_1, X_2) = -0.0392. This implies that X_1 and X_2 can be thought to be mutually independent. Using numerical integration yields the distribution of T, which also can be approximated by a normal distribution with parameters $(\mu, \sigma) = (9.903, 0.7546)$. The tradeoff Bx life of T is 8.573, which corresponds to a reliability of 96.1%.

Finally, the reliability functions estimated from the pseudo-failure time approach and two-phase degradation modeling approach are also displayed in figure 2. From the figure, the following remarks can be made:

- The reliability function obtained from the pseudo-failure time approach is close to the reliability function of [15] under the independent assumption. This confirm the fact that Y_1 and Y_2 are independent.
- For $t < 10$, the reliability functions obtained from the pseudo-failure time approach and two-phase degradation model are close to each other. This confirm the appropriateness of the pseudo-failure time approach.
- The reliability function obtained from the two-phase degradation model has the smallest life dispersion while a small dispersion implies a good failure prediction ability. This validates the appropriateness of the proposed modeling approach.

4. CONCLUSIONS

This paper has dealt with modeling of multidimensional degradation processes. A hybrid modeling approach has been proposed, which is generally applicable to multidimensional degradation processes. This is the main contribution of the paper. Another contribution of the paper is a weighted LSM that combines with a linear model to estimate pseudo-failure time, which is applicable for the situation where extrapolation is needed.

A real-world example that involves two degradation indicators has been analyzed in detail, and a two-phase degradation model has been developed to fit the data. The model in the first phase is similar to the first part of the delay time model, and the model in the second phase is an age-state-dependent stochastic process model. The contribution and finding obtained from this example have been:

- An asymptotic-relation-based method is developed to estimate the change point, which is applicable for the situations where the degradation rate curve in the best coordinate system is S-shaped or inverse S-shaped.
- The distribution of time to the failure threshold generally has a light left tail so that the normal and inverse Weibull distributions can be appropriate for approximating this distribution.

A topic for future research is to examine the influence of parameter p of the generalized weighted mean model on the resulting model.

ACKNOWLEDGMENT

The research was supported by the National Natural Science Foundation of China (No. 71771029).

REFERENCES

[1] Liu K, Gebraeel N Z and Shi J 2013 A Data-Level Fusion Model for Developing Composite Health Indices for Degradation Modeling and Prognostic Analysis *IEEE Trans. Autom. Sci. Eng.* **10** 652-64.

[2] Chehade A, Song C, Liu K, Saxena A and Zhang X 2018 A data-level fusion approach for degradation modeling and prognostic analysis under multiple failure modes *Journal of Quality Technology* **50** 150-65.

[3] Song C and Liu K 2018 Statistical degradation modeling and prognostics of multiple sensor signals via data fusion: A composite health index approach *IIE Trans.* **50** 853-67.

[4] Jiang R and Chen Z 2018 A standard-based approach for multi-criteria performance evaluation of engineered systems *Reliab. Eng. Syst. Saf.* **202** 107001.

[5] Jiang R 2020 Multidimensional performance degradation data fusion and residual life prediction for machine tools *2020 Global Reliability and Prognostics and Health Management(PHM-Shanghai)* pp 1-8.

[6] Jiang R 2020 A novel approach for selecting underlying variables of a stochastic degradation process *11th International Conference on Prognostics and System Health Management* pp 74–9.

[7] Wang W 2008 Delay Time Modelling *Complex System Maintenance Handbook* (Springer vol 2) ed Obbacy K A H and Murthy D N P (London: Springer-Verlag) Chapter 14 pp 345-370.
[8] Fisher R A 1992 Statistical methods for research workers *Breakthroughs in statistics* (New York: Springer) pp 66-70.
[9] Jiang R 2019 A novel normalization method for using in multiple criteria decision analysis *2019 IEEE International Conference on Industrial Engineering and Engineering Management (IEEM)* pp 268-72.
[10] Paradowski B,Wieckowski J and Dobryakova L 2020 Why TOPSIS does not always give correct results? *Procedia Computer Science*, **176** 3591-600.
[11] Son Y K 2011 Reliability prediction of engineering systems with competing failure modes due to component degradation *J. Mech. Sci. Technol.* **25** 1717-25.
[12] Jiang R 2015 Estimating residual life distribution from fractal curves of a condition variable *2015 Prognostics and System Health Management Conference* pp 1-6.
[13] Jiang R and Huang C 2018 Predicting distribution of time to degradation limit using a weighted approach *J. Mech. Sci. Technol.* **32** 5133-38.
[14] Bullen P S 2003 *Handbook of Means and Their Inequalities*(Springer Netherlands).
[15] Xu Z, Liu C, Tang J and Tan Q 2020 Reliability modeling for multi-performance degraded data based on mixed correlated competition failure *Journal of Chongqing Institute of Technology* **34** 253-62.
[16] Jiang R 2013 A tradeoff Bx life and its applications *Reliab. Eng. Syst. Saf.* **113** 1-6.

Classification of helicopter's typical flight state based on threshold

Li Sheng-nan[1,2], Wang Jing-lin[1,2], Yang Le[1,2], ZhangShang-tian[1,2]

1. Aviation Industry Group Corporation Shanghai Institute of Aeronautical Measurement and Control Technology, Shanghai 201601, China.
2. Key Laboratory of Fault Diagnosis and Health Management Science and Technology. Shanghai 201601, China.

E-mail: 1475291570@qq.com

Abstract. Dividing the 37 flying state of a certain line number helicopter. Firstly, dividing the helicopter rotation and single-engine flight. Secondly, performing preliminary state division for the remaining samlpes, the specific division of yaw angle, helicopter flight altitude and indicated air speed are different states, the least squares polynomial method is used for smoothing respectively. Calculating the extreme value of each parameter data, with the difference value of the extreme value of the parameter data being less than 10 as the limiting condition, dividing the original data segment into non-turning, level flight and steady speed state. The remaining sampling points are in the state of unsteady turning and non-level flight. Taking the difference value 0 as the limiting condition, further divide the non-steady speed and non-level flight state. Dividing the state of turning and non-turning, level flight, ascent and descent, steady speed, increase speed and deceleration state, which is the preliminary division state. Finally, dividing the near-ground and non-near-ground, classifying the helicopter status according to the height threshold, and analyze the accuracy of the classification results. The results show that this method is versatile, can quickly divide helicopters with different flight complexity, and has high accuracy.

1. Introduction

Helicopters for sightseeing, entertainment and material transportation when disasters occur, environmental monitoring, combat, training, etc. [1-2]. The influence of complex factors such as external airflow, vibration and noise in flight changes the load carrying capacity of mobile components under different flight conditions, causing damage to the mobile components [3]. In the event of an accident, it may cause casualties and damage to the helicopter, Therefore, the classification of helicopter flight status has important practical significance for the fault diagnosis and life prediction of helicopter structural components [4-7].

Liu Yu et al. proposed a method based on support vector machines for state division [89].But in actual operation, the time complexity is not been combined with the research. In practical applications, the use value is low. Wang Jin-sheng uses random forest method to achieve flight status division [9], However, the accuracy of the pre-classification method is uncertain, needing to verify the accuracy of the actual flight status classification results. This paper uses the least squares polynomial fitting method to smooth the helicopter state parameters on the scientific research flight test data of a certain type of helicopter. Obtaining the extreme value of each parameter and performing the difference to obtain the extreme value difference sequence. When the difference value of the extreme value is less than 10, it is the limiting condition, considering the time continuity, and finally the corresponding original data segment that meets the above conditions is divided into non-turning, the remaining sampling points are in the state of turning, unstable speed and non-level flight. The remaining yaw angle, altitude, and indicated airspeed are differentiated, and the difference value is 0 as the limiting condition, and the unsteady speed and the non-level flight state are further divided. If the value is greater than 0, the speed is increasing, and the state is ascending. Less than 0 means deceleration and descending state. Combining the engine torque, main rotor speed, pitch angle and other parameter thresholds to divide the helicopter's 37 flying states. Dividing the flying states of data with different levels of complexity, and the accuracy rate is high. Dividing actual scientific research flight test data, it has certain practical engineering application value.

2. Method: Least squares polynomial

Statistical research and scientific research, Building an approximate expression $y = P(x)$, This requires a lot of experimental data $(x_k, y_k)(k = 0,1,,...,m)$ to find its functional relationship[10],Measuring the deviation $P(x)$ of a function from the given data, It can be measured by the sum of squares of deviations often used in curve fitting.

For the given data $(x_k, y_k)(k = 0,1,,...,m)$, in the selected function class ϕ ,Solve $P(x) \in \phi$, Minimize the sum of squares of the deviation $r_i = P(x_i) - y_i (i = 0,1,...,m)$.

$$\sum_{i=0}^{m} r_k^2 = \sum_{i=0}^{m} [P(x_k) - y_k]^2 = min \tag{1}$$

Geometrically, the so-called least squares curve $y = P(x)$ fitting problem is the problem of finding the minimum curve at a given point in the sum of the square $(x_0, y_0), (x_1, y_1), ..., (x_m, y_m)$ of the distance from the point $x_0, x_1, ..., x_m$ [11].

Let the fitting polynomial be

$$P(x) = a_0 + a_1 x + ... + a_1 x^n, n < m \tag{2}$$

Getting the result by finding the partial derivative of the coefficients:

Firstly, finding the partial derivative of $S(a_0, a_1, ..., a_n)$ with respect to the coefficient $(a_0, a_1, ..., a_n)$ [12], Easy to get

$$\frac{\partial S}{\partial a_j} = 2[\sum_{i=0}^{n}(\sum x_k^{i+j})a_i - \sum_{i=0}^{m} y_k x_k^j], \quad j = 0,1,...,n \tag{3}$$

get:

$$\sum_{i=0}^{n}(\sum x_k^{i+j})a_i - \sum_{i=0}^{m} y_k x_k^j = 0, \quad j = 0,1,...,n \tag{4}$$

Equivalent to:

$$\sum_{i=0}^{n}(P(x_k) - y_k)x_k^j = 0, \quad j = 0,1,...,n \tag{5}$$

Regarding the linear equations of coefficients $a_0, a_1, ..., a_n$, Getting each coefficient by solving the above equation.

The title is set 17 point Times Bold, flush left, unjustified. The first letter of the title should be capitalized with the rest in lower case. It should not be indented. Leave 28 mm of space above the title and 10 mm after the title.

3. Helicopter status classification

3.1Classification of helicopter hovering, single-engine flight and rotation status

Single-engine flight: When the helicopter is off the ground and one of the left engine torque or the right engine torque is 0, the helicopter is in a single-engine flying state.

Rotation decline: The helicopter has a single-phase clutch, which can rotate on its mechanical structure. When the main rotor system and the engine are disconnected, the engine no longer provides

IOP Conf. Series: Materials Science and Engineering 1207 (2021) 012024 doi:10.1088/1757-899X/1207/1/012024

power to the main rotor. The relative airflow drives the main rotor system, and only the airflow passing upwards through the rotor drives the rotor blades.

When the helicopter is off the ground, the left engine torque and the right engine torque are 0 at the same time, and the altitude drops, that is, if the altitude difference value is less than or equal to 0 within a continuous period of time, it is classified as autorotation.

The list of authors should be indented 25 mm to match the abstract. The style for the names is initials then surname, with a comma after all but the last two names, which are separated by 'and'. Initials should not have full stops—for example A J Smith and *not* A. J. Smith. First names in full may be used if desired. If an author has additional information to appear as a footnote, such as a permanent address or to indicate that they are the corresponding author, the footnote should be entered after the surname.

3.2. Classification of helicopter other status

After the rotation and single-engine flight status are divided, the remaining sampling points will be further divided into the flight status. In the helicopter scientific research flight test data, the state is divided according to the state parameters yaw angle, indicated air speed, and altitude, and the helicopter is preliminarily divided into turning non-turning, level flight non-level flight, steady speed and unsteady speed respectively. In the process of non-level flight and unstable speed, the state is further divided into altitude rise and fall, increase speed and deceleration state.

The process of dividing the initial state of helicopter altitude is as follows：

Step 1: Polynomial fitting for smoothing

Through the analysis of the height data, the height data has volatility. Use least squares polynomial to fit height data, and convert the data to matrix form. The polynomial fits each row of data. Due to the volatility in the tail of the fitted data, in order to eliminate the influence of the tail fitting data fluctuation, each row adds m more data for fitting, the fitting order is n, the matrix data after the change is shown below.

$$D_{n\times(r+m)} = \begin{bmatrix} t(1) & t(2) & \cdots & t(r) & t(r+1) & \cdots & t(r+m) \\ t(r+1) & t(r+2) & \cdots & t(2r) & t(2r+1) & \cdots & t(2r+m) \\ \vdots & \vdots & \vdots & & \vdots & \vdots & \vdots \\ t(\mathrm{n}r+1) & t(\mathrm{n}r+2) & \cdots & & 0 & 0 \cdots & 0 \end{bmatrix} \tag{6}$$

The height data is transformed into a matrix form, each row is 200, in order to eliminate the influence of the end of the fitting data fluctuation, each row is added m to 60 data for fitting. Take the top 200 data after fitting each row. The best effect is when the order of each line fitting is n to 25.

Step 2: Take extreme values

Take the maximum value and minimum value for the height data after the least squares polynomial smoothing process. Sort the position numbers of the maximum and minimum sampling points according to the order of the height of the original data sampling points to obtain the extreme point data sequence.

Step 3: Obtain the extreme difference sequence

Difference the extreme value data sequence to obtain the difference value sequence and add 0 in front of the sequence.

Step 4: Classification of non-level flight status

Considering the time continuity of the flight state combined with the characteristics of the altitude data sequence change, when the altitude is within continuous time, the altitude value fluctuates within a stable range, and the fluctuation range is within 10, which is classified as a level flight state.

That is to say, a segment of extreme value difference sequence that meets at least two consecutive points and the absolute value of the extreme value difference sequence is less than 10 is divided into the level flight range in the extreme value difference sequence. The original altitude data corresponding to the extreme difference sequence belonging to the level flight range is further divided into level flight state data, and the remaining altitude sampling points are divided into non-level flight state data.

Step 5: Preliminary classification of helicopter ascent and descent

Divide the ascent and descent status for non-level flight data, performing difference calculation on the non-level flight altitude data of step 4, The divided rising state is the height data corresponding to the position where the difference value is greater than 0, and the rest is the falling state data. The ascent, level flight and descent state divided by the original height and the fitted height are shown in Figure1.

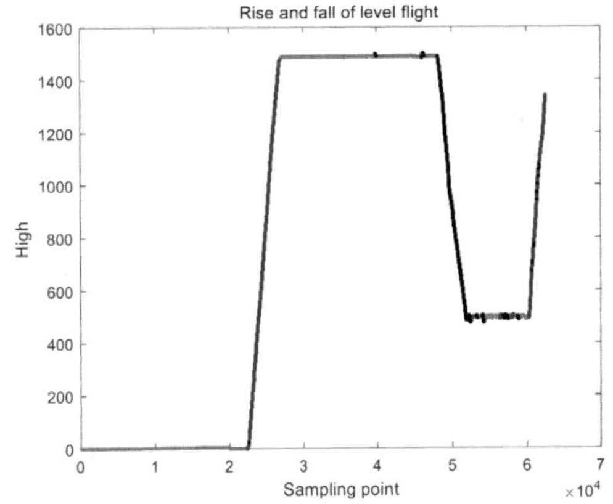

Figure 1. Helicopter ascent and descent and level flight state diagram

In the picture, blue is height rise, black is height drop, and red is level flight.

According to the division process from step 1 to step 5, the preliminary state division results of helicopter speed and yaw angle are shown in Figure 2 and Figure 3 below.

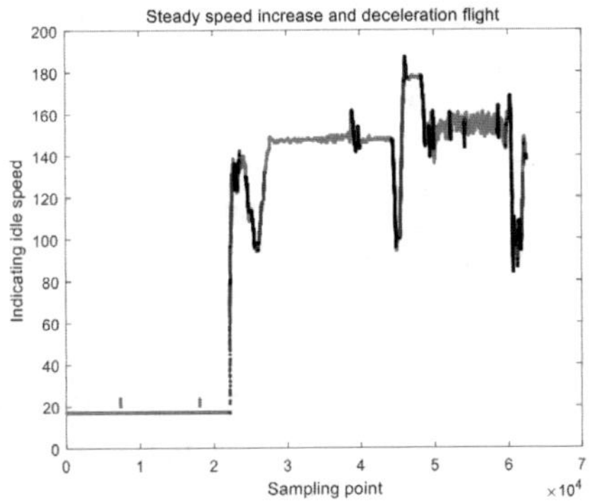

Figure 2. Helicopter speed increase, deceleration and steady speed flight status diagram

In the figure, blue is increasing speed, black is decelerating, and red is steady speed.

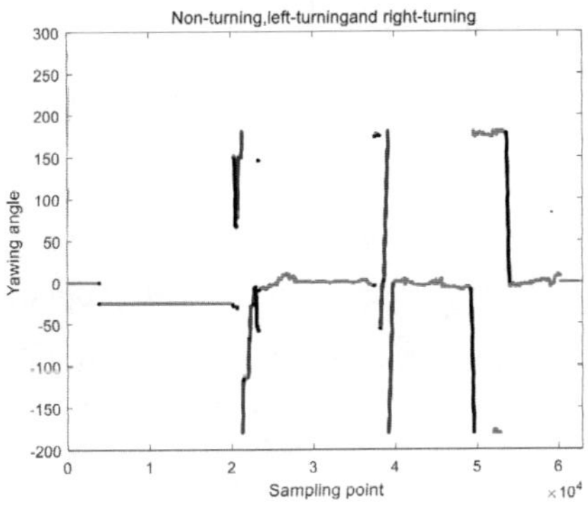

Figure 3. Helicopter non-turning left turn right turn flight state diagram

In the picture, blue means left turn, black means right turn, red means non-turning. The near-ground non-near-ground threshold is 50 meters. According to the speed threshold, the speed is divided into different ranges. Excessive speed: 70~90km/h; speed for maximum endurance: 90~130km/h; Speed for maximum endurance: Between Sailing speed: 130~190km/h; Long Sailing speed: 190~215km/ h. There are two speed ranges between 130~190km/h cruising speed, 130~170km/h cruising speed and 170~190km/h cruising speed.

3.3.Helicopter status classification rules

The total number of sampling points for a helicopter sampling point is 77334, and each 13 sampling points is 1 second. According to the state parameter threshold and the following rules, the helicopter is divided into 37 flying states, 37 flying states shown in Table 1 below.

Table 1. Flight status classification criteria

status	Flight status name	Status label
Helicopter ascends and turns in flight	Ascend to turn or circling ascend	10
	Helicopter turns sharply	13
Helicopter turning and flying	(90-130)Helicopter long-range speed horizontal turn	17
	(130-170)Helicopter cruise speed horizontal turn	21
	(170-190)Helicopter cruise speed horizontal turn	23
	(190-215)Helicopter voyage speed horizontal turn	25
	Helicopter hovering without ground effect	28
	Helicopter speed increase and turn flight	12
Helicopter turning and descending flight	Helicopter descending and turning flight	32
	Helicopter descends sharply and turns in flight	31
Helicopter non-turning ascending flight	Helicopter climbs obliquely	6
	Helicopter flying vertically	14
Helicopter non-turning level flight	Helicopter level acceleration flight	18
	(70-90)Helicopter cruising speed level flight	15
	(90-130Helicopter long-range speed level flight	16
	(130-170)Helicopter cruising speed level flight	20
	(170-190)Helicopter cruising speed level flight	22
	(190-215)Helicopter voyage speed level flight	24
	Helicopter maximum cruising speed level flight	26
	Helicopter decelerates horizontally	19
Helicopter non-turning down flight	Helicopter unsteady descending flight	8
	Helicopter dive flight	29
	The helicopter glide down at a steady speed	7
	Helicopter vertical descending flight	35
The flight speed is>270km/h	The helicopter flies at an insurmountable speed	27
Rise near the ground	Helicopter taking off and increasing speed	5
Descent near the ground	Helicopter vertical landing flight	11
Level flight near the ground	Helicopter flattened landing flight	37
	Helicopter approach/glide/deceleration flight	36
	Helicopter idle status on the ground	1
	Helicopter flying slowly	2
	Helicopter flying at low speed near the ground	9
	Helicopter has ground effect hovering flight	4

Helicopter hovering and slewing flight	34
Helicopter taxiing and taxiing state	3
Helicopter rotates down to fly	33
Helicopter single-engine flight	30

- Hover: When the helicopter is at a certain height, it can maintain the same state of direction and position. The helicopter is off the ground, but the yaw angle and altitude are fixed, and the speed is about 0, it is hovering.
- Helicopter idle status on the ground: When the rotation speed of the main rotor is 127~129, and the speed of the helicopter is about 0 without leaving the ground, it is a slow vehicle on the ground.
- Helicopter air slow state: For a certain type of helicopter, when the height of every two sampling points rises by about 5m, and when the height of every two sampling points drops by about 5m, it is dropping sharply, and the yaw angle is turning.
- Helicopter ascends, descends and turns to fly: For a certain type of helicopter, when the height of every two sampling points rises by about 5m and when the height of every two sampling points drops by about 5m, it is a sharp drop, and the yaw angle is in a turning state.
- Helicopter flattened landing: When the helicopter is landing at altitude, the pitch angle increases monotonously for 4 to 8 seconds.

According to the above rules, the flight process of this sortie divided the 37 flight states of the helicopter, and the flight state changes are shown in Figure 4.

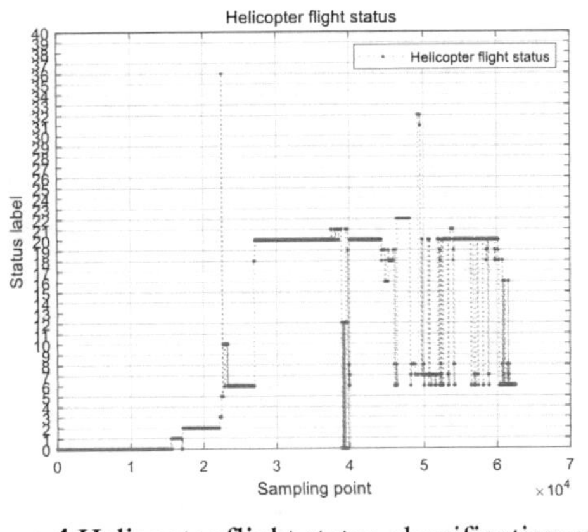

Figure 4. Helicopter flight status classification results

4. Result

In the above figure, the abscissa is the sampling point number, and the ordinate is the flight status number.

In the result of divided the helicopter flying status obtained, judging the helicopter altitude, speed, and yaw angle according to different status and corresponding limited conditions. When the corresponding yaw angle standard deviation, speed standard deviation, and altitude standard deviation in this segment of data are less than 1, it is non-turning, level flight, and steady speed; otherwise, turning, non-level flight, and unstable speed. Judging each segment of data after segmentation, and marking the sampling start point and ending point of each segment that does not meet the corresponding state determination criteria. Comparing the number of each sampling point divided by the helicopter with the corresponding actual flight status result. The wrong sampling points are as follows.

Table 2. Analysis of the accuracy of flight status division

Divide the error state	Divide the range of the serial number of the wrong sampling point	Divide the error	Divide the range of the serial number of the wrong sampling point
State 6	46310-46390		46310-46220
	50610-50590	State 8	52270-52240
	50940-50900	State 9	39080-39100
	49930-49980	State 12	38920-38930
	52380-52340	State 18	52210-52240
	53320-53370		60190-60250
	56850-56880	State 19	54020-54070
State7	60700-60730	State21	38140-38190

Then the number of wrong sampling points is 728. Among them, the number of sampling points in state 0 is 16,207, and state 0 means that the helicopter is not started. The accuracy of the final classification of helicopter flight status is 98.34%.

5. Conclusion

The least squares polynomial fitting method is used to smooth the helicopter state parameters. The extreme value of each parameter data is obtained, and the difference value of the extreme value of the parameter data is less than 10 as the limiting condition, and the corresponding original data segment is divided into non-turning, level flight and steady speed state. The remaining sampling points are turning, unstable speed, and non-level flight. The parameter difference value 0 is used as the limiting condition, and the non-steady speed and non-level flight state are further divided, and finally combined with engine torque, main rotor speed, pitch angle, etc. The parameter threshold divides the 37 flight states of the helicopter. Due to the large fluctuations of the state parameter data, after smoothing the polynomial fitting, there are still large fluctuations, and there is a certain error in the polynomial fitting. The division result is compared with the actual state, and the division accuracy rate reaches 98.34%, which can divide the flight state of helicopters with different flight complexity levels, and has a high accuracy rate

6. References

[1] Shanhani A R,2016 Mohammadi S.Damage tolerance and classic fatigue life predicion of a helicopter main rotor blade. *J.* MECCANICA **51** 1869-1886.

[2] Davies D P,Jenkins S L, Belben F R, 2013,Survey of fatiguegailures in helicopter componments and some lessons-leanrt. *J.* Engineering-FailureAnalysis,**32** 134-151.

[3] Wang Jinsheng.2017 Research on Helicopter Flight Status Recognition Technology Based on Machine Learning D. Jiangxi: Nanchang Hangkong University.

[4] Widodo A,Yang B S. 2011 Machine health prognostics using survival probability and support vector machine . *J.* Expert Systems with Applications, **38** 8430-8437.

[5] Qiu Xiaomei. 2019 Research on Fault Diagnosis and Remaining Life Prediction Methods of Rolling Bearings. D. Zibo: Shandong University of Technology.

[6] L Ren,J Cui,Y Sun,et al. 2017 Multi-bearing remaining useful life collaborative prediction：a deep learning approach. *J.* Journal of Manufacturing Systems, ,**43** 248-256.

[7] Meng K, Dong Z Y, Wang D H, et al. 2010 A self-Adaptive RBF neural network classifier fortransformer fault analysis. *J.* IEEE Transactions on Power Systems, **25** 1350-1360.

[8] Liu Yu. 2016 Research on Helicopter Flight Status Recognition Method and Application Based on SVM .D. Nanchang Hangkong University.

[9] Xiong Bangshu, et al. 2016 Helicopter flight status recognition based on SVM .*J.* Journal of Applied Science, **34** 469-474.

[10] Zhao Haisen, Du Zhonglan, Liu Xiaofang, et al.2014 On-line identification method of squirrel-cage asynchronous motor rotor resistance based on recursive least square method and model reference adaptive method. *J.* Proceedings of the Chinese Society of Electrical Engineering,,**34** 30.

[11] Li Ping,2019 Application of least squares polynomial fitting algorithm in managing high consumption to use low-value consumables .*J.* Chinese Health Economics. **38** 72-75

[12] Tang Yue,2017 Pump performance test curve piecewise least squares polynomial fitting algorithm .*J.* Journal of Drainage and Irrigation Machinery Engin,**35** 744-748.

Aircraft System State Recognition and Fault Prediction Based on a Test Diagnostic Model

Yan Su[1,2], Hongcai Chen[2,3], Chenxuan Gu[1], Xiangyu Xing[1], Xuerui Liang[1]

[1]Civil Aviation College, Nanjing University of Aeronautics and Astronautics, Nanjing, China
[2]State Key Laboratory of Smart Grid Protection and Control, Nanjing, China
[3]Nari Group Corporation, Nanjing, China

E-mail: chenghongcai@sgepri.sgcc.com.cn, suyannj@nuaa.edu.cn

Abstract. The existing testability models for fault prognosis of aircraft systems limit the implementation of prognosis and health management systems. This paper develops a test diagnosis modeling method and relevant algorithms to support dynamic testing and to evaluate fault prognostic ability during aircraft system design. According to the system principles and the complex function structure of aircraft systems, a test diagnostic model is established by integrating testing and prognostic information with a test diagnostic skeleton model using multi-signal flow. New test indexes are identified to assess the testability and prognostic ability of aircraft systems. Relevant state recognition and fault prediction algorithms are established by fusing the improved particle swarm optimization algorithm and Hidden Semi-Markov Model. The feasibility and validity of the test diagnostic modeling method and relevant algorithms are verified in an aircraft's engine bleed air system. Training and test show that the model can support analysis and estimation, and the algorithms can ensure accurate results after training the HSMM using improved PSO algorithm.

1. Introduction

Flight safety is a common goal of the global civil aviation industry. Airlines strive to minimize operating costs and increase economic efficiency while safeguarding aircraft operation for passengers and crew. Aircraft economic efficiency can be improved by reducing diagnostic and maintenance costs [1, 2]. Manufacturers use design for testability (DFT) to reduce the cost of a product during its design and life cycle and to limit additional investment in test equipment. Maintenance efficiency is improved by using built-in test (BIT) coverage methods [3]. DFTs can improve system fault diagnosis ability and fault coverage by adding sensors and tests in the design stage of the system. Most of the test designs focus on professional microcosmic fields, such as test problems of large-scale integrated circuits and self-built test problems [4].

DFT uses two methods [5, 6] based on either empirical engineering weights or models. Model-based DFT has been widely used in engineering and have resulted in many testability models. The most widely used models are logic models, hybrid diagnosis models, information flow models, and multi-signal (flow graphs) [7]. In the 1980s, Data Sciences International (DSI) developed a logic model based on correlation, which is the earliest applied testability model [8]. Later, DSI researched the hybrid diagnostic modeling technology that expressed the function and the failure mode in the same correlation model [9]. In the 1990s, express, test design software based on a hybrid diagnostic model, came out [10]. In the multi-signal flow model, the fault and the test are not directly related, and the correlation information is expressed through the signal propagation relationship [11]. Hybrid diagnostic models and later information flow and multi-signal flow models are all improvements of the logic model, the essence of which can be called the correlation model [12, 13, 14, 15]. At present, multi-signal flow models are commonly used to improve the testability of the product in the design stage [16, 17, 18, 19].

The existing testability models obtain the test parameters such as the fault detection rate and the fault isolation rate, and can evaluate and optimize the fault detection ability and fault isolation ability of the product [20]. However, the existing testability models have the following limitations: (a)Much of the test and diagnostic information is lost during the modeling process. The diagnosis ability and the dynamic development progress of faults cannot be described in detail. (b)The established model cannot evaluate the prediction ability of the system. The information cannot meet the requirements of system dynamic test and fault prognosis.

With the development and application of new technology and new equipment, the causes of aircraft faults are increasing. The parameters, which need to be monitored to ensure the operational safety of the aircraft, are also increasing. The testability of aircraft systems has greatly increased the complexity of aircraft diagnosis and repair, and has become a critical factor that affects the entire life cycle cost of aircraft systems. Aircraft maintenance costs are high primarily because aircraft system designs lack fault prediction capability [21].

Considering these limitations, this paper develops a new test diagnostic modeling method based on the system principles and multi-signal flow to test aircraft systems and predict faults. The model integrates system test and fault prognosis information. The relevant state recognition and fault prediction algorithms are used for the developed test diagnostic model. An aircraft engine's bleed air system is used as an example to demonstrate the developed modeling method and verify the test analysis and evaluation algorithms.

2. Test diagnosis modeling

The test diagnosis modeling process is shown in figure 1. Firstly, a skeleton model was established based on aircraft system principles and multi-signal flow by applying the functional structure of the hierarchy analysis method. The test, fault, and status information was then obtained based on testability design data and the predictable properties through failure modes and effects analysis. Finally, the test, fault, status, and other information was dynamically related with the skeleton model to establish a test diagnostic model.

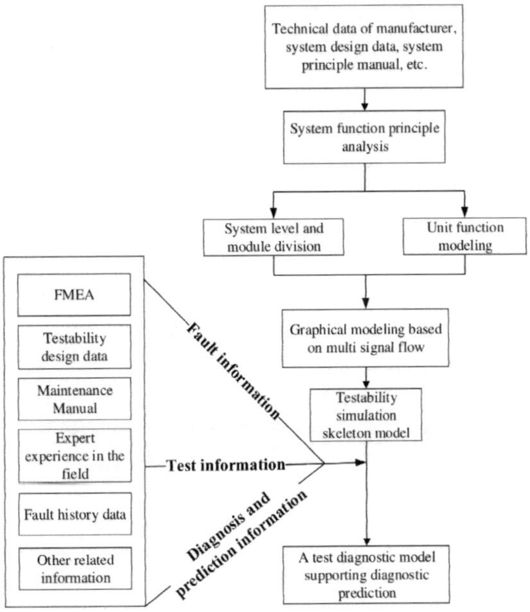

Figure 1. Test diagnosis modeling process

2.1. Skeleton modeling based on system theory and multi-signal flow

The test diagnostic skeleton model is defined as FS:

$$FS = (M, E, D, I, O) \quad (1)$$

where M is the module and can be divided into different levels of system, subsystem, least replaceable unit (LRU), and other layers, expressed as:

$$M = [m_1, m_2, \cdots, m_l] \quad (2)$$

E represents the set of directed edges of each node, indicating the connection between each node and the propagation direction of each variable, expressed as $E = [e_{ij}]$. D is the functional description to specify the function of the unit in the system. I represents the input variable, describing all input variables of the functional unit. O represents the output variable, describes all the output variables of the functional unit, and is also the input variable of the subsequent functional unit. The model can be optimized to improve system-level diagnosis and prediction by adding and deleting test points and optimizing test methods based on the testability skeleton model.

2.2. Information modeling based on test diagnostic skeleton model

In order to support fault prognosis, relevant fault prognosis information and test information were added to the test diagnostic skeleton model. The failure mode and failure mode attributes of each module are related to the nodes of the skeleton model.

The fault prognosis information model is defined as follows:

$$Fa = (F_l, F_m, F_r, F_{fs}, F_P, F_S) \quad (3)$$

where F_l indicates the location of the fault, F_m is the failure mode, F_r indicates the fault occurrence rate, F_{fs} represents the function of each failure mode function, F_P is the predictability of the fault, and F_s expresses the fault status.

The test information model is defined as follows:

$$Te = (T_l, T_d, T_i, T_f, T_{dp}, T_p, T_t, T_c) \quad (4)$$

where T_l is the location of the test point, T_d is the functional text description, T_i is the input signal, T_f is the signal feature extraction method, T_{dp} is the output of the test, T_p is the ability to predict the test, and T_t and T_c are the test time and cost of testing, respectively.

The fault prognosis information Fa and test information Te were set as the attributes of the relevant test nodes of the skeleton model according to the correlation between information and nodes. The relationship between the signal and the behavior of the test node was described by the directed edge. Finally, the test prognostic model with test information and prognosis information was established to satisfy the requirements of dynamic fault testing by changing the attributes of the nodes.

2.3. Test analysis and evaluation

Hypothesis fault f_i is related to test t_j, and the occurrence of fault f_i will cause test t_j to fail. This relationship is the correlation between failure and testing and can be expressed by a binary matrix $FT_{m\times n}$ [19, 22].

$$FT_{m\times n} = \begin{bmatrix} ft_{11} & \cdots & ft_{1n} \\ \vdots & \ddots & \vdots \\ ft_{ml} & \cdots & ft_{mn} \end{bmatrix} \quad (5)$$

The value of ft_{ij} is 0 or 1. If $ft_{ij}= 1$, then fault f_i is related to test t_j, that is, fault f_i can be detected by test t_j. If fault fi has nothing to do with test t_j, the fault fi cannot be detected by test t_j, assuming there are m failure modes and n tests in the model.

The developed test diagnostic model adds fault prediction properties to describe the development process of the fault. In order to evaluate the model's testability and prognosis, this paper develops a new concept of predictable fault coverage.

Definition 1. Predictable fault

Predictable fault means that the fault mode itself supports fault prediction, and the existing testing methods can meet the requirements of its fault prediction.

Definition 2. Predictable fault coverage.

Predictable fault coverage indicates the ratio of the number of faults that can be predicted by the system's existing methods to the total number of predictable faults under the given conditions.

$$\gamma_{FS_i} = \frac{N_{PS_i}}{N_{PS_i}} \tag{6}$$

The predictable fault and predictable fault coverage are combined with traditional testability parameters, i.e., fault detection rate, fault isolation rate, unmeasured fault, and fuzzy group, and are set as test indexes. The model's testability and prognostic ability are evaluated using these indexes to identify design deficiencies.

3. State recognition and fault prediction based on PSO and HSMM

Normally, the function of aircraft system from normal to failure does not happen instantaneously. It will go through a series of different fault states before failure. There is a gradual transition relationship between different fault states, and the fault states are hidden. The specific state must be determined by the collected test data. In the developed test diagnosis modeling process, the predictable attributes of the fault are defined, and different fault states of the same failure mode are divided, so that the test diagnosis model can well support the identification and prediction of fault states. Considering that the Hidden Semi-Markov Model (HSMM) focuses on fault development and evolution, and has the advantages of strong state classification capabilities, it is suitable for system state identification and prediction0[25]. The hidden semi-Markov model is used to identify and effectively evaluate the fault state of aircraft systems, and to predict the remaining service life of the system.

According to the relevant definitions of HSMM, the HSMM model parameters are defined as follows [24]:

$$\lambda = (\pi, A, B, D, N, M) \tag{7}$$

where D represents the state dwell time probability distribution. Assuming that the system will experience failure states from the initial normal state to the final complete fault, is the expected dwell time of the system in the fault state (i=1, 2, ..., n-1). Set as observed variable, RUL as the left life of system, then:

$$\mathrm{RUL} = d_i^* + \sum d_i \tag{8}$$

$$d_i = \mu_i + \rho\sigma_i^2$$

$$\rho = (T - \sum_{i=1}^{N} \mu_i) / \sum_{i=1}^{N} \sigma_i^2$$

$$d_i^* = p(i|v)d_i \tag{9}$$

The recognition and fault prediction steps are as following:

Step 1. According the life cycle test data of all fault states of the system, estimate the parameters of HSMM model, such as the transition probability matrix, the dwell time of the fault states, and the variance so on. A model library including all fault states of the system is established.

Step 2. Input the real-time characteristic data of the object system into the model library established by HSMM for fault status identification.

Step 3. Calculate the dwell time and the left dwell time of the system state.

Step 4. Calculate the RUL.

In above steps, parameter estimation is a very important and difficult step of solving with HSMM. Baum-Welch algorithm [27] is used to solve the parameter estimation of HSMM model at present. Baum-Welch algorithm trains the HSMM model converging to a local optimum. In order to converge to system's global optimal, this paper combines the improved particle swarm optimization (PSO) algorithm [28] with the original training algorithm to train HSMM. The process of HSMM training algorithm combined with improved particle swarm is as follows:

Step 1. Initialize the position and velocity of each particle, and set the iteration termination conditions. Let the original number of iterations t=0.

The dynamic adjustment of inertia weight variables ω according to the exponential decrease is introduced into the standard particle swarm algorithm.

$$\omega[t] = \exp(\frac{-t}{maxt}) \tag{10}$$

Step 2. Estimate the HSMM parameters using the sample data and Baum-Welch algorithm, and let the parameters as the initial positions of the particles in the particle swarm.

Step 3. Calculate observation sequence probability $P(O|\lambda)$ of the model parameter λ using following formula:

Herein, $(O_1 O_2 \cdots O_{t-d})$ is the observed variable of the state i at time t-d, and $(O_{t-d} O_{t-d+1} \cdots O_t)$ the observed variable of the state j at time t.

$$P(O|\lambda) = \sum_{i=1}^{N}\sum_{j=1}^{N}\sum_{d=1}^{t} \alpha_{t-d}(i)\alpha_{ij}P_{j(d)} \prod_{j=i-d+1}^{t} b_j(o_s)\beta_{(t)}(j)$$

$$\alpha_i(t) = P(O_1 O_2 \cdots O_t|\lambda)$$

$$\beta_t(i) = \sum_{j=1}^{N}\sum_{d=1}^{D} \beta_{t+d}(j)\alpha_{ij}P_j(d)b_j(O_{t+1}^{t+d}) \tag{11}$$

Step 4. Update the local and global optimal positions of each particle according to the formula of the improved particle swarm algorithm.

$$V_i(t) = \omega(t)V_i(t-1) + c_1 r_1\big(P_i(t-1) - X_i(t-1)\big) + c_2 r_2(P_g(t-1) - X_i(t-1)) \tag{12}$$

$$c_1, c_2 \in (0,2)$$

$$r_1, r_2 \in (0,1)$$

In the above algorithm, c_1, c_2, r_1, r_2 are random number, and $\omega(t)$ is weight of inertia. Learning factor c_1 and c_2 delegate the acceleration weight of particles advancing to their own extreme value and global extreme value.

Step 5. Determine whether the above parameter estimate meet the requirements. If not, set $t = t + 1$, go back to step 2.

4. Application

An engine bleed air system of certain type aircraft is selected as the typical application to verify the test diagnostic modeling method. According to the system principle of the engine bleed air system, its components can be divided into several LRU-level components, namely one-way control valve IPVC, high pressure valve HPV, pressure regulating shut-off valve PRSOV, pre-cooler PCE, fan control valve FAV, Bleed-air Temperature sensor BTS and PESOV downstream pipe. Through analysis, the engine bleed air system has 7 modules, expressed using set of M, which is defined as$\{m_1, m_2, \cdots, m_7\}$.

Further analysis of input and output variables, the connection between the engine bleed air system nodes and the propagation of each variable direction, expressed as $E = [e_{ij}]$. Input and output variables of engine bleed air system are shown in table 1. The engine bleed air system testability skeleton model is established in figure 2.

Where m_1 is the medium pressure one way valve IPCV, m_2 is the high pressure valve HPV, m_3is the pressure regulating shut off valve PRSOV, m_4 is the PRSOV downstream piping, m_5 is the pre-cooler PCE, m_6 is the temperature sensor BTS, m_7 is the fan air valve F.

Table 1. Formatting sections, subsections and subsubsections.

Modules(D)	Input variable (I)	Output variable (O)
Intermediate pressure control valve (IPVC)	Engine bleed air selection command/ C_{HPV}	IPCV opening degree/V_{IPVC}

High pressure valve(HPV)	Engine bleed air selection command/CHPV High pressure upstream pressure/PHPV-IN	HPV opening degree/V_{HPV}
Pressure regulation shut-off valve (PRSOV)	PRSOV inlet pressure/PPRV-IN Bleed air pressure adjustment value/PPIPS	PRSOV opening degree/V_{PRV} PRSOV outlet pressure/$P_{PRV-OUT}$
PRSOV downstream pipeline	PRSOV outlet pressure/$P_{PRV-OUT}$	Precooling module inlet pressure /P_{PIPS}
Fan air valve (FAV)	FAV Torque current/I_{PRV} FAV powered by/W_{FAV}	FAV opening degree /V_{FAV}
Pre-cooler engine(PCE)	Precooling module inlet pressure/P_{PIPS} FAV opening degree /V_{FAV}	PCE hot side outlet temperature/T_{PCE}
Bleed-air Temperature Sensor (BTS)	PCE hot edge outlet temperature/T_{PCE}	Engine bleed air temperature/T_{BTS}

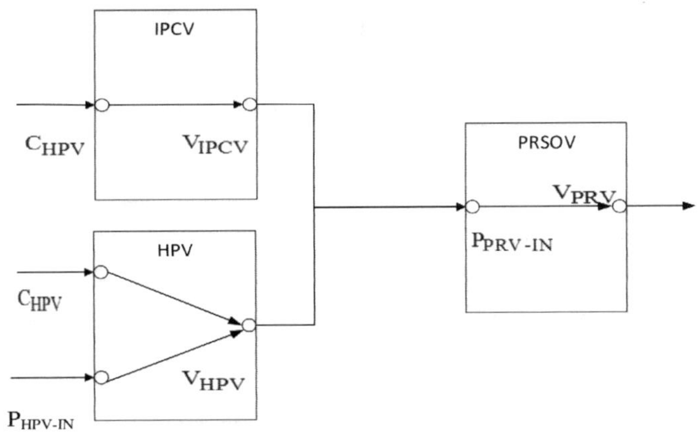

Figure 2. Test diagnosis skeleton model of engine bleed air system

According to relevant manuals of engine bleed air system, production materials, test design manual, maintenance manual and other technical data, the fault modes of the engine bleed air subsystem, the associated characteristic signals are got, seeing table 2. Engine bleed air system has eight test $T = \{t_1, t_2, \cdots, t_8\}$. The relevant test information is shown in table 3. Finally, the establishment of a certain type of aircraft engine air injection subsystem test diagnostic model shown in figure 3.

Table 2. Engine bleed air system fault mode and associated signals.

Fault location F_l	Fault mode F_m	Fault characteristic parameters F_{fs}	Whether to support forecasting F_p
Intermediate pressure control valve IPCV	IPCV On position failure	IPCV opening degree/V_{IPCV}	Yes
	IPCV Off position failure	HPV outlet pressure/$P_{HPV-OUT}$, IPCV opening degree/ V_{IPCV}	Yes
HPV	HPV On/Off position failure	HPV opening degree/ V_{HPV}	Yes

PRSOV	PRSOV On/Off bit failure	PRSOV opening degree/ V_{PRV}, PRSOV outlet pressure /$P_{PRV\text{-}OUT}$	Yes
	PRSOV internal leakage	PRSOV outlet pressure /$P_{PRV\text{-}OUT}$	No
PRSOV downstream pipeline	Cracked pipe weld	PRSOV outlet pressure /$P_{PRV\text{-}OUT}$	No
PCE	PCE internal leakage/efficiency reduction	PCE hot side outlet temperature /T_{PCE}	No
BTS	Engine bleed air over/under temperature	Engine bleed air temperature /T_{BTS}	No
FAV	FAV On/Off bit failure	FAV opening degree/V_{FAV}	No

Table 3. Engine bleed air system fault mode and associated signals.

Test number	Test name	Detection signal	Pass condition	Whether to support forecasting
t_1	IPCV open view	V_{IPCV}	C_{HPV}=1, V_{IPCV} =0/ C_{HPV}=0, V_{IPCV} =1	No
t_2	HPV open view	V_{HPV}	C_{HPV}=1, V_{IPCV} =1/ C_{HPV}=0, V_{IPCV} =0	No
t_3	PRSOV open view	V_{PRV}	C_{HPV}=1, V_{PRV} =1/ C_{HPV}=0, V_{PRV} =0	No
t_4	PRSOV outlet pressure monitoring	$P_{PRV-OUT}$	[2.8, 3.3] bar gauge	No
t_5	PRV downstream pressure monitoring	P_{PIPS}	[2.8, 3.3] bar gauge	No
t_6	PCE hot side outlet temperature monitoring	T_{PCE}	[190,260] ° C	No
t_7	Engine bleed air temperature monitoring	T_{BTS}	[190,260] ° C	No
t_8	FAV open view	V_{FAV}	+I_{FAV} ->+V_{FAV}	No

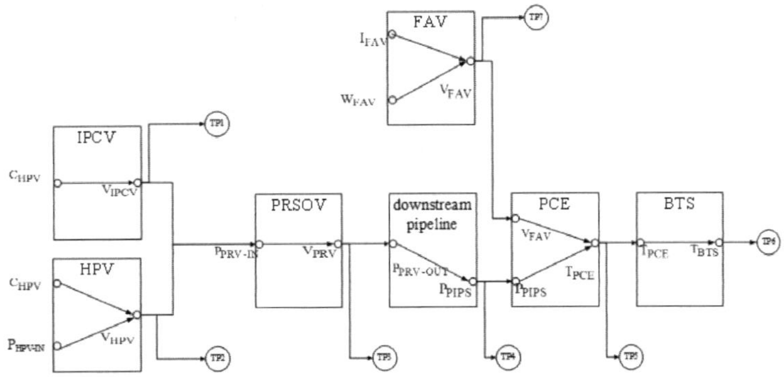

Figure 3. Test diagnosis skeleton model of engine air bleed subsystem

Through testing and diagnosis analysis of the engine bleed air system, the fault detection rate and fault isolation rate of the established model all equal to 100%, the predictable fault coverage is just 16.667%. The pressure regulating shut-off valve PRSOV has four open position fault status, i.e. normal status S_0 at open degree 0° position, degenerating status S_1 at open degree 60° position, deterioration status S_2 at open degree 120° position, failure status S_0 at open degree 180° position. In order to verify the developed state recognition and fault prediction method based on PSO and HSMM, training and test samples of different states are collected, shown in figure 4. The test samples are put into the HSMM model library, and test data recognition results under different fault conditions are calculated in table 5.

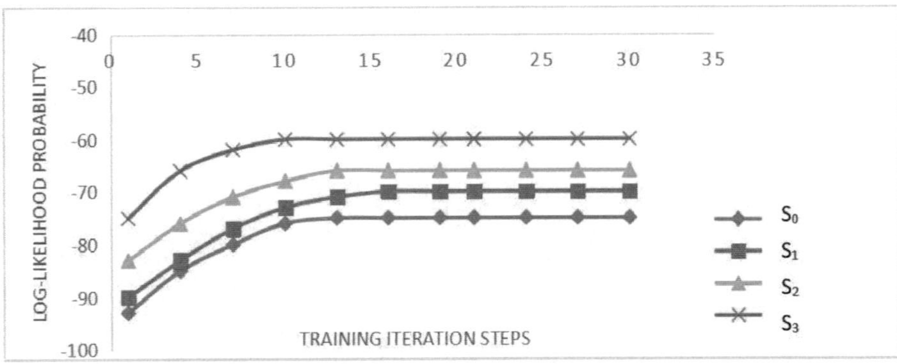

Figure 4. HSMM training process based on improved PS.

Table 4. HSMM train and test samples of PRSOV.

Fault status	S_0		S_1		S_2		S_3	
	Train	Test	Train	Test	Train	Test	Train	Test
Samples	10	2	12	2	12	2	10	2

Table 5. Fault status recognition results.

Log-likelihood probability	S_0		S_1		S_2		S_3	
	T_1	T_2	T_1	T_2	T_1	T_2	T_1	T_2
$HSMM_0$	-17.647	-17.042	-325.156	-341.164	-378.655	-370.164	-	-
$HSMM_1$	-359.249	-364.189	-23.975	-347.165	-347.165	-342.785	-389.154	-394.454
$HSMM_2$	-425.189	-465.187	-378.156	-379.145	-34.265	-36.131	-278.487	-285.486
$HSMM_3$	-	-	-398.154	-400.564	-301.165	-298.575	-12.545	-14.543

According to the magnitude of the log-likelihood probability, the fault state with the largest output value is the current fault state. It can be seen from table 5 that all test samples have the largest output value in the HSMM in its original state, so the fault state recognition rate based on HSMM is 100%. HSMM is trained based on an improved PSO algorithm training method. The maximum number of iterative steps in the training process is set to 50, and the algorithm's convergence error is 0.0001. Therefore, the fault state recognition method based on HSMM is accurate. Full-life history data of PRSOV is used as a training sample for HSMM. Using the HSMM training method in this paper, the failure state transition probability, the expected time, mean and variance of the fault status are obtained, shown in table 6 and table 7. Therefore the system fault state identification and remaining life prediction can be obtained on developed method. Taking the data of a certain observation point in S1 as an example, the fault state recognition probability can be calculated. The probability that it is in S1 is 0.8536. According to the remaining life calculation equation (8), the remaining life of the system is 11.6121.

Table 6. Transition matrix for four fault states of PRSOV.

Fault state	S_0	S_1	S_2	S_3
S_0	0.9063	0.0937	0.0000	0.0000
S_1	0	0.6955	0.3045	0.0000
S_2	0	0	0.9145	0.0855
S_3	0	0	0	1

Table 7. Expected dwell time for four fault states of PRSOV.

Fault state	S_0	S_1	S_2	S_3
mean	10.4515	7.1256	5.5483	5.4275
variance	1.2864	0.7256	0.6384	0.2548
d_i	10.4764	7.1456	5.5617	5.4924

5. Conclusions

This paper presents a test diagnosis modeling method and relevant algorithms to support dynamic testing and evaluate the ability of fault prognosis. A test diagnostic model was developed to support analyzing and evaluating the ability of test and prognosis simultaneously during system testability design. Considering the requirements of test and fault prognosis, the test information and prognosis information were added to a skeleton model as attributes. New testability indicators based on test diagnostic models were developed to evaluate system fault diagnosis and prediction capabilities.

In order to achieve effective identification of system fault states and estimate the remaining service life of the system during testability design, state recognition and fault prediction algorithms on HSMM are established. Because HSMM training is easily trapped in local extreme values, the improved PSO algorithm is used to converge to system's global optimal in the training process of HSMM. Finally, a test diagnostic model of an aircraft's engine bleed air system was established to verify the developed model.

Through test and diagnosis analysis of the engine bleed air system, the fault detection rate and fault isolation rate of the established model all equal to 100%, the predictable fault coverage is only 16.667%. The system fault state identification and remaining life prediction of the engine bleed air system can be obtained using the developed state recognition and fault prediction method. The algorithm's convergence error is just 0.0001. Therefore, the fault state recognition and fault prediction algorithms based on PSO and HSMM are accurate. The developed modeling method is helpful for optimizing system testability design and implementing prognosis and health management for improved system design.

6. References

[1] Liu C 2010 Analysis and optimization of civil aircraft maintenance cost based on Reliability. *Nanjing:Nanjing University of Aeronautics and Astronautics*
[2] Zhang R 2012 Research on Test Analysis and Evaluation Technology of Fault Diagnosis for Civil Aircraft System, N*anjing:Nanjing University of Aeronautics and Astronautics*
[3] Zhang B and Zeng T 2000 Intelligent BIT Technology, *Measurement and Control Technology* **19(11)** pp 1-4
[4] Veillette B R and Roberts G W 1995 A built-in self-test strategy for wireless communication systems. IEEE International Test Conference on Driving Down the Cost of Test. *IEEE Computer Society* pp 930-939
[5] Parsan F A, Smith S C and Al-Assadi W K 2016 Design for Testability of Sleep Convention Logic. *J. IEEE Transactions on Very Large Scale Integration (VLSI) Systems* **24(2)** pp 743-753
[6] Petrovic M, Milic M 2017 Analog Device Design for Testability in the Case of Oscillation Based Testing, 2017 IEEE 30th International Conference on Microelectronics (MIEL). Proceedings. **10** pp 283-286
[7] Nair R, Lin C and Haynes L 1996 Automatic dependency model generation using SPICE event driven simulation. *Autotestcon 96. Test Technology and Commercialization* pp 318-328
[8] Fairfax R K, Lancaster T C and Hurlburt G F 1981 Information technology and its impact on test and evaluation at the Naval Air Test Center. *1st Flight Test Conference*
[9] Gould E 2004 Modeling it both ways: hybrid diagnostic modeling and its application to hierarchical system designs, *Autotestcon . Proceedings. IEEE*, pp 576-582
[10] Simpson W R and Sheppard J W 1990 A hierarchical approach to system-level diagnosis(avionics). *Digital Avionics System Conference. IEEE.* pp 254-259
[11] Simpson W R and Sheppard J W 1992 System Testability Assessment for Integrated Diagnostics. *J. IEEE Design & Test* **9(1)** pp 40-54
[12] Tsai Y and Hsu Y 2012 A study of function-based diagnosis strategy and testability analysis for a system. *J. Proceedings of the Institution of Mechanical Engineers Part C Journal of Mechanical Engineering science*. **226(1)** pp 273-282
[13] Su Y and Zhang R 2014 Test Strategy Optimization Method for Aircraft System Fault Diagnosis. *International Conference on Simulation and Modelling Methodologies*. pp 741-748
[14] Zheng Y 2014 Testability Modelingand Analysis of Hierarchical System. *Harbin Institute of Technology*
[15] Qiu J 2012 Equipment Testability Modeling and Design Technology. *Beijing: China Environmental Science Press*
[16] Yang C, Zhang S, Tong C and Gu D 2013 Research on testability modeling with Bayesian network based on multi-signal flow model. 2013 IEEE 8th Conference on Industrial Electronics and Applications (ICIEA). *Piscataway. NJ:IEEE*. 1870-73
[17] Su Y, Liang X, Gu C and Pecht M 2019 A Multivalued Test and Diagnostic Strategy Optimization Method for Aircraft System Fault Diagnosis, 2019 IEEE International Conference on Prognostics and Health Management. *USA. San Francisco*. June 17-19, pp 1-6
[18] Tian L, Lu C, Yu J, Wu D, Lu M and Hu Y 2015 Test point optimization of control moment gyro (CMG) based on multi-signal flow model and improved genetic algorithm, *Piscataway. NJ:IEEE*. pp 122-125
[19] Su Y, Liang X R, Wang H, Wang J J and Pecht M G 2019 A Maintenance and Troubleshooting Method Based on Integrated Information and System Principles, *IEEE Access*, **7** 70513–24
[20] Zhang Y, Qiu J and Liu G 2011 Comparison and Prospect of Testability Models, *Journal of Test and Measurement Technology* **6** pp 504-514
[21] Janasak K M and Beshears R R 2007 Diagnostics To Prognostics - A Product Availability Technology Evolution, *Reliability and Maintainability Symposium* pp 113-118

[22] IEEE Standards Board. IEEE Std 1522-2004. IEEE Trial-Use Standard for Testability and Diagnose Ability Characteristics and Metrics, *Piscataway: IEEE Standards Press* 2004

[23] Heather S, Brian C, James P and Lindquist Martin A 2019 Improved state change estimation in dynamic functional connectivity using hidden semi-Markov models. *NeuroImage* pp 243-257

[24] Ferguson J D 1980 Application of Hidden Markov Models to Text and Speech, Variable duration models for speech, in: Symp.Institute for Defense Analyses. *Princeton NJ* Oct pp 143–179.

[25] Salzberg S L, Searls D B and Kasif S 1998 Computational Methods in Molecular Biology. *Elsevier Science B.V*

[26] Yu S 2010 Hidden semi-Markov models. *Artificial Intelligence* **174** pp 215–243

[27] Rezaei V and Pezeshk H 2013 HoracioPérez-Sa'nchez.Generalized Baum-Welch algorithm based on the similarity between sequences. *PLOS ONE* **8(12)** pp 1-8

[28] Zheng B, Huang H, Guo W, Li Y and Mi J 2018 Fault diagnosis method based on supervised particle swarm optimization classification algorithm. *Intelligent Data Analysis* **22(1)** pp 191-210

AUTHOR INDEX